BAYESIAN PHYLOGENETICS

Methods, Algorithms, and Applications

CHAPMAN & HALL/CRC
Mathematical and Computational Biology Series

Aims and scope:
This series aims to capture new developments and summarize what is known over the entire spectrum of mathematical and computational biology and medicine. It seeks to encourage the integration of mathematical, statistical, and computational methods into biology by publishing a broad range of textbooks, reference works, and handbooks. The titles included in the series are meant to appeal to students, researchers, and professionals in the mathematical, statistical and computational sciences, fundamental biology and bioengineering, as well as interdisciplinary researchers involved in the field. The inclusion of concrete examples and applications, and programming techniques and examples, is highly encouraged.

Series Editors

Proposals for the series should be submitted to one of the series editors above or directly to:
CRC Press, Taylor & Francis Group
3 Park Square, Milton Park
Abingdon, Oxfordshire OX14 4RN
UK

Published Titles

An Introduction to Systems Biology: Design Principles of Biological Circuits
Uri Alon

Glycome Informatics: Methods and Applications
Kiyoko F. Aoki-Kinoshita

Computational Systems Biology of Cancer
Emmanuel Barillot, Laurence Calzone, Philippe Hupé, Jean-Philippe Vert, and Andrei Zinovyev

Python for Bioinformatics
Sebastian Bassi

Quantitative Biology: From Molecular to Cellular Systems
Sebastian Bassi

Methods in Medical Informatics: Fundamentals of Healthcare Programming in Perl, Python, and Ruby
Jules J. Berman

Computational Biology: A Statistical Mechanics Perspective
Ralf Blossey

Game-Theoretical Models in Biology
Mark Broom and Jan Rychtář

Structural Bioinformatics: An Algorithmic Approach
Forbes J. Burkowski

Spatial Ecology
Stephen Cantrell, Chris Cosner, and Shigui Ruan

Cell Mechanics: From Single Scale-Based Models to Multiscale Modeling
Arnaud Chauvière, Luigi Preziosi, and Claude Verdier

Bayesian Phylogenetics: Methods, Algorithms, and Applications
Ming-Hui Chen, Lynn Kuo, and Paul O. Lewis

Statistical Methods for QTL Mapping
Zehua Chen

Normal Mode Analysis: Theory and Applications to Biological and Chemical Systems
Qiang Cui and Ivet Bahar

Kinetic Modelling in Systems Biology
Oleg Demin and Igor Goryanin

Data Analysis Tools for DNA Microarrays
Sorin Draghici

Statistics and Data Analysis for Microarrays Using R and Bioconductor, Second Edition
Sorin Drăghici

Computational Neuroscience: A Comprehensive Approach
Jianfeng Feng

Biological Sequence Analysis Using the SeqAn C++ Library
Andreas Gogol-Döring and Knut Reinert

Gene Expression Studies Using Affymetrix Microarrays
Hinrich Göhlmann and Willem Talloen

Handbook of Hidden Markov Models in Bioinformatics
Martin Gollery

Meta-analysis and Combining Information in Genetics and Genomics
Rudy Guerra and Darlene R. Goldstein

Differential Equations and Mathematical Biology, Second Edition
D.S. Jones, M.J. Plank, and B.D. Sleeman

Knowledge Discovery in Proteomics
Igor Jurisica and Dennis Wigle

Introduction to Proteins: Structure, Function, and Motion
Amit Kessel and Nir Ben-Tal

Biological Computation
Ehud Lamm and Ron Unger

Optimal Control Applied to Biological Models
Suzanne Lenhart and John T. Workman

Clustering in Bioinformatics and Drug Discovery
John D. MacCuish and Norah E. MacCuish

Spatiotemporal Patterns in Ecology and Epidemiology: Theory, Models, and Simulation
Horst Malchow, Sergei V. Petrovskii, and Ezio Venturino

Published Titles (continued)

Stochastic Dynamics for Systems Biology
Christian Mazza and Michel Benaïm

Engineering Genetic Circuits
Chris J. Myers

Pattern Discovery in Bioinformatics: Theory & Algorithms
Laxmi Parida

Exactly Solvable Models of Biological Invasion
Sergei V. Petrovskii and Bai-Lian Li

Computational Hydrodynamics of Capsules and Biological Cells
C. Pozrikidis

Modeling and Simulation of Capsules and Biological Cells
C. Pozrikidis

Cancer Modelling and Simulation
Luigi Preziosi

Introduction to Bio-Ontologies
Peter N. Robinson and Sebastian Bauer

Dynamics of Biological Systems
Michael Small

Genome Annotation
Jung Soh, Paul M.K. Gordon, and Christoph W. Sensen

Niche Modeling: Predictions from Statistical Distributions
David Stockwell

Algorithms in Bioinformatics: A Practical Introduction
Wing-Kin Sung

Introduction to Bioinformatics
Anna Tramontano

The Ten Most Wanted Solutions in Protein Bioinformatics
Anna Tramontano

Combinatorial Pattern Matching Algorithms in Computational Biology Using Perl and R
Gabriel Valiente

Managing Your Biological Data with Python
Allegra Via, Kristian Rother, and Anna Tramontano

Cancer Systems Biology
Edwin Wang

Stochastic Modelling for Systems Biology, Second Edition
Darren J. Wilkinson

Bioinformatics: A Practical Approach
Shui Qing Ye

Introduction to Computational Proteomics
Golan Yona

Chapman & Hall/CRC Mathematical and Computational Biology Series

BAYESIAN PHYLOGENETICS

Methods, Algorithms, and Applications

Edited by

Ming-Hui Chen, Lynn Kuo, and Paul O. Lewis

University of Connecticut
Storrs, USA

CRC Press is an imprint of the
Taylor & Francis Group, an **informa** business
A CHAPMAN & HALL BOOK

CRC Press
Taylor & Francis Group
6000 Broken Sound Parkway NW, Suite 300
Boca Raton, FL 33487-2742

First issued in paperback 2022

ISBN-13: 978-1-466-50079-2 (hbk)
ISBN-13: 978-1-03-234023-4 (pbk)
DOI: 10.1201/b16965

Library of Congress Cataloging-in-Publication Data

Bayesian phylogenetics : methods, algorithms, and applications / editors, Ming-Hui Chen, Lynn Kuo, Paul O. Lewis.
pages cm. -- (Chapman & Hall/CRC mathematical and computational biology series)
"A CRC title."
Includes bibliographical references and index.
ISBN 978-1-4665-0079-2 (hardcover : alk. paper)
1. Phylogeny. 2. Biometry. 3. Molecular genetics. 4. Bayesian statistical decision theory. I. Chen, Ming-Hui, 1961- II. Kuo, Lynn, 1949- III. Lewis, Paul O., 1961-

QH367.5.B39 2014
576.8'8--dc23 2014013188

Visit the Taylor & Francis Web site at
http://www.taylorandfrancis.com

and the CRC Press Web site at
http://www.crcpress.com

Contents

List of Figures

List of Tables

Preface

Since Felsenstein's groundbreaking 1981 paper (Felsenstein, 1981), molecular phylogenetics has closely tracked developments in statistics, more so than many other subdisciplines of biology. This trend has continued as statistics has become increasingly Bayesian, and the gap between phylogenetics and coalescent-based population genetics slowly disappears. Phylogenetics differs from typical statistical modeling efforts in its combination of discrete (tree topology) and continuous (substitution model) components. Because of its theoretical and computational challenges, phylogenetics is now proving to be an attractive area of research for a widening diversity of young Bayesian statisticians.

This project grew out of a successful collaboration between Paul Lewis, a biologist in the Department of Ecology and Evolutionary Biology long interested in model based phylogenetic methods, and Ming-Hui Chen and Lynn Kuo, professors in the Department of Statistics at the same institution (The University of Connecticut). This book provides a snapshot of current research in Bayesian phylogenetics, and was envisioned as a way of bringing state-of-the-art phylogenetics to the attention of the Bayesian statistical community, and state-of-the-art Bayesian statistics to the attention of the phylogenetics community, with the ultimate goal of encouraging further successful interdisciplinary research. The book is suitable for researchers at the graduate level or above in either statistics or biology. Topics addressed include model selection, sampling efficiency, priors, recombination, phylodynamics, evolutionary dependence, and divergence time estimation.

We would like to sincerely thank the authors for the uniformly high quality of their contributions, which made the job of editing straightforward. We thank them also for their patience as the book project evolved over time. Our hope is that this book will stimulate more successful collaborations between researchers in phylogenetics, population genetics and statistics.

Ming-Hui Chen
Lynn Kuo
Paul O. Lewis
Storrs, Connecticut

Editors

Ming-Hui Chen is a Professor of Statistics and director of the Statistical Consulting Services at the University of Connecticut. He was formerly an associate professor at Worcester Polytechnic Institute. He was the recipient of the 2013 American Association of the University Professors Research Excellence Award and the 2013 College of Liberal Arts and Sciences (CLAS) Excellence in Research Award in the Physical Sciences Division at University of Connecticut, the 2011 International Chinese Statisticians Association (ICSA) Outstanding Service Award, the 1998-2000 Harold J. Gay Assistant Professorship in Mathematical Sciences at Worcester Polytechnic Institute, and the 1993 I.W. Burr Award in Statistics at Purdue University. He is an elected fellow of the ASA and the IMS. He has served on numerous professional committees, including the 2013 president of International Chinese Statisticians Association, the Board of Directors of the International Society for Bayesian Analysis for 2011-2013, the Executive Director of the ICSA for 2007-2010, and the ICSA Board of Directors for 2004-2006. He has also served on editorial boards for *Bayesian Analysis*, *Journal of the American Statistical Association*, *Journal of Computational and Graphical Statistics*, *Lifetime Data Analysis*, *Sankhyā*, and *Statistics and Its Interface.* His research interests include Bayesian statistical methodology, Bayesian computation, Bayesian phylogenetics, categorical data analysis, design of Bayesian clinical trials, DNA microarray data analysis, meta-analysis, missing data analysis (EM, MCEM, and Bayesian), Monte Carlo methodology, prior elicitation, statistical methodology and analysis for prostate cancer data, and survival data analysis.

Lynn Kuo is a Professor of Statistics at the University of Connecticut (Storrs, Connecticut). She was formerly a visiting assistant professor at The University of Michigan and UC Davis, and an assistant professor at SUNY Stony Brook. She received all her three degrees BS, MA, and Ph.D. in Mathematics from UCLA. She is an elected fellow of the American Statistical Association (ASA). She worked as a research fellow at the Statistical Survey Institute at USDA, and later at the Statistical and Applied Mathematical Sciences Institute (SAMSI). She received an outstanding service award from ICSA in 2013. She has served as an Associate Editor for the Journal of American Statistical Association and Naval Research Logistics. She has also served on numerous review panels for CDC, NIH, and NSF. She was elected to be the Secretary and Treasurer of the Section of Bayesian Statistics of ASA in 1998-1999. She has published more than 80 papers in mainstream statistical journals including

Annals of Statistics, Journal of the American Statistical Association, Biometrics, Biometrika, The American Statistician, Canadian Journal of Statistics, Journal of Computational and Graphical Statistics, Sankhyā, Statistica Sinica, and *Statistics in Medicine.* Since her collaborations with the other two editors and other biologists, she has also published in *Systematic Biology, Molecular Biology and Evolution, Nature Genetics,* and *Statistics in Biosciences.* Her research areas include nonparametric Bayesian statistics, survey sampling, survival analysis, longitudinal data analysis, Bayesian phylogenetics, and 'omics' data analysis. She has supervised 10 Ph.D. students who have current positions in academia, government, and industry.

Paul O. Lewis is an Associate Professor of Ecology and Evolutionary Biology and co-director of the Bioinformatics Facility within the Biotechnology/Bioservices Center at the University of Connecticut (Storrs, Connecticut). He was formerly an assistant professor of Biology at the University of New Mexico (Albuquerque, New Mexico). He received a B.S. degree in Biology and Mathematics from Georgetown College (Georgetown, Kentucky) in 1982, an M.S. degree in Biology from the University of Memphis (Memphis, Tennessee) in 1984, and his Ph.D. in Plant Biology from The Ohio State University (Columbus, Ohio) in 1991. He had postdoctoral training under Bruce S. Weir in the Department of Statistics, North Carolina State University (Raleigh, North Carolina) and David L. Swofford at the Smithsonian Institution Laboratory of Molecular Systematics (Suitland, Maryland). He has served as an associate editor for the journal Systematic Biology, and has been elected president of the Society of Systematic Biologists for 2015. His research interests include maximum likelihood and Bayesian methods in phylogenetics, and the systematics and evolution of green plants from green algae to angiosperms.

Contributors

Serena Arima
Department of Methods and Models for Economy, Territory and Finance
Sapienza Universitá di Roma
Roma, Italy

Guy Baele
Department of Microbiology and Immunology
Rega Institute, KU Leuven
Leuven, Belgium

Peter Beerli
Department of Scientific Computing
Florida State University
Tallahassee, Florida, USA

Alexandre Bouchard-Côté
Department of Statistics
University of British Columbia
Vancouver, British Columbia, Canada

David Bryant
Department of Mathematics and Statistics
University of Otago
Dunedin, New Zealand

Ming-Hui Chen
Department of Statistics
University of Connecticut
Storrs, Connecticut, USA

Sooyoung Cheon
Department of Informational Statistics
Korea University
Sejong-city, South Korea

Mandev S. Gill
Department of Biostatistics
University of California
Los Angeles, California, USA

Tracy A. Heath
Department of Integrative Biology
University of California
Berkeley, California, USA

Asger Hobolth
Bioinformatics Research Center
Aarhus University
C.F. Møllers Alle, Aarhus C, Denmark

Mark T. Holder
Department of Ecology and Evolutionary Biology
University of Kansas
Lawrence, Kansas, USA

Mary K. Kuhner
Department of Genome Sciences
University of Washington
Seattle, Washington, USA

Lynn Kuo
Department of Statistics
University of Connecticut
Storrs, Connecticut, USA

Philippe Lemey
Department of Microbiology and Immunology
Rega Institute
KU Leuven, Leuven, Belgium

Paul O. Lewis
Department of Ecology and Evolutionary Biology
University of Connecticut
Storrs, Connecticut, USA

Faming Liang
Department of Statistics
Texas A&M University
College Station, Texas, USA

Vladimir N. Minin
Department of Statistics
University of Washington
Seattle, Washington, USA

Brian R. Moore
Department of Evolution and Ecology
University of California
Davis, California, USA

Julia A. Palacios
Department of Statistics
University of Washington
Seattle, Washington, USA

Michal Palczewski
Department of Scientific Computing
Florida State University
Tallahassee, Florida, USA

Marc A. Suchard
Departments of Biomathematics, Human Genetics, and Biostatistics
University of California
Los Angeles, California, USA

David L. Swofford
Department of Biology and Institute for Genome Sciences and Policy
Duke University
Durham, North Carolina, USA

Luca Tardella
Department of Statistical Sciences
Sapienza Universitá di Roma
Roma, Italy

Jeffrey L. Thorne
Bioinformatics Research Center
North Carolina State University
Raleigh, North Carolina, USA

Ying Wang
Beijing Institute of Genomics
Chinese Academy of Science
Chaoyang District, Beijing, PR China

Rui Wu
Department of Statistics
University of Connecticut
Storrs, Connecticut, USA

Ziheng Yang
Department of Biology
University College London
London, United Kingdom

1

Bayesian phylogenetics: methods, computational algorithms, and applications

CONTENTS

1.1 Introduction

Bayesian statistics experienced a Renaissance with the advent of Markov chain Monte Carlo simulation, which made possible the analysis of posterior distributions that are analytically intractable (Geman and Geman, 1984; Gelfand and Smith, 1990). Phylogenetics transitioned to a likelihood-based framework with Felsenstein's seminal paper (Felsenstein, 1981), and Bayesian approaches to phylogenetics were introduced in the mid-1990s with papers by Rannala and Yang (1996) and Mau and Newton (1997). Bayesian phylogenetics then quickly took off with the introduction of several software packages, the most popular to date being MRBAYES (Huelsenbeck and Ronquist, 2001; Ronquist et al., 2012b) and BEAST (Drummond and Rambaut, 2007; Drummond et al., 2012), which made it possible for empirical practitioners to carry out analyses on their own data. These programs pointed out the power of the Bayesian approach, and subsequent publications illustrated how the Bayesian paradigm allowed practical evaluation of innovative models exhibiting unprecedented biological detail. Examples include relaxed-clock divergence time estimation (Thorne et al., 1998; Kishino et al., 2001; Thorne and Kishino, 2002a; Drummond et al., 2006; Drummond and Suchard, 2010), analyses of co-speciation (Huelsenbeck et al., 2000c), ancestral state estimation (Huelsenbeck and Bollback, 2001), population history from serial sampled sequences (Drummond et al., 2002; Heled and Drummond, 2008), character correlation (Huelsenbeck and Rannala, 2003; Pagel and Meade, 2006), stochastic character mapping (Huelsenbeck et al., 2003), evolutionary dependence among sites (Robinson et al., 2003), detection of positively selected sites (Huelsenbeck and Dyer, 2004; Huelsenbeck et al., 2006), allowing substitution model heterogeneity across sites (Lartillot and Philippe, 2004; Pagel and Meade, 2004), combined analyses of morphological and sequence data (Nylander et al., 2004), joint estimation of phylogeny and alignment (Redelings and Suchard, 2005), and species tree estimation (Heled and Drummond, 2010), to name only a few.

Although recently published books have included chapters on Bayesian phylogenetics (Bininda-Emonds, 2004; Felsenstein, 2004; Hall, 2004; Nielsen, 2005; Yang, 2006; Lemey et al., 2009; Dey et al., 2011; Knowles and Kubatko, 2011; Wiley and Lieberman, 2011), the field of Bayesian phylogenetics has never before had a book all its own, and the time is right for a volume that highlights the latest innovations in the field. The remainder of this introductory chapter provides an overview of the organization of the book and a brief summary of each contributed chapter.

1.2 Overview

This book provides a snapshot of current trends in Bayesian phylogenetic research. There is a heavy emphasis on model selection, reflecting the recent active interest in accurately estimating marginal likelihoods. Other important themes include new approaches to improve mixing in Bayesian phylogenetic analyses in which the tree topology varies, divergence time estimation, accommodation of biologically realistic models, and the burgeoning interface between phylogenetics and population genetics.

Wang and Yang (Chapter 2) present a comprehensive discussion on prior specification in different settings of Bayesian phylogenetics and its impact on the posterior distribution. They also point out the difficulties of specifying high-dimensional priors.

Arima and Tardella (Chapter 3) review harmonic mean, generalized harmonic mean (GHM), inflated density ratio (IDR), generalized stepping-stone (GSS), path sampling (PS), and generalized path sampling (GPS) estimators for computing marginal likelihoods. They evaluate the performance of these estimators with RMSE using both synthetic and real datasets fitted by JC69 and GTR+Γ models. GHM is comparable to GSS and GPS when they rely on the same reference distribution. The IDR method compares well in accuracy to GSS and GPS while having a computational efficiency close to that of GHM.

Baele and Lemey (Chapter 4) re-examine the same estimators considered by Arima and Tardella for model selection. They also review the AICM method (Raftery et al., 2007a) which estimates the AIC criterion using posterior samples of the MCMC method. They conduct a simulation study on relaxed molecular clock models to assess the performance of these estimators. They show that path sampling, stepping-stone, and maximum a posteriori (MAP) methods are comparable and clearly outperform the harmonic mean, stabilized/smoothed harmonic mean, and AICM methods.

Holder, Lewis, Swofford, and Bryant (Chapter 5) extend the GSS method for a fixed tree topology to the case of varying tree topology. They propose a reference distribution for tree topologies and edge lengths where topologies containing splits judged important by the posterior are selected with high probability and edge length distributions are kept separately for high posterior probability splits to improve the match between reference distribution and posterior distribution. They show that their method compared well to a brute-force method when applied to binary unrooted trees.

Wu, Chen, Kuo, and Lewis (Chapter 6) extend the IDR method to accommodate varying tree topology. They show the method is consistent in estimating the marginal likelihood that marginalizes over tree topologies. In addition, proof of statistical consistency of marginal likelihood estimation is provided for the harmonic mean, stepping-stone, and IDR methods.

Cheon and Liang (Chapter 7) review many MCMC algorithms used in phylogenetics. They point out the difficulties of these algorithms with high-dimensional summations and integrals and local trapping. So they propose a sequential stochastic approximation Monte Carlo (SSAMC) algorithm which is based on SAMC by Liang et al. (2007) and Cheon and Liang (2008). They show that SSAMC can avoid the local trapping problem because of its self-adjusting mechanism. They further show that SSAMC can be computationally efficient by comparing its CPU time per iteration to that of BAMBE (Simon and Larget, 2001) and MRBAYES (Huelsenbeck and Ronquist, 2001).

Alexandre Bouchard-Côté (Chapter 8) reviews sequential Monte Carlo (SMC) methods that are also known as particle filter methods. SMC offers the potential to greatly improve efficiency in analyses in which the tree topology is of interest. He discusses the computational gains of the SMC and also shows how to implement multi-core parallelization of SMC.

Palczewski and Beerli (Chapter 9) derive a useful identity establishing that the combined marginal likelihood over all independent data blocks can be calculated as a product of independently calculated marginal likelihoods for each data block and a term that depends on the model and data under the assumption that the parameters under different data blocks are independent. This facilitates the analysis for large-scale biographic or population genetic datasets on computer clusters.

Kuhner (Chapter 10) reviews the ancestral recombination graph (ARG), and discusses inference, sampling methods, and software used for inferring the ARG. She points out that inference and analysis of ARG are not only wide open for future innovation, but also become more important as recombination, reassortment, gene transfer, and other reticulate processes in evolution are increasingly appreciated.

Palacios, Gill, Suchard, and Minin (Chapter 11) review modern Bayesian nonparametric methods for inference of effective population size trajectories. They show that coalescent-based estimation of population size trajectories can be thought of as estimation of an intensity function of a temporal point process. Consequently, statistical and computational techniques for point process literature will be germane for further development of Bayesian nonparametric phylodynamics models.

Hobolth and Thorne (Chapter 12) present an attractive alternative approach for studying molecular evolution in a likelihood-based framework, especially when the evolutionary models are too biologically rich to be computational feasible by the conventional approaches. Their approach infers evolutionary trajectories with Markov models conditional upon the endpoints of a branch. They consider both independent-site models and dependent-site models and develop EM and MCMC algorithms for them, respectively.

Heath and Moore (Chapter 13) review Bayesian inference of species divergence times. For branch rates, they discuss priors on autocorrelated-rate, independent-rate, local molecular-clock, and mixture models. For node times, they discuss generic and branching-process priors on node ages. They also discuss priors for calibrating divergence times and practical issues in estimating divergence times.

2

Priors in Bayesian phylogenetics

Ying Wang

Beijing Institute of Genomics, Chinese Academy of Science, Beijing, PR China

Ziheng Yang

Department of Biology, University College London, London, United Kingdom

CONTENTS

2.1 Introduction

2.1.1 The Bayesian paradigm

The key feature of Bayesian statistics is its use of probability distributions to represent the uncertainty in the parameters of the model. The distribution of the parameters before the collection and analysis of the data is called the prior, while the distribution incorporating the information in the data is called the posterior. In contrast, in Frequentist statistics, a parameter is an unknown but fixed constant and cannot have a distribution. When prior knowledge of the parameters is available, Bayesian analysis provides a natural way to incorporate such information. When no such information is available, a vague or diffuse prior has to be used for Bayesian inference to proceed. The posterior distribution combines the information from the prior and the information from the data and forms the basis for all Bayesian inferences.

In this chapter, we will start by introducing some basic notations, and use a simple example in phylogenetic analysis, that is, estimation of the distance between two sequences, to illustrate the different priors. We then discuss prior specification in different aspects of Bayesian phylogenetics, highlight the difficulties of specifying high-dimensional priors, and discuss the impact of the prior on the posterior.

Let $\mathbf{x}$ denote aligned DNA or protein sequences sampled from s species. We are interested in the phylogeny of the species and possibly also in the parameters in the evolutionary model. Let τ be the phylogeny (tree topology), $\mathbf{b}$ be the branch lengths of tree τ, and θ be the parameters in the evolutionary model, such as the transition/transversion rate ratio κ. Let $\pi(\tau, \mathbf{b}, \theta)$ denote the prior and $p(\mathbf{x}|\tau, \mathbf{b}, \theta)$ the likelihood. The posterior distribution is then

$$p(\tau, \mathbf{b}, \theta|\mathbf{x}) = \frac{p(\mathbf{x}|\tau, \mathbf{b}, \theta)\, \pi(\tau, \mathbf{b}, \theta)}{m(x)}, \tag{2.1}$$

where the marginal probability distribution of the data is

$$m(\mathbf{x}) = \sum_{\tau} \int_{B} \int_{\Theta} p(\mathbf{x}|\tau, \mathbf{b}, \theta)\, \pi(\tau, \mathbf{b}, \theta)\, \mathrm{d}\theta\, \mathrm{d}\mathbf{b}. \tag{2.2}$$

This is a normalizing constant, to ensure that the posterior distribution of (2.1) is a proper density. In most models used in Bayesian phylogenetics, θ is independent of τ and $\mathbf{b}$.

2.1.2 Objective versus subjective priors

There are two philosophies of Bayesian statistics, that is, the objective and subjective Bayesian, which differ in their interpretation of the prior. In subjective Bayesian, the prior represents the investigator's personal belief concerning the parameters. Evolutionary biologists, in general, are dismayed that prior solicitation may require a psychoanalytic and psychiatric assessment of their personal beliefs and that a statistical analysis of their data should be influenced by their subjective beliefs. In objective Bayesian, the prior represents our objective assessment of the information available on the parameter prior to the experiment. Much effort (by Laplace, Jeffreys, etc.) was taken to derive priors to represent a total lack of information on the parameter. For a time, uniform or flat priors were thought to serve this purpose, but it is now generally acknowledged that uniform (flat) priors are not "non-informative" and that truly "non-informative" priors do not exist. Later effort instead focused on use of formal rules to construct vague or diffuse priors which are expected to have minimal influence on the posterior. Criteria proposed for specifying an objective prior include the principle of insufficient reason, which underlies the uniform prior used by Bayes (1763) and Laplace (1812); invariance to reparameterization, which underlies the Jeffreys prior (Jeffreys, 1961b); and maximization of missing information, which underlies the reference prior of Bernardo (1979). See Kass and Wasserman (1996) and Ghosh (2011) for reviews.

It appears that subjective Bayesian is now the more popular version among Bayesian statisticians, apparently because objective Bayesians were unable to produce one coherent theory, and, in particular, to construct a prior of total ignorance. Nevertheless, where prior information is lacking concerning the parameters, the objective Bayesian idea of letting the data dominate the inference is appealing to biologists. Here we are not interested in this distinction in the interpretation of the prior, and instead emphasize the importance of evaluating the impact of the prior on the posterior in any Bayesian analysis.

Whether we adopt the objective or subjective views, a prior should reflect our information on the parameter before collection and analysis of the data. Thus we can construct the prior by making use of information gained in past experiments under similar conditions or other independent evidence. The prior can also be specified by modeling the physical biological process. For example, in population genetics, the genealogy underlying a random sample of DNA sequences has a probability distribution described by the coalescent process and this distribution may be considered a prior on the gene tree. Recombination, population demography, migration, etc., will alter the prior distribution of gene genealogies. In this context, the prior offers a convenient framework to incorporate such information. Experimental studies have demonstrated the distributions of human recombination hotspots over several genomic regions (e.g., the average width of hotspots and the average distance between adjacent hotspots), so it is natural to use such information in Bayesian inference through the prior. For phylogeny reconstruction, the birth–death process can be used to model speciation and extinction and to generate a prior distribution of phylogenetic trees.

2.2 Estimation of distance between two sequences

2.2.1 The maximum likelihood estimate (MLE)

We consider estimation of the distance between two DNA sequences under the JC69 model (Jukes and Cantor, 1969b). The single parameter θ is the sequence distance, measured by the expected number of nucleotide substitutions per site. The data ($\mathbf{x}$) can be summarized as the number of variable sites (x) out of a total of n sites in the alignment. The likelihood is

$$p(x|\theta) \propto \phi^x (1-\phi)^{n-x}, \tag{2.3}$$

where

$$\phi = \frac{3}{4} - \frac{3}{4}\mathrm{e}^{-\frac{4}{3}\theta} \tag{2.4}$$

is the probability of observing a difference at a site. If $\frac{x}{n} < \frac{3}{4}$, the MLE of ϕ is $\hat{\phi} = \frac{x}{n}$, and the MLE of θ is $\hat{\theta} = -\frac{3}{4}\log(1 - \frac{4}{3}\frac{x}{n})$. If $\frac{x}{n} \geq \frac{3}{4}$, then $\hat{\phi} = \frac{3}{4}$, $\hat{\theta} = \infty$. Because for any $\theta > 0$, $\Pr(x \geq \frac{3}{4}n) > 0$, we have $\mathrm{E}[\hat{\theta}] = \infty$, and $\hat{\theta}$ in general does not have finite expectation or variance. In practical data analysis, ad hoc decisions are taken when $\frac{x}{n} \geq \frac{3}{4}$, for example, by setting $\hat{\theta}$ to a pre-specified arbitrary large number such as 10 or 100.

As a specific example, for the human-orangutan mitochondrial 12s rRNA, $n = 948$, $x = 90$, so that $\hat{\theta} = 0.1015$. The standard deviation is 0.0109, so the 95% confidence interval is $\hat{\theta} \pm 1.96 \times SD$ or (0.0801, 0.1229) (Yang, 2006, page 9). Alternatively, by lowering the log likelihood from its peak by $\frac{1}{2}\chi^2_{1,5\%} = \frac{1}{2} \times 3.841$, we get the 95% likelihood interval to be (0.0817, 0.1245) (Yang, 2006, page 25).

In a Bayesian analysis, suppose the prior is $\pi(\theta)$. Then the posterior is

$$p(\theta|x) = \frac{p(x|\theta)\pi(\theta)}{m(x)}, \quad 0 \leq \theta < \infty,$$

where the normalizing constant

$$m(x) = \int_0^\infty p(x|\theta)\pi(\theta)\,\mathrm{d}\theta.$$

2.2.2 Uniform or flat priors

Bayes (1763) used a uniform prior $p \sim \mathrm{U}(0,1)$ on the binomial probability parameter p, where the data are a binomial sample $\mathrm{Bin}(x|p)$, and provided a long argument justifying his choice. It is since then common to assign a uniform prior on p when no other information is available. When the data are a sample from the normal distribution $x_i \sim \mathrm{N}(\mu, \sigma^2)$, with $-\infty < \mu < \infty$, $\sigma > 0$, one can specify a flat prior on μ and a flat prior on $\log\sigma$, so that $\pi(\mu) \propto 1$, and $\pi(\sigma) \propto \frac{1}{\sigma}$. Those priors do not integrate to 1 and are called improper priors, but they lead to proper posteriors. For a time, such flat or uniform priors are considered to be "noninformative." However, flat priors are not flat anymore after a reparameterization and are thus not noninformative.

For our example of distance estimation, if we let $\pi(\theta) = c$ with c a constant, the marginal distribution of the data

$$m(x) = c\int_0^\infty p(x|\theta)\mathrm{d}\theta = \infty,$$

as the likelihood $p(x|\theta)$ goes to a constant $\left(\frac{3}{4}\right)^x\left(\frac{1}{4}\right)^{n-x}$ as $\theta \to \infty$. There is thus no proper posterior. If we assign a uniform prior on ϕ instead, $\phi \sim \mathrm{U}(0, \frac{3}{4})$, we will have an exponential prior on θ with mean 3/4: $\pi(\theta) = \pi_\phi(\phi)|\frac{\mathrm{d}\phi}{\mathrm{d}\theta}| = \frac{4}{3}\mathrm{e}^{-4\theta/3}$, $0 \le \theta < \infty$. For the example data $n = 948$, $x = 90$, the posterior mean and 95% equal-tail credibility interval (CI) for θ under this prior are 0.1026 and (0.0823, 0.1252).

Alternatively we can assign a proper uniform prior on $\theta \sim \mathrm{U}(0, A)$. The posterior mean is found to be 0.1027 and 95% CI to be (0.0824, 0.1254) for $\theta \sim \mathrm{U}(0, 10)$ or $\theta \sim \mathrm{U}(0, 100)$.

2.2.3 The Jeffreys priors

The Jeffreys prior is based on the Fisher (expected) information and is invariant to reparameterization. Let $\mathrm{I}(\theta)$ be the Fisher information of θ with respect to the likelihood function $p(x|\theta)$

$$\mathrm{I}(\theta) = -\mathrm{E}_\theta\left[\frac{\mathrm{d}^2 \log p(x|\theta)}{\mathrm{d}\theta^2}\right].$$

The Jeffreys prior is then given as

$$\pi(\theta) \propto \mathrm{I}(\theta)^{\frac{1}{2}}.$$

For the multiple-parameter case, $\theta = (\theta_1, ...\theta_k)$, the Jeffreys prior is given as

$$\pi(\theta) \propto [\det \mathbf{I}(\theta)]^{\frac{1}{2}},$$

where $\det \mathbf{I}(\theta)$ is the determinant of the Fisher information matrix with elements

$$\mathbf{I}_{ij}(\theta) = -\mathrm{E}_\theta\left[\frac{\partial^2 \log p(x|\theta)}{\partial\theta_i\, \partial\theta_j}\right].$$

An advantage of the Jeffreys prior is its invariance under reparameterization, but the prior may not perform well in the case of multiple parameters and appeared to have problems when there are nuisance parameters in the model (Datta and Ghosh, 1996).

In our distance example, we have

$$\log p(x|\theta) = x\log(\frac{3}{4} - \frac{3}{4}e^{-\frac{4}{3}\theta}) + (n-x)\log(\frac{1}{4} + \frac{3}{4}e^{-\frac{4}{3}\theta}),$$

$$\frac{d^2 \log p(x|\theta)}{d\theta^2}$$
$$= -\left[\frac{e^{-\frac{8}{3}\theta}}{(\frac{3}{4} - \frac{3}{4}e^{-\frac{4}{3}\theta})^2} + \frac{\frac{4}{3}e^{-\frac{4}{3}\theta}}{\frac{3}{4} - \frac{3}{4}e^{-\frac{4}{3}\theta}}\right]x - \left[\frac{e^{-\frac{8}{3}\theta}}{(\frac{1}{4} + \frac{3}{4}e^{-\frac{4}{3}\theta})^2} - \frac{\frac{4}{3}e^{-\frac{4}{3}\theta}}{\frac{1}{4} + \frac{3}{4}e^{-\frac{4}{3}\theta}}\right](n-x).$$

As $E(x) = n\phi = n(\frac{3}{4} - \frac{3}{4}e^{-4/3\,\theta})$,

$$I(\theta) = \frac{16\,n}{3(e^{\frac{8}{3}\theta} + 2e^{\frac{4}{3}\theta} - 3)}.$$

Thus the Jeffreys prior is

$$\pi(\theta) \propto (e^{\frac{8}{3}\theta} + 2e^{\frac{4}{3}\theta} - 3)^{-\frac{1}{2}}, \quad 0 < \theta < \infty. \tag{2.5}$$

This is a proper density. This prior is derived by Ferreira and Suchard (2008, page 361), although their formula is incorrect (their α should be 4/3 instead of 1/3).

If instead we use ϕ in (2.4) as the parameter (parameters ϕ and θ form a one-to-one correspondence), we have

$$\begin{aligned} I(\phi) &= \frac{n}{\phi(1-\phi)}, \\ \pi(\phi) &\propto \phi^{-\frac{1}{2}}(1-\phi)^{-\frac{1}{2}}, \quad 0 \le \phi < \frac{3}{4}. \end{aligned} \tag{2.6}$$

Thus ϕ has a truncated beta distribution. We can see that (2.5) and (2.6) are consistent in that $\pi_\theta(\theta) = \pi_\phi(\phi(\theta))|\frac{d\phi}{d\theta}|$.

With $n = 948$, $x = 90$, the posterior mean $\tilde{\theta} = 0.1021$, and the 95% equal-tail CI is $(0.0818, 0.1247)$. This interval is nearly identical to the likelihood interval which is also invariant to reparameterization.

2.2.4 Reference priors

The reference prior was first proposed by Bernardo (1979), and was further developed by Berger and Bernardo (1992) and Berger et al. (2009). For a comprehensive review, refer to Bernardo (2005). The reference prior maximizes the expected Kullback-Leibler divergence between the prior and the posterior. Suppose the prior is $q(\theta)$, the likelihood is $p(x|\theta)$, with $\theta \in \Theta, x \in \mathcal{X}$, and the posterior is

$$p(\theta|x) = \frac{q(\theta)p(x|\theta)}{p(x)},$$

where $p(x) = \int_\Theta q(\theta)p(x|\theta)\,\mathrm{d}\theta$. Then the Kullback-Leibler divergence between the prior and the posterior for the given dataset x is

$$I_x(q(\theta), p(\theta|x)) = \int_\Theta p(\theta|x) \log \frac{p(\theta|x)}{q(\theta)} \mathrm{d}\theta.$$

The expected Kullback-Leibler divergence is

$$I(q) = \int_\mathcal{X} p(x) \int_\Theta p(\theta|x) \log \frac{p(\theta|x)}{q(\theta)} \mathrm{d}\theta \mathrm{d}x.$$

The formula for deriving the reference prior was given in Berger et al. (2009). If there is only one parameter in the model, and if the model is regular (in the sense that the MLE has an asymptotic normal distribution when the data size approaches infinity), the reference prior is known to be the Jeffreys prior (Berger et al., 2009).

While the Jeffreys and reference priors are important in objective Bayesian statistics, they are rarely used in phylogenetics. Indeed, the only study we are aware of appears to be Ferreira and Suchard (2008), who discussed the use of reference priors for parameters in continuous-time Markov Chain models of nucleotide substitution. (Note, however, that their likelihood function, their equation 4, is incorrectly defined.) The discussion is limited to comparison of two sequences only. We suspect that two factors may have contributed to this lack of enthusiasm. First, construction of Jeffreys and reference priors require analytical manipulation of the likelihood function, the Fisher information etc., which may not be possible for complex models or multiple sequences. Second, pairwise sequence comparison, such as distance estimation under JC69, is often not of direct interest: often the distances are calculated for reconstruction of the phylogeny. It is unclear whether the reference or Jeffreys priors may be made a better use of when our interest is in the parameters in the evolutionary models. One such example is the nonsynonymous/synonymous rate ratio ($\omega = d_N/d_S$) in comparison of two protein-coding gene sequences. There is much interest in ω as the parameter can inform us of the direction and magnitude of natural selection on the protein. The MLE of ω (Goldman and Yang, 1994) can be 0 or ∞ in some datasets, so that $\hat{\omega}$ does not have finite mean or variance. Use of a prior in Bayesian analysis will have a shrinkage effect, producing an estimator with nice (Frequentist) properties.

2.3 Priors on model parameters in Bayesian phylogenetics

2.3.1 Priors on branch lengths

The estimation of branch lengths has many implications as branch lengths contain information about species divergence times and about sequence evolutionary rate. Even if our interest is in the tree topology and branch lengths are not of direct interest and are thus nuisance parameters, inaccurate branch length estimates may cause problems to the estimation of the tree topology.

An binary unrooted tree for s species has $2s - 3$ branches, so the branch lengths may be represented by $\mathbf{b} = \{b_1, ..., b_{2s-3}\}$. Each is measured by the expected number of substitutions per site, with $b_i > 0, i = 1, ..., 2s-3$. Standard implementations of Bayesian phylogenetic methods (e.g., Larget and Simon (1999) and Huelsenbeck and Ronquist (2001)) have used independently and identically distributed (i.i.d.) priors on the $2s - 3$ branch lengths, with each branch length having either a uniform or exponential distribution. Both are implemented in the program MrBayes. The uniform prior $b_i \sim \mathrm{U}(0, A)$ has an upper bound A specified by the user (the default is 100, so that the prior mean of b_i is 50). The exponential prior $b_i \sim \exp(\lambda)$ has a rate parameter λ specified by the user (the default is 10, so that the prior mean of b_i is 0.1).

It has recently been realized that those i.i.d. priors are problematic as they make highly informative but unreasonable statements about the tree length, the sum of branch lengths, $T = \sum_{i=1}^{2s-3} b_i$. In several studies, it has been observed that MrBayes generates exceptionally large trees with the 95% CI of the tree length in the Bayesian analysis excluding its MLE under the same model (Marshall, 2010; Brown et al., 2010). Rannala et al. (2012) pointed out that the prior on branch lengths is to blame. The default prior on branch lengths in MrBayes is the independent exponential densities with parameter $\lambda = 10$ (and mean 0.1),

$$\pi(\mathbf{b}) = \prod_{i=1}^{2s-3} \pi(b_i) = \prod_{i=1}^{2s-3} \lambda e^{-\lambda b_i}. \tag{2.7}$$

The tree length thus has an induced gamma distribution:

$$\pi(T) = \mathrm{Gamma}(2s - 3, \lambda), \tag{2.8}$$

with mean $\mathrm{E}(T) = (2s - 3)/\lambda$, and variance $\mathrm{V}(T) = (2s - 3)/\lambda^2$. With s large, the gamma is nearly normal. The 95% prior CI is approximately $(2s - 3)/\lambda \pm 1.96\sqrt{2s - 3}/\lambda$, so that this induced prior on tree length can be extremely informative and favors unreasonably large values. For example, with $s = 100$, the 95% prior interval for tree length is 19.7 ± 2.75 or (16.9, 19.7), but the MLE may be ~ 1 or 2.

The situation is worse for the i.i.d. uniform prior $b_i \sim \mathrm{U}(0, A)$. While most branch lengths in phylogenetic trees are small (< 1 or even < 0.1), large values (such as 5 or 10) occasionally occur. To avoid precluding large values, a large bound A (the default value is 100) is used. With

$$\pi(\mathbf{b}) = \prod_{i=1}^{2s-3} \pi(b_i) = A^{-(2s-3)}, \tag{2.9}$$

the induced prior on $T = \sum_{i=1}^{2s-3} b_i$ is

$$\pi(T) = \frac{1}{(z-1)!A^z} \sum_{i=0}^{z} (-1)^i \binom{z}{i} \left[(T - i\,A)^+\right]^{z-1}, \tag{2.10}$$

where $z = 2s-3$, $[x]^+ = \max\{0, x\}$ (Buonocore et al., 2009). With a large s, this is also approximately normal, with $\mathrm{E}(T) = (2s-3)\frac{A}{2}$, and $\mathrm{V}(T) = (2s-3)\frac{A^2}{12}$. With s=100, A=100, the 95% prior interval will be $9850 \pm 1.96 \times 405.2$ or (9055, 10644). Such long trees (with an average $\sim$ 10000 substitution per site) are never encountered or expected in phylogenetic analysis of real datasets and are unreasonable. It is indeed ironical that uniform priors are sometimes advertised as being noninformative in the statistics literature. The problem is illustrated in a dataset in Table 2.1 and Figure 2.1.

Besides generating extremely long and unreasonable trees in the posterior, the i.i.d. priors may potentially create a local peak in the posterior (because the likelihood is flat for very large branch lengths and tree length), and cause convergence and mixing problems for the Bayesian MCMC algorithms (Rannala et al., 2012).

One modification of the i.i.d. priors is the hierarchical prior suggested by Suchard et al. (2001). While $b_i \sim \mathrm{Exp}(\lambda)$, λ is assigned an inverse-gamma distribution with hyper parameters. This was found to perform slightly better than the two i.i.d. priors by introducing more variance in the induced prior on T.

A compound Dirichlet prior on branch lengths was recently suggested by Rannala et al. (2012). This assigns a gamma or inverse gamma prior on the

Table 2.1: Prior and posterior means and 95% equal-tail credibility intervals for the tree length in the analysis of the clams data (Figure 2.1).

Branch length prior	Prior	Posterior
i.i.d. exponential		
exp(10)	18.3 (15.745, 21.045)	16.163 (13.630, 18.919)
exp(1)	183 (157.45, 210.45)	182.00 (156.48, 209.32)
exp(0.1)	1830 (1574.5, 2104.5)	1168.7 (1023.3, 1323.2)
i.i.d. uniform		
uniform(0, 1)	91.500 (83.847, 99.137)	32.661 (26.659, 39.376)
uniform(0, 10)	915.00 (838.47, 991.37)	344.26 (287.38, 405.43)
uniform(0, 100)	9150.0 (8384.7, 9913.7)	3469.7 (2897.6, 4143.5)
GammaDirichlet		
gammadir(1, 0.1, 1, 1)	10 (0.2532, 36.889)	1.0840 (0.9323, 1.2504)
gammadir(1, 0.01, 1, 1)	100 (2.5318, 368.89)	1.0842 (0.9344, 1.2513)
gammadir(1, 0.0001, 1, 1)	10000 (253.18, 36889)	1.0851 (0.9383, 1.2541)

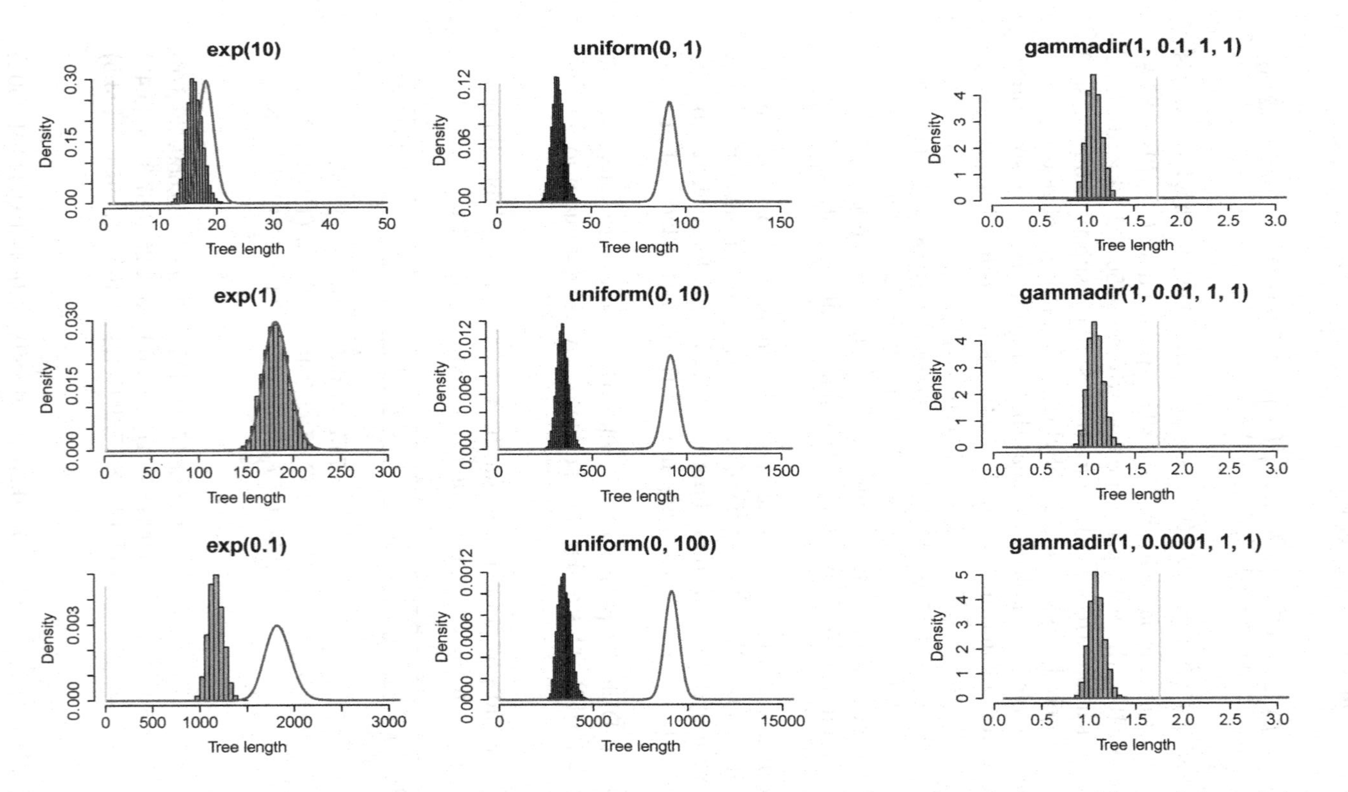

Figure 2.1: The effect of the branch length prior on the posterior distribution of tree length (sum of branch lengths) in the analysis of the clams data of Hedtke et al. (2008). The solid lines are the prior distributions, while the histograms are the estimated posterior distributions of tree length. The MLE of tree length under the same model is 1.75, indicated on the plots by vertical lines (Zhang et al., 2012). Note that the x-axis ranges are different for some plots.

tree length, and then partitions the tree length into branch lengths according to a Dirichlet distribution. With a gamma distribution on T:

$$T \sim \text{Gamma}(T|\alpha_T, \beta_T), \tag{2.11}$$

and a uniform Dirichlet on b_1/T, b_2/T, ..., the prior on branch lengths is specified as

$$f(b) = \frac{\beta_T^{\alpha_T}}{\Gamma(\alpha_T)} e^{-\beta_T T} T^{\alpha_T - 1} (2s-4)!\, T^{-2s+4} \tag{2.12}$$

(Rannala et al., 2012, equation 37). Zhang et al. (2012) implemented the compound Dirichlet priors on branch lengths in MrBayes 3.1.2 and 3.2.1 and used them to analyze six problematic datasets of Brown et al. (2010). They found that the new priors are effective in overcoming the problem of extremely long trees.

We analyzed one of the six datasets of Brown et al. (2010) using MrBayes 3.2.1 modified by Zhang et al. (2012) to compare the independent exponential priors ((2.7) and (2.8)), the independent uniform priors ((2.9) and (2.10)), and the compound Dirichlet priors ((2.11) and (2.12)). The sequence length ($n = 584$) in the clams dataset (Hedtke et al., 2008) is relative small given that the number of species is $s = 93$. The GTR+Γ_4 substitution model (Yang, 1994b,a) is assumed. The number of iterations is 10^7 with the number of chains equal to 4 and a temperature of 0.2. The sample frequency is 1000, and the last 8000 samples were used for estimating the posteriors. Different parameters in the prior densities were assumed (see Table 2.1) to examine how robust the posteriors are. Figure 2.1 shows the results of using the clams data. We can see that with the exponential or uniform priors, the parameter in the prior has a profound effect on the posterior of tree length. The Gamma-Dirichlet prior is robust to different values of the parameter in the Gamma distribution (e.g., when the prior mean of T changes by three orders of magnitude). The results highlight the complexity and importance of specifying high-dimensional priors in Bayesian analysis.

2.3.2 Priors on parameters in substitution models

Nucleotide substitutions over time are described by a continuous-time Markov model with instantaneous rate matrix $Q = \{q_{ij}\}$, where q_{ij}, $i \neq j$, is the rate of change from nucleotide i to nucleotide j. Many time-reversible models of nucleotide substitution have been developed, ranging from the simple Jukes-Cantor's model (Jukes and Cantor, 1969b) to the general time-reversible model (GTR) (Tavaré, 1986; Yang, 1994a). The JC69 model contains no free parameters, assuming equal rate between any two nucleotides and equal equilibrium base frequencies. The K80 model (Kimura, 1980) involves the transition/transversion rate ratio (κ), with $0 < \kappa < \infty$. Parameter κ can be assigned a gamma distribution, or $\kappa/(\kappa+1)$ can be assigned a beta distribution. The HKY85 model (Hasegawa et al., 1984, 1985) accommodates the transition/transversion rate ratio (κ) as well as different frequencies for the four nucleotides when the process is in equilibrium. The base frequencies $(\pi_T, \pi_C, \pi_A, \pi_G)$ are typically assigned a Dirichlet prior, $\text{Dir}(\alpha_T, \alpha_C, \alpha_A, \alpha_G)$, or simply Dir(1,1,1,1). The GTR model is parameterized by five substitution rate parameters and four equilibrium base frequency parameters, with the substitution rate matrix

$$Q = \{q_{ij}\} = \begin{pmatrix} - & r_{TC}\pi_C & r_{TA}\pi_A & r_{TG}\pi_G \\ r_{TC}\pi_T & - & r_{CA}\pi_A & r_{CG}\pi_G \\ r_{TA}\pi_T & r_{CA}\pi_C & - & r_{AG}\pi_G \\ r_{TG}\pi_T & r_{CG}\pi_C & r_{AG}\pi_A & - \end{pmatrix}.$$

The diagonal of the matrix is specified such that the sum of each row equals 0. For example, the dash on the first row represents $-(r_{TC}\pi_C + r_{TA}\pi_A + r_{TG}\pi_G)$. The row and column shows entries that are ordered T, C, A, G. Because time and rate are confounded, the Q matrix is usually scaled so that the mean rate equals 1. Thus, the branch length is defined as the expected number of substitutions per site. Consequently, only the relative substitution rates matter. Usually a Dirichlet(1,1,1,1) prior is assigned on the equilibrium nucleotide frequencies (e.g., in Suchard et al. (2001)). Two alternative ways to parameterize the five substitution rates have been suggested (Zwickl and Holder, 2004, e.g., in). One was referred to as 5RR, in which one of the rates (e.g., the rate for GT substitution) is fixed at 1, and the other rates are measured relative to it. An exponential distribution with rate parameter λ is assigned with λ having an exponential hyperparameter:

$$r_{ij} \sim \text{Exp}(\lambda),$$
$$\lambda \sim \text{Exp}(1).$$

A uniform prior $r_{ij} \sim \text{U}(0, 100)$ is found not to work well and lead to upwardly biased posterior estimates (Zwickl and Holder, 2004).

The other approach was referred to as ST1, in which the sum of all six rates is forced to be 1; that is, $\sum_{i,j} r_{ij} = 1, i, j \in \{T, C, A, G\}$ and $i \neq j$., while the relative rates r_{ij} are assigned a Dirichlet prior, $\text{Dir}(\alpha_{TC}, \alpha_{TA}, \alpha_{TG}, \alpha_{CA}, \alpha_{CG}, \alpha_{AG})$. By far the most commonly used version is the uniform Dirichlet, $\text{Dir}(1, 1, 1, 1, 1, 1)$, but other parameters may be used as well to reflect our knowledge of the sequence evolutionary process. For example,

$$\{r_{TC}, r_{TA}, r_{TG}, r_{CA}, r_{CG}, r_{AG}\} \sim \text{Dir}(4, 8, 4, 4, 8, 4)$$

may reflect the transition–transversion rate difference (Zwickl and Holder, 2004).

2.3.3 Priors for heterogeneous substitution rates among sites and over time

The substitution rate depends on the mutation and fixation process, and often varies among sites of the DNA or protein sequence. In protein coding genes, the second codon position typically harbors the smallest number of substations, while the third codon position contains the highest number due to the redundancy of the genetic code. In protein sequences, different sites or domains evolve at different rates due to different pressures from the protein structure and function. Relaxing the assumption of a homogeneous substitution rate over sites in phylogenetic analysis is thus essential. The among-site heterogeneity in overall substitution rate, in stationary frequencies, and in exchange rates between nucleotides have been considered in the literature.

The assumption of homogeneity in substitution rates across sites can be relaxed by assuming that the rates at sites are independent gamma variables (Yang, 1993, 1994b). The branch lengths at site i are multiplied by a random variable $r_i \sim \text{Gamma}(\alpha, \alpha)$. As the rates are relative, the mean is fixed at 1. If the rate multiplier is close to 0, the site is close to being invariable. A large rate multiplier suggests that the site is a substitution hotspot. In Bayesian implementations, the gamma shape parameter α can be assigned a uniform prior $\text{U}(0, A)$ with $A = 200$ in MrBayes, although an exponential prior may be preferable since most estimates of α from real datasets are small (< 1) (Yang, 1996).

Instead of the parametric gamma distribution, several authors suggested using nonparametric Dirichlet process to model variable rates at sites (Lartillot and Philippe, 2004; Huelsenbeck and Suchard, 2007), with both the number of rate classes and the assignments of sites to classes treated as random variables (Ferguson, 1973a; Antoniak, 1974a). The Dirichlet process is characterized by a concentration parameter, denoted by α, and a base distribution, denoted by $G_0(\cdot)$. Parameter α determines the expected number of classes in the prior, with larger α favoring more classes. In Huelsenbeck and Suchard's implementation (2007), only the relative rate (or tree length) varies among sites, while parameters in the Q matrix, such as the equilibrium base frequencies, are shared across all sites. In Lartillot and Philippe's implementation (2004), equilibrium base frequencies are allowed to vary among sites as well. Similar mixture models have been used to account for "pattern-heterogeneity" across sites, using multiple rate matrices (Pagel and Meade, 2004).

One of the surprising findings from the nonparametric Bayesian analysis of real datasets is that the Bayesian Dirichlet prior favored many rate classes (30 or 50) (Lartillot and Philippe, 2004). In contrast, maximum likelihood analysis of similar datasets under finite-mixture models can fit only a few (3 or 4, say) site classes: use of 6 or 7 site classes in the model, say, would simply lead to collapse of the model into one of few classes. The difference is apparently due to the impact of the Dirichlet process prior. In effect, one gets out in the posterior what one puts in in the prior, although one may be under the illusion that the Dirichlet process prior allows one to estimate the number of rate classes from the data. The Dirichlet process favors more clusters (rate classes) in larger datasets and it favors partitions with very large and very small clusters (as opposed to medium-sized clusters) (Green and Richardson, 2001). Those features may be undesirable if our objective is to estimate the number of clusters (the number of rate classes).

For distantly related species, different sequences may have different proportions of nucleotides or amino acids, indicating that the process of nucleotide or amino acid substitution is not stationary. Non-stationary Markov models of sequence evolution incorporating composition heterogeneity among lineages have been proposed by several authors (Yang and Roberts, 1995; Blanquart and Lartillot, 2006, 2008, for example,). Yang and Roberts (1995) and Galtier and Gouy (1998) implemented ML models that allow the base frequencies to drift in different directions on the tree. The general model assumes a set of base frequency parameters for every branch and involves too many parameters. Blanquart and Lartillot (2006, 2008) implemented the model in a Bayesian MCMC program, with a Poisson process used to describe the distribution of change points along the tree while at each change point, a new set of base frequency parameters $\pi = \{\pi_T, \pi_C, \pi_A, \pi_G\}$ is proposed from a uniform Dirichlet distribution. In this model, both the number of change points (compositional shift events) and their positions on the tree are random variables and are estimated from the data.

2.3.4 Partition and mixture models for analysis of large genomic datasets

Modern phylogenetic analysis often involves many genes or proteins. In such analysis, one may use a random-effects model to deal with heterogeneity of the substitution process among sites in the same gene or partition of genes. Examples of such models include the parametric gamma or the nonparametric Dirichlet process models of rates among sites discussed above.

One can then use a fixed-effects model (partition model) to deal with further heterogeneity among partitions. For simplicity, assume one gene is one partition, although we recognize that it may be more appropriate to have many genes in one partition and that it may be more appropriate to partition the sites according to some other features such as the codon position rather than by gene. In the ML framework, such methods were implemented by Yang (1996). Bayesian versions of such models are available in the program MrBayes, which allows the user to "link" or "unlink" parameters across partitions, such as the tree topology, rate, base compositions, etc. Some authors (e.g., Pagel and Meade, 2004) suggested treating all sites as one partition and using a mixture model to deal with the heterogeneity among sites. For the present it is not clear whether partition models or mixture models should be preferred, or how the hundreds or even thousands of genes should be partitioned in phylogenetic analysis. Such decisions may depend on the particular datasets being analyzed (Yang and Rannala, 2012).

2.4 Priors on the tree topology

2.4.1 Prior on rooted trees

If the trees are rooted and the nodes are associated with times (either absolute geological times or relative times), the probabilistic distribution of trees can be generated from a stochastic process of cladogenesis, that is, the process of speciation, extinction, and random sampling of extant species. Such a process often generates a distribution of divergence times (node ages) as well. Yang and Rannala (1997a) used a birth–death sampling process as a prior for phylogenetic trees, specified by a per-lineage birth rate λ, a per-lineage death rate μ, and a sampling fraction ρ. The joint probability of node times $\mathbf{t}$ and the tree topology τ, given two lineages that arose at time t_1 (the time of the root node), is

$$\pi(\tau, \mathbf{t}|t_1, \lambda, \mu, \rho) = \frac{2^{s-1}}{s!(s-1)} \prod_{j=2}^{s-1} \frac{\lambda\, p_1(t_j)}{1 - \frac{1}{\rho}\, \mathrm{P}(0, t_1)\, \mathrm{e}^{(\mu-\lambda)t_1}},$$

where $\mathrm{P}(0, t)$ is the probability that a lineage arising at time t in the past leaves one or more descendants in the present-day sample, and $p_1(t)$ is the probability that a lineage arising at time t in the past leaves one descendant in the sample. And

$$\begin{aligned} \mathrm{P}(0, t) &= \frac{\rho\,(\lambda - \mu)}{\rho\lambda + (\lambda(1-\rho) - \mu)\, \mathrm{e}^{(\mu-\lambda)t}}, \\ p_1(t) &= \frac{1}{\rho}\, \mathrm{P}(0, t)^2 \mathrm{e}^{(\mu-\lambda)t}. \end{aligned}$$

If $\mu = 0$, the process reduces to a pure birth process (a Yule process), and if $\rho = 1$, it corresponds to a birth–death process with complete sampling. The birth–death sampling model has been extended by Stadler (2010) to accommodate sampling of ancestral lineages through time.

In population genetics, the genealogy of a sample of DNA sequences is usually modeled by the coalescent process (Kingman, 1982b,a). The waiting times between nodes on the genealogy are exponentially distributed random variables with the rate parameter determined by the population size and the number of lineages ancestral to the sample. If other population genetic forces are considered, for example mutation, recombination, and migration, the waiting times will also depend on other parameters such as the recombination rate, migration rate, etc. Let N denote the population size, and n denote the number of chromosomes in a sample. The joint distribution of waiting times $\mathbf{t}$ and the tree topology τ is

$$\begin{aligned}\pi(\tau, \mathbf{t}|N) &= \frac{2^{n-1}}{n!(n-1)!}\prod_{j=2}^{n}\binom{j}{2}\mathrm{e}^{-\binom{j}{2}t_j}, \quad \text{or} \\ &= \prod_{j=2}^{n}\mathrm{e}^{-\binom{j}{2}t_j},\end{aligned}$$

where t_j is measured in $2N$ generations (Hudson, 1990), and $\frac{n!(n-1)!}{2^{n-1}}$ is the total number of tree topologies (or more precisely, the total number of labeled histories, that is, rooted tree topologies with interior nodes ordered by age).

The single-population neutral coalescent was extended to the case of multiple species, connected by a species phylogeny (Rannala and Yang, 2003). Basically, coalescent events occur independently in different extant and extinct species, and at different rates because different species have different population sizes. This so-called multi-species coalescent model forms the basis of a class of methods in statistical phylogeography and of the so-called species-tree methods, which attempt to estimate the species tree from multiple genes despite gene tree–species tree conflict (Edwards, 2009).

2.4.2 Priors on unrooted trees and impossibility of equal clade probabilities

Most phylogenetic analyses do not assume the molecular clock and infer unrooted trees. The most commonly used prior on unrooted tree topology is the uniform distribution. The total number of unrooted trees is $\prod_{i=3}^{s}(2i-5)$, where s is the number of species (Felsenstein, 1978b), so the prior probability for each tree topology is $1/\prod_{i=3}^{s}(2i-5)$.

Pickett and Randle (2005) pointed out that when applying equal prior probability of all possible trees, the induced prior probability on clades is not equal. A clade is a monophyletic group including all closely related species, so that species within a clade are more closely related to each other than they are to those outside the clade. The probability of a clade is calculated by summing over the probabilities of all trees containing the clade. Under the prior of equal probability for all possible trees, small and large clades (clades containing few or many species) tend to have higher probabilities than middle-sized clades. The unequal clade probability is due to the fact that combinatorially there are many clades of small or large sizes than of intermediate sizes. While different phylogenetic trees are different models (Yang, 2006), so that posterior tree (model) probabilities are the most appropriate measure of accuracy, posterior tree probabilities are often low, and it is common to calculate posterior clade probabilities as summaries. The concern thus is that if the prior clade probabilities vary greatly depending on the clade size, the posterior clade probabilities may not be comparable. Steel and Pickett

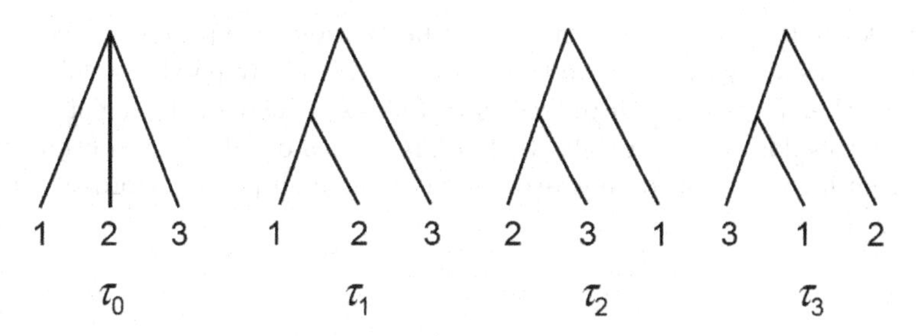

Figure 2.2: The star tree and the three binary tree topologies for three species.

(2006) pointed out that no matter what reasonable priors are assigned to trees, it is impossible to achieve equal clade probabilities. Brandley et al. (2006) argue that if the data are informative, the likelihood will dominate the prior, and the effect of unequal clade probabilities will not be a concern for practical data analysis.

2.4.3 Prior on trees and high posterior probabilities

It has been observed that the Bayesian method tends to produce extremely high posterior probabilities for trees or clades in real data analysis. Sometimes nearly every clade receives a posterior probability of $\sim 100\%$ while the tree is clearly incorrect. For example, the first and second codon positions strongly support one tree, while the third position supports another, but it is biologically unimaginable that the first two codon positions should have a different history from the third positions.

In the so-called star-tree paradox, sequence data are generated using the star tree (τ_0 in Figure 2.2), and analyzed by assigning prior probabilities 1/3 each for the three binary trees (τ_1, τ_2, τ_3). It was found that in large datasets, the posterior probabilities for the three binary trees fluctuate a lot, often strongly rejecting or supporting one of them, while biologists expect them to be close to 1/3 each. The problem is well studied but poorly understood (Suzuki et al., 2002; Lewis et al., 2005; Yang and Rannala, 2005; Yang, 2007; Susko, 2008).

Two priors were suggested to alleviate the problem of very high posterior probabilities, by what Alfaro and Holder (2006) called "conservative Bayesian inference." Lewis et al. (2005) suggested assigning nonzero probabilities to the star tree and other degenerate multifurcating trees. Such a prior clearly leads to reduced posterior probabilities for the binary trees. This has the conceptual difficulty of assigning prior probabilities to models (trees) that are known to be untrue. A second prior was suggested by Yang and Rannala (2005). Only binary trees are considered, but the internal branch lengths on the tree are assigned exponential priors with extremely small means: indeed when the data size $n \to \infty$, the prior mean approaches zero at a rate faster than $1/\sqrt{n}$ but more slowly than $1/n$ (Yang, 2007). For both priors, the posterior tree or clade probabilities are very sensitive to the prior, which may partly explain the fact that neither is used much in real data analysis. Instead, biologists do what they always do: take the results with a large pinch of salt: they continue to use Bayesian inference but do not really believe the high posterior probabilities for trees or clades.

2.5 Priors on times and rates for estimation of divergence times using fossil calibrations

Under the molecular clock hypothesis (Zuckerkandl and Pauling, 1962), the evolutionary rate remains constant over time or across a phylogenetic tree, so that substitutions accumulate over time in a clock-like fashion. Consequently, if we know the number of substitutions that occurred along a branch since the speciation event, with the incorporation of fossil information, we can estimate divergence times for those nodes that do not have a fossil calibration. The molecular clock thus provides a powerful methodology for dating species divergences. There are several complications to this simple strategy.

First, the molecular clock assumption is often unrealistic, and concerns about the impact of the violation of the clock has prompted the development of many relaxed-clock models (Thorne et al., 1998; Drummond et al., 2006; Rannala and Yang, 2007). The rate variation is usually modeled by a stochastic process, for example, a geometric Brownian motion process, running over time or across branches of the tree (Thorne et al., 1998; Kishino et al., 2001; Thorne and Kishino, 2002b; Rannala and Yang, 2007) or a Ornstein-Uhlenbeck process (Aris-Brosou and Yang, 2002). Alternative models include the Poisson process model of rate evolution, which assumes that the rate changes on the branches according to a Poisson process (Huelsenbeck et al., 2000a), and the independent-rates model, which assumes that branch rates are independent random variables from a common distribution (Drummond et al., 2006; Rannala and Yang, 2007). A recent study implemented an interesting Bayesian version of early likelihood-based local-clock models (Hasegawa et al., 1989; Kishino and Hasegawa, 1990; Rambaut and Bromham, 1998; Yoder and Yang, 2000b; Yang and Yoder, 2003a), which partition the branches on the tree into several classes with different rate parameters. The Bayesian versions (Heath et al. 2012; see also Drummond and Suchard, 2010) rely on the nonparametric Dirichlet process to partition the branches, so that the number of branch (rate) classes and the assignment of branches to classes are estimated from the data.

Second, in phylogenetic analysis, the sequence data allow us to estimate only the expected number of substitutions along branches, the product of the substitution rate (r) and divergence time (t), but not r and t separately. The likelihood depends on rt and stays unchanged if we multiply r by a constant c

and divide t by c. This statistical nonidentifiability is "resolved" through the use of the prior on times and the prior on rates. The time and rate priors will thus remain important no matter how much sequence data is available (Yang and Rannala, 2006a). The prior on times also incorporates fossil calibration information, which typically involves a lot of uncertainties (Thorne et al., 1998; Thorne and Kishino, 2002b; Yang and Rannala, 2006a; Ronquist et al., 2012b). Lepage et al. (2007) implemented three different priors on times (node ages) and used the Bayes factor to compare how well the prior model fits the data. They found that the fit of different priors are data dependent. For some datasets, the uniform or Dirichlet prior was favored over the birth–death prior. For other datasets, the birth–death process prior was favored.

2.6 Summary

Bayesian inference requires the specification of prior probability distributions. This may be a benefit when we have prior knowledge and hope to incorporate it in our analysis, but can be a burden otherwise. In Bayesian phylogenetics, the high dimension and complex parameter space of the inference problem adds considerable complexity to prior specification. A typical phylogenetic analysis may include dozens or hundreds of parameters in the DNA or amino acid substitution model, but there are a near-infinite number of tree topologies (which are different statistical models), and each of them involves hundreds or thousands of branch-length parameters. Essentially, all real-world Bayesian phylogenetic analyses have been conducted using "default" priors implemented in the computer programs, and we may expect this trend to persist. While there do not exist truly "noninformative" priors, diffuse or minimally informative priors that do not have undue influence on the posterior will be desirable and merit careful consideration when new models are implemented. This is particularly the case as the phylogenetic model assumed is increasingly complex and parameter-rich (partly to accommodate the ever-increasing sequence datasets), so that strong correlations in the model are inevitable and sensitivity to the prior of some parameters may be hard to avoid. We emphasize the importance of examining the impact of the prior on the posterior inference. It remains to be seen whether use of priors favoring short branch lengths, or polytomies, may prove to be a useful approach to alleviating the problem of spuriously high posterior probabilities for trees and clades.

Acknowledgment

ZY is supported by a BBSRC grant.

3

Inflated density ratio (IDR) method for estimating marginal likelihoods in Bayesian phylogenetics

Serena Arima

Department of Methods and Models for Economy, Territory and Finance, Sapienza Universitá di Roma, Roma, Italy

Luca Tardella

Department of Statistical Sciences, Sapienza Universitá di Roma, Roma, Italy

CONTENTS

3.1 Introduction

It is widely accepted that species diversified in a tree-like pattern from a common descendant and that the diversification is mainly due to changes in the genetic codes of the species accumulating during the centuries. The main aim of phylogenetics is to investigate the evolutionary relationships among species, studying similarities and differences of aligned genomic sequences. From a statistical point of view, the problem of analyzing phylogenetic sequences is often formalized as follows: given a set of DNA sequences of different species, we aim at inferring the tree that best represents the evolutionary relationships. Alternative tree estimation methods such as parsimony methods (Felsenstein (2004), chapter 7) and distance methods (Fitch and Margoliash, 1967; Cavalli-Sforza and Edwards, 1967) have been proposed. We consider stochastic models for substitution rates in a fully Bayesian framework. We focus on model selection issues and several estimation procedures of the Bayesian model evidence will be reviewed. We address model choice within a fully Bayesian framework proposing alternative model evidence estimation procedures.

3.2 Substitution models: a brief overview

Phylogenetic data consist of homologous DNA strands or protein sequences of related species. Observed data consist of a nucleotide matrix X with n rows representing species and k columns representing sites. Comparing DNA sequences of two related species, we define *substitution* to be the replacement, at the same site, of one nucleotide by another one along the path between two species. The stochastic models describing this replacement process are called *substitution models.* A phylogeny or a phylogenetic tree is a representation of the genealogical relationships among species, also called *taxa.* Tips (leaves or external nodes) represent the present-day species, while internal nodes usually represent extinct ancestors for which genomic sequences are no longer available. The ancestor of all sequences is the root of the tree. The branching pattern of a tree is called the *topology,* and is denoted with τ, while the lengths ν_τ of the branches of the tree τ represent the time periods elapsed until a new substitution occurs.

DNA substitution models are probabilistic models which aim at modeling comprehensively changes in nucleotide state at homologous DNA sites. Changes at each site occur at random times. Nucleotides at different sites are usually assumed to evolve independently of each other. For a fixed site, nucleotide replacements over time are modeled by a 4-state Markov process, in which each state represents a nucleotide. The Markov process indexed with time t is completely specified by a substitution rate matrix $Q(t) = r_{ij}(t)$: each element $r_{ij}(t)$, $i \neq j$, represents the instantaneous rate of substitution from nucleotide i to nucleotide j. The diagonal elements of the rate matrix are defined as $r_{ii}(t) = -\sum_{j\neq i} r_{ij}(t)$, so that $\sum_{j=1}^{4} r_{ij}(t) = 0$, $\forall i$. The transition probability matrix is $P(t) = \{p_{ij}(t)\}$, which gives, for each couple of nucleotides (i, j), the probability that state i is replaced by state j at time t. The substitution process is assumed to be homogeneous over time t so that $Q(t) = Q$ and $P(t) = P$. It is also commonly assumed that the substitution process at each site is stationary with equilibrium distribution $\Pi = (\pi_A, \pi_C, \pi_G, \pi_T)$ and time-reversible, that is

$$\pi_i r_{ij} = \pi_j r_{ji}, \tag{3.1}$$

where π_i is the proportion of time the Markov chain spends in state i and $\pi_i r_{ij}$ is the amount of flow from state i to j. Equation (3.1) is known as the *detailed-balance condition* and means that flow between any two states in the opposite time direction is the same. Following the notation in Hudelot et al. (2003), we define $r_{ij}(t) = r_{ij} = \rho_{ij}\pi_j$, $\forall i \neq j$, where ρ_{ij} is the transition rate from nucleotide i to nucleotide j. This reparameterization is particularly useful for the specification of substitution models, since it makes clear the distinction between the nucleotide frequencies $\pi_A, \pi_G, \pi_C, \pi_T$ and substitution rates ρ_{ij}, allowing as to spell out different assumptions on evolutionary patterns. The most general time-reversible nucleotide substitution model is the so-called **GTR** defined by the following rate matrix

$$Q = \begin{pmatrix} - & \rho_{AC}\pi_C & \rho_{AG}\pi_G & \rho_{AT}\pi_T \\ \rho_{AC}\pi_A & - & \rho_{CG}\pi_G & \rho_{CT}\pi_T \\ \rho_{AG}\pi_A & \rho_{CG}\pi_C & - & \rho_{GT}\pi_T \\ \rho_{AT}\pi_A & \rho_{CT}\pi_C & \rho_{GT}\pi_G & - \end{pmatrix} \tag{3.2}$$

and more thoroughly illustrated in Lanave et al. (1984). Several substitution models can be obtained by simplifying the Q matrix reflecting specific biological assumptions: the simplest one is the **JC69** model, originally proposed in Jukes and Cantor (1969a), which assumes that all nucleotides are interchangeable and have the same rate of change, that is, $\rho_{ij} = \rho \;\; \forall i, j$ and $\pi_A = \pi_C = \pi_G = \pi_T$. In this chapter, for illustrative purposes, we will consider only instances of GTR and JC69 models. One can look at Felsenstein (2004) and Yang (2006) for a wider range of alternative substitution models.

3.2.1 Bayesian inference for substitution models

The parameters of a phylogenetic model can be represented as

$$(\tau, \nu_\tau, \theta),$$

where $\tau \in \mathcal{T}$ is the tree topology, $\nu_\tau \in \mathcal{V}_\tau$ is the set of branch lengths of topology τ, and $\theta = (\rho, \pi) \in \Theta$ are the parameters of the rate matrix. We denote $K_\mathcal{T}$ the cardinality of $\mathcal{T}$. Notice that $K_\mathcal{T}$ is a huge number even for few species. For instance, with $n = 10$ species there are about $K_\mathcal{T} \approx 2 \cdot 10^6$ different trees.

Observed data form a nucleotide matrix X; once the substitution model M is specified, the likelihood $p(X|\tau, \nu_\tau, \theta, M)$ can be computed using the pruning algorithm, a practical and efficient recursive algorithm proposed in Felsenstein (1981). One can then make inferences on the unknown model parameters looking for the values which maximize the likelihood. Alternatively, one can adopt a Bayesian approach where the parameter space is endowed with a joint prior distribution $\pi(\tau, \nu_\tau, \theta)$ on the unknown parameters and the likelihood is used to update the prior uncertainty about (τ, ν_τ, θ) following the Bayes rule:

$$p(\theta, \nu_\tau, \tau|X, M) = \frac{p(X|\tau, \nu_\tau, \theta, M)\pi(\tau, \nu_\tau, \theta)}{m(X|M)},$$

where

$$m(X|M) = \sum_{\tau \in \mathcal{T}} \int_{\mathcal{V}_\tau} \int_{\Theta} p(X|\tau, \nu_\tau, \theta, M)\pi(\tau, \nu_\tau, \theta) d\nu_\tau d\theta.$$

The resulting distribution $p(\tau, \nu_\tau, \theta|X, M)$ is the posterior distribution which coherently combines prior beliefs and data information. Prior beliefs are usually conveyed as $\pi(\tau, \nu_\tau, \theta) = \pi_{\mathcal{T}}(\tau)\pi_{\mathcal{V}_\tau}(\nu_\tau)\pi_{\Theta}(\theta)$. The denominator of Bayes' rule $m(X|M)$ is the normalizing constant of the posterior distribution $p(\theta, \nu_\tau, \tau|X, M)$. It is also called the marginal likelihood of model M (or predictive distribution) and it plays a key role in discriminating among alternative models. In fact, the posterior distribution $p(\theta, \nu_\tau, \tau|X, M)$ as such neither allows criticism of a model in light of the observed data, nor permits comparison among models since different models may present different sets of unobservable parameters. On the other hand, the marginal likelihood can be interpreted as the likelihood of the model M, given the data, regardless of the parameters of the model, and is a natural tool for comparing different models. The ratio of the marginal likelihood of competing models is called the *Bayes factor*. The Bayes factor can be interpreted as the strength of the evidence favoring one of the models for the given data and it is the most popular criterion in selecting competing models in a Bayesian framework (Jeffreys, 1961a). However, the computation of the marginal likelihood and, consequently, of the Bayes factor, is still a challenging problem even in simple models since it involves summation and integration over the whole parameter space. Several approximation techniques have been developed (see Robert and Wraith (2009)). In the next section, we briefly introduce the Bayesian model choice issue and describe some computational techniques developed in order to approximate the marginal likelihood and the Bayes factor.

3.3 Bayesian model choice

Computing the normalizing constant of a probability density function $g(\theta)$, known only up to a multiplicative constant, is a crucial point of many statistical problems. However, the analytic computation is not possible in many statistical models since it involves the solution of awkward multiple integrals.

The normalizing constant of the unnormalized non-negative function $g(\theta)$ can be denoted as

$$c = \int_{\Omega} g(\theta) d\theta, \tag{3.3}$$

where θ is a high-dimensional variable defined in the parameter space Ω. The normalized density function will be hereafter named $\tilde{g}(\theta)$. It can be easily understood that such integration is not trivial to carry out, especially when θ is defined in a high-dimensional space. The computation of (3.3) is not just a theoretical statistical/mathematical matter, but it can be an important issue for several fields. Gelman and Meng (1998) give an extensive account of such fields, ranging from computational chemistry to genetics, in which, despite the different terminology, the problem of the computation of c is a crucial point. For phylogenetic models, the computation of the normalizing constant involves not only the integration of continuous model parameters but also the summation over all possible topologies. The specification of the possible topologies is unfeasible even for small n.

Computational methods for estimating normalizing constants have been developed not only in statistical literature, but also in physical, mathematical, and computational chemical literatures. Several alternative estimation procedures have been introduced. As we will see in the next sections, most of these methods rely on powerful Monte Carlo (MC) and Markov chain Monte Carlo (MCMC) methods, which allow us to simulate from complex probability models without the knowledge of the normalizing constant.

We will quickly illustrate basic concepts of Bayesian inference in which the normalizing constant plays a role. Bayesian inference, in general, aims at exploring the posterior probability distribution over the parameters of interest. Given a model M, with parameter vector $\theta \in \Omega$ and observed data X, the Bayes theorem states that the posterior probability distribution is defined as

$$\pi(\theta|X,M) = \frac{\pi(\theta|M)f(X|\theta,M)}{\int_\Omega \pi(\theta|M)f(X|\theta,M)d\theta}, \tag{3.4}$$

where the denominator of (3.4),

$$m(X|M) = \int_\Omega \pi(\theta|M)f(X|\theta,M)d\theta, \tag{3.5}$$

is the normalizing constant of the posterior distribution $\pi(\theta|X,M)$, also called *marginal likelihood* or *predictive distribution*. It follows from (3.5) that the evaluation of the posterior distribution involves also the computation of $m(X|M)$.

When two models M_0 and M_1 have to be compared, the *Bayes factor* in favor of M_1 over M_0 is defined as the ratio of the marginal likelihoods $m(X|M_1)$ and $m(X|M_0)$, that is,

$$BF_{10} = \frac{m(X|M_1)}{m(X|M_0)}, \tag{3.6}$$

which provides the relative weight of evidence of the model M_1 compared to the model M_0. The Bayes factor corresponds to the ratio of the posterior odds for M_1 versus M_0 to the prior odds for M_1 versus M_0. A rough calibration

Table 3.1: Calibration of the Bayes factor BF_{10} as reported in Jeffreys (1961a).

BF_{10}	Evidence for M_1
< 1	negative (supports M_1)
1 to 3	barely worth mentioning
3 to 12	positive
12 to 150	strong
> 150	very strong

of the Bayes factor has been proposed in Jeffreys (1961a) and is reported in Table 3.1.

For many purists, model adequacy requires nothing but the calculation of the marginal likelihood and model selection requires nothing but the Bayes factor (Kass and Raftery (1995)). However, discussion about the computation of the Bayes factor and its interpretability in the presence of improper priors is still open (for a review, see, for example, Gilks et al. (1996a)). The marginal likelihood is central also to the so-called Bayesian model averaging procedure, which is based on the construction of the average predictive distribution: this predictive distribution is built up by averaging the model predictives over all models, that is, by averaging their marginal likelihoods (Hoeting et al., 1999).

Model comparison of phylogenetic models is clearly a difficult task since the Bayes factor of two competing models M_0 and M_1 is

$$BF_{10} = \frac{m(X|M_1)}{m(X|M_0)},$$

where, for $i = 0, 1$

$$\begin{aligned} c^{(i)} &= m(X|M_i) \\ &= \sum_{\tau \in \mathcal{T}} \int_{\nu_\tau^{(i)}} \int_{\theta^{(i)}} p(X|\theta, \tau, M_i)\pi_\Theta(\theta^{(i)})d\theta^{(i)}\pi_\mathcal{V}(\nu_\tau^{(i)}|\tau)d\nu_\tau^{(i)}\pi_\mathcal{T}(\tau). \end{aligned} \quad (3.7)$$

The marginal likelihood $c^{(i)}$ cannot be computed analytically even in very simple cases: it involves integrating the likelihood over k-dimensional subspaces for the branch length parameters ν_τ and the substitution rate matrix $\theta = (\rho, \pi)$ and eventually summing over all possible topologies.

Most of the marginal likelihood estimation methods proposed in the literature have also been applied extensively in molecular phylogenetics (Minin et al., 2003; Lartillot et al., 2007; Suchard et al., 2001). Among these methods, many are valid only under very specific conditions. For instance, the Dickey-Savage ratio (Verdinelli and Wasserman, 1995) applied in phylogenetics in Suchard et al. (2001), assumes nested models. Laplace approximation (Kass and Raftery, 1995) and BIC (Schwarz, 1978), first applied in phylogenetics in Minin et al. (2003), require large sample approximations around the maximum likelihood, which can sometimes be difficult to compute or inaccurate for very complex models. A recent appealing variation of the Laplace approximation has been proposed in Rodrigue et al. (2007): however, its applicability and performance are endangered when the posterior distribution deviates from normality and maximization of the likelihood is neither straightforward nor accurate.

The reversible jump approach (Green, 1995; Bartolucci et al., 2006) is another MCMC option applied to phylogenetic model selection in Huelsenbeck et al. (2004). Unfortunately, the implementation of this algorithm is not straightforward for the end user and it often requires appropriate delicate tuning of the Metropolis Hastings proposal. Moreover, its implementation suffers extra difficulties when comparing models based on an entirely different parametric rationale (Lartillot et al., 2007). An efficient tool for computing the marginal likelihood has been proposed in Ardia et al. (2012): the method is mainly based on the adaptive mixture of Student-t (AdMit) approach proposed in Hoogerheide et al. (2007). The method consists of two steps. First, it constructs a mixture of Student-t distributions which approximates a target distribution of interest. The fitting procedure relies only on a kernel of the target density, so that the normalizing constant is not required. In a second step, this approximation is used as an importance function in an importance sampling method to estimate characteristics of the target density. The estimation procedure is fully automatic and avoids difficult tasks. Ardia et al. (2012) show that AdMit performs accurately even with very skewed and multimodal distributions. However, when the target distribution lies in a high-dimensional space, AdMit shows some numerical instability and it is computationally demanding. This fact is, probably, one of the main motivations preventing the use of AdMit for phylogenetic models. However, we believe that it is a promising tool for computing the marginal likelihood and its performance in a phylogenetic context will be investigated in future research.

As recently pointed out in Ronquist and Deans (2010), among the most widely used methods for estimating the marginal likelihood of phylogenetic models are thermodynamic integration, also known as path sampling, and the harmonic mean approach. Thermodynamic integration, reviewed in Gelman and Meng (1998) and first applied in a phylogenetic context in Lartillot and Philippe (2006a), produces reliable estimates of Bayes factors of phylogenetic models in a large variety of models. Although this method has the advantage of general applicability, it can incur high computational costs and may require specific adjustments. For certain model comparisons, a full thermodynamic integration may take weeks on a modern desktop computer, even under a fixed tree topology for small single protein datasets (Rodrigue et al., 2007; Ronquist and Deans, 2010). Alternative approaches, based on a similar rationale and resembling the power prior approach of Friel and Pettitt (2008) are the stepping-stone and the generalized stepping-stone (GSS) methods, recently proposed in two consecutive papers by Xie et al. (2011) and Fan et al. (2011): similar to thermodynamic integration, these methods are computationally intensive since they need to sample from a series of distributions in addition to the posterior.

Such dedicated path-based Markov chain Monte Carlo (MCMC) analyses appear to be the price for estimating marginal likelihoods more accurately than more direct standard methods. In Fan et al. (2011), it is shown that GSS produces very stable estimates of the marginal likelihood and in Xie et al. (2011) it appears to be slightly more precise than the plain thermodynamic integration based on power priors. The thermodynamic integration method and GSS method will be described in more detail in Section 3.4.3.

The harmonic mean (HM) estimator can be easily computed and it does not demand further computational efforts other than those already made to draw inference on model parameters, since it only needs simulations from the posterior distributions. However, it is well known that the HM estimator is unstable and biased (Raftery et al., 2007a; Lenk, 2009), since it can end up with an infinite variance.

Although several solutions have been proposed, the computation of the marginal likelihood is still an open problem: none of the proposed methods can be considered intrinsically fail-safe nor will the proposed methods always outperform the others and the comparison among different methods can be useful in order to accurately compare competing models and more thoroughly validate the final estimate. In the next section we focus on an alternative generalized harmonic mean estimator, the IDR estimator, which shares the computational simplicity of HM estimator but, unlike it, better copes with the infinite variance issue. Its simple implementation makes the IDR estimator a useful tool to easily provide a reliable method for comparing competing substitution models. We will argue how IDR can be useful as a confirmatory tool even in those models for which more complex estimation methods, such as the thermodynamic integration and GSS, can be applied.

3.4 Computational tools for Bayesian model evidence

3.4.1 Harmonic mean estimators

We introduce the basic ideas and formulas for the class of estimators known as generalized harmonic mean (GHM). Since the marginal likelihood is simply the normalizing constant of the unnormalized posterior density resulting from the joint density of parameters and data, we illustrate the GHM estimator as a general solution for estimating the normalizing constant of a non-negative, integrable density g defined as

$$c = \int_{\Omega} g(\theta) d\theta, \tag{3.8}$$

where $\theta \in \Omega \subset \Re^k$ and $g(\theta)$ is the unnormalized version of the probability distribution $\tilde{g}(\theta)$. The GHM estimator of c is based on the following identity

$$c = \frac{1}{E_{\tilde{g}}\left[\left(\frac{g(\theta)}{f(\theta)}\right)^{-1}\right]}, \tag{3.9}$$

where f is a convenient instrumental Lebesgue integrable function, which is only required to have a support which is contained in that of g and to satisfy

$$\int_{\Omega} f(\theta) d\theta = 1. \tag{3.10}$$

The GHM estimator, denoted as $\hat{c}_{GHM}$ is the empirical counterpart of (3.9), namely

$$\hat{c}_{GHM} = \frac{1}{\frac{1}{T}\sum_{t=1}^{T} \frac{f(\theta_t)}{g(\theta_t)}}, \tag{3.11}$$

where $\theta_1, \theta_2, ..., \theta_T$ are sampled from $\tilde{g}$. In Bayesian inference, the very first instance of such a GHM estimator was introduced in Gelfand and Dey (1994) to estimate the marginal likelihood considered as the normalizing constant of the unnormalized posterior density $g(\theta) = \pi_\Theta(\theta)L(\theta)$. Hence, taking $f(\theta) = \pi_\Theta(\theta)$ one obtains as a special case of (3.11), the harmonic mean estimator

$$\hat{c}_{HM} = \frac{1}{\frac{1}{T}\sum_{t=1}^{T} \frac{1}{L(\theta_t)}}, \tag{3.12}$$

which can be easily computed by recycling simulations $\theta_1, ..., \theta_T$ from the target posterior distribution $\tilde{g}(\theta)$ available from the MC or MCMC sampling scheme. This probably explains the original enthusiasm in favor of $\hat{c}_{HM}$ which indeed was considered a potential convenient competitor of the standard Monte Carlo importance sampling estimate given by the (prior) arithmetic mean (AM) estimator. The implementation of (3.12), and, more generally, (3.11), requires a lighter computational burden, hence it can reduce computing time with respect to alternative more complex estimators such as those based on thermodynamic integration. The simplicity of the computation has then favored the widespread use of the harmonic mean estimator with respect to more complex methods. In fact, the harmonic mean estimator is implemented in several Bayesian phylogenetic software programs and recent biological papers (Yamanoue et al., 2008; Wang et al., 2009) report the HM as a routinely used model selection tool.

However, $\hat{c}_{HM}$ can end up with a very large variance and unstable behavior. This fact cannot be considered as an unusual exception, but it often occurs, and the reason for that can be argued on a theoretical ground. Indeed, starting from the original paper (Newton and Raftery, 1994a) (see in particular R. Neal's discussion), it has been shown that even in very simple and standard Gaussian models, the HM estimator can end up having an infinite variance, hence yielding unreliable approximations. This fact sometimes raises the question whether it is an effective tool and certainly has encouraged researchers to look for alternative solutions. Several generalizations and improved alternatives have been proposed and recently reviewed in Raftery et al. (2007a).

In the following subsections we will consider the performance of alternative implementations of GHM method. The first one is directly based on (3.11), where g is taken as the posterior density up to the unknown proportionality constant, while f is taken as an auxiliary probability density which is possibly a convenient fully available probability density approximating the posterior. Such auxiliary density is not easy to build up appropriately in general, but in principle, setting it as close as possible to the posterior could yield a very efficient estimator. The second one is derived from a different perspective and does not require any additional calibration of an auxiliary density, although this prevents it from achieving in principle arbitrary precision. It has been originally proposed in Petris and Tardella (2007) and it is called the inflated density ratio (IDR) estimator, although it can also be seen as a particular instance of the generalized harmonic mean (GHM) approach.

These two estimators basically share the original simplicity and the computational feasibility of the HM estimator but, unlike it, they can guarantee greater stability and important theoretical properties, such as a bounded variance.

3.4.2 IDR: inflated density ratio estimator

The inflated density ratio approach proposed in Petris and Tardella (2007) is an alternative tool for evaluating the normalizing constant of a density function. The new IDR estimator is obtained via a basic identity involving the ratio of two functions: the unnormalized target density and a perturbed version of it. The estimator, which can be seen as an alternative formulation of the generalized harmonic mean estimator, can be easily computed by recycling MC or MCMC simulations from the target distribution. It is a consistent estimator and, under fairly general conditions, the variance is proved to be finite.

Under some general conditions, such as the integrability of $g(\cdot)$ and the absolute continuity of $g(\cdot)$ with respect to $f(\cdot)$, (3.11) defines a biased, consistent, and easy-to-compute estimator of the normalizing constant c. However, as shown in Newton and Raftery (1994a), $\hat{c}_{GHM}$ can end up with a very large variance and sometimes infinite even for simple models when $f(\theta)$ is not carefully chosen, as in the case of the original HM estimator where $f(\theta)$ corresponds to the prior.

Petris and Tardella (2007) propose an alternative formulation of the generalized harmonic mean estimator, based on a particular choice of $f(\theta)$. The original idea is to consider a different function obtained as a perturbation of the original target density function $g(\theta)$: this perturbed version of g, which we will denote as g_{P_k}, is defined such that its total mass has some known functional relation to the total mass c of the target density g. In particular, as described in Petris and Tardella (2003), they propose to inflate g parametrically so that

$$\int_{\Omega} g_{P_k}(\theta)d\theta = c + k, \tag{3.13}$$

where k is a known inflation constant which can be arbitrarily fixed; however, as we will see in the following section, the choice of k will be crucial since it is strongly related to the error of the estimate.

The perturbation g_{P_k} of the density g can be obtained using the idea of the *hyperplane inflation* method (Petris and Tardella, 2003): in one-dimensional space, when $g(\theta)$ is fully supported on the real line, that is, $g(\theta) > 0$ for any $\theta \in \Omega = \mathbb{R}$, the perturbed density g_{P_k} can be constructed as follows:

$$g_{P_k}(\theta) = \begin{cases} g(\theta + r_k) & \text{if } \theta < -r_k, \\ g(0) & \text{if } -r_k \leq \theta \leq r_k, \\ g(\theta - r_k) & \text{if } \theta > r_k, \end{cases} \tag{3.14}$$

where

$$2r_k = \frac{k}{g(0)}$$

is the length of the interval centered around the origin. The transformation in (3.14) consists of a perturbation of the original function g through a horizontal shift of the original g around the origin. In practice, the original density is shifted by opening a rectangle of side $2r_k$ and area equal to k around the origin: the area of the perturbed density g_{P_k} is then equal to $c + k$. Figure 3.1 can help in understanding how the perturbation acts, showing the original density and its perturbed version in the one-dimensional case.

In the multidimensional case, that is when $\Omega \in \mathbb{R}^n$, the perturbation is induced by opening an n dimensional closed ball of radius r_k around the origin: more formally, the perturbed density g_{P_k} is defined as

$$g_{P_k}(\theta) = \begin{cases} g(0) & \text{if } \theta \in B_{r_k}, \\ g(\phi_n(\theta, r_k)) & \text{otherwise,} \end{cases} \tag{3.15}$$

where

$$\phi_n(\theta; r_k) = \frac{\theta}{|\theta|}(|\theta|^n - r_k^n)^{1/n}$$

and

$$B_{r_k} = \{\theta \in \mathbb{R}^n : |\theta| \leq r_k\}$$

is the n-dimensional closed ball of radius r_k centered at the origin. Since Petris and Tardella (2007) prove that the transformation $\phi_n(\theta, r_k)$ preserves the Lebesgue measure λ, the perturbed density in (3.15) has total mass $c + k$, and $k = g(0)\lambda(B^n_{r_k})$.

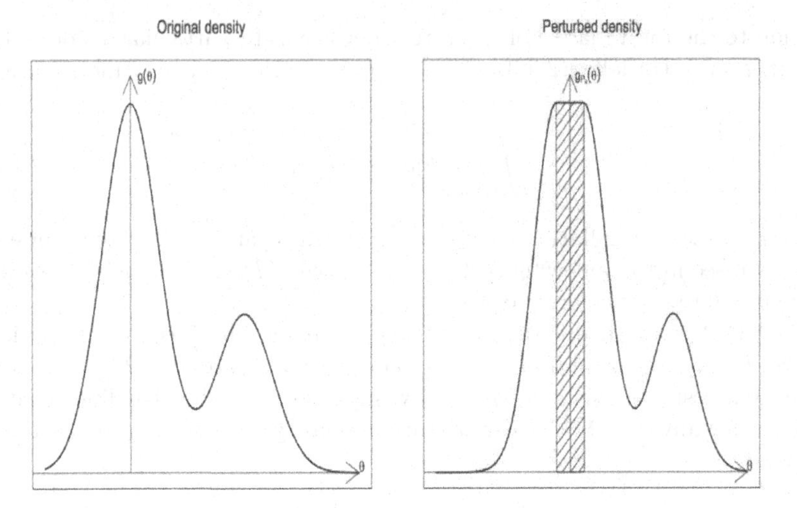

Figure 3.1: Plots of original density $g(\cdot)$ (left) and radially expanded perturbed density $g_{P_k}(\cdot)$ (right). The area of the shaded region is equal to the perturbation parameter k corresponding to the inflated mass.

Using this inflated density idea, in the same fashion as the generalized harmonic mean estimator, we can express the unknown normalizing constant c as a function of the original and of the perturbed density as follows:

$$c + k = \int g_{P_k}(\theta) d\theta = c \int \frac{g_{P_k}(\theta)}{g(\theta)} \frac{g(\theta)}{c} d\theta = c E_{\tilde{g}} \left[\frac{g_{P_k}(\theta)}{g(\theta)} \right], \tag{3.16}$$

where $\tilde{g}(\theta) = \frac{g(\theta)}{c}$ is the normalized target density. From (3.16), it follows that the normalizing constant c can be written as

$$c = \frac{k}{E_{\tilde{g}} \left[\frac{g_{P_k}(\theta)}{g(\theta)} \right] - 1}. \tag{3.17}$$

This alternative formulation of the normalizing constant involves the computation of the expected value with respect to the target distribution, which is, clearly, unknown. However, since we can obtain a sample from $\tilde{g}$, the expression in (3.17) can be estimated by the inflated density ratio estimator $\hat{c}_{IDR}$ defined as

$$\hat{c}_{IDR} = \frac{k}{\frac{1}{T} \sum_{t=1}^{T} \frac{g_{P_k}(\theta_t)}{g(\theta_t)} - 1}, \tag{3.18}$$

where $\theta_1, ..., \theta_T$ is a sample from the normalized target density $\tilde{g}$.

In order to better understand the connection of this approach with the generalized harmonic mean and (3.11), we can rewrite the inflated density ratio estimator as

$$\hat{c}_{IDR} = \frac{1}{\frac{1}{T}\sum_{t=1}^{T}\left[\left(\frac{kg(\theta_t)}{g_{P_k}(\theta_t)-g(\theta_t)}\right)^{-1}\right]},$$

which can be easily seen as a special case of the generalized harmonic mean estimator $\hat{c}_{GHM}$ in (3.11) substituting $f_k(\theta_t) = \frac{g_{P_k}(\theta_t)-g(\theta_t)}{k}$. However, although the method is similar and closely related to the generalized harmonic mean approach, the function $f_k(\theta) = \frac{g_{P_k}(\theta)-g(\theta)}{k}$ is not necessarily positive. Petris and Tardella (2007) prove that the use of the perturbed density as an importance function has some advantages: unlike $\hat{c}_{GHM}$, the new estimator $\hat{c}_{IDR}$ has finite variance under general conditions, such that $g(\theta)$ is a continuous positive log-Lipschitz density. They provide a sufficient condition for the finiteness of the variance, as well as for densities for which the log-Lipschitz condition is not satisfied (Petris and Tardella (2007)). Moreover, the use of a parametric perturbation makes the method more flexible and applicable to different parametric families.

As for all methods based on an importance sampling strategy, the properties of the estimator $\hat{c}_{IDR}$ strongly depend on the ratio $\frac{g_{P_k}(\theta)}{g(\theta)}$: in particular, the asymptotic Relative Mean Square Error of the estimator $\text{RMSE}_{\hat{c}_{IDR}}$ is defined as

$$\text{RMSE}_{\hat{c}_{IDR}} = \sqrt{E_{\tilde{g}}\left[\left(\frac{\hat{c}_{IDR}-c}{c}\right)^2\right]} \tag{3.19}$$

and it can be approximated as

$$\text{RMSE}_{\hat{c}_{IDR}} \approx \frac{1}{\sqrt{n}}\frac{c}{k}\sqrt{\text{Var}_{\tilde{g}}\left[\frac{g_{P_k}(\theta)}{g(\theta)}\right]}. \tag{3.20}$$

A sufficient and necessary condition for the finiteness of the $\text{RMSE}_{c_{I\hat{D}R}}$ is the finiteness of the second moment of the ratio $\frac{g_{P_k}(\theta)}{g(\theta)}$, that is:

$$\text{RMSE}_{\hat{c}_{IDR}} \approx \frac{1}{\sqrt{n}}\frac{c}{k}\sqrt{\text{Var}_{\tilde{g}}\left[\frac{g_{P_k}(\theta)}{g(\theta)}\right]} < \infty \Longleftrightarrow E_{\tilde{g}}\left[\left(\frac{g_{P_k}(\theta)}{g(\theta)}\right)^2\right] < \infty.$$

The relative mean square error in (3.20) can be estimated as follows:

$$\widehat{\text{RMSE}}_{\hat{c}_{IDR}} \approx \frac{1}{\sqrt{n}}\frac{\hat{c}_{IDR}}{k}\sqrt{\widehat{\text{Var}}_{\tilde{g}}\left[\frac{g_{P_k}(\theta)}{g(\theta)}\right]}, \tag{3.21}$$

where $\widehat{\text{Var}}_{\tilde{g}}$ is the sample variance of the ratio.

The expression in (3.20) clarifies the key trade-off of the choice of k with respect to the error of the estimator: in fact, as $k \to 0$, the inflated density tends to approximate the original density such that $\frac{g_{P_k}(\theta)}{g(\theta)} \approx 1$. It means that the smaller k is, the smaller the variance of the empirical mean of the ratio $\frac{g_{P_k}(\theta)}{g(\theta)}$ will be. However, the same empirical mean is an unbiased estimator of $\frac{c}{c+k}$, which approaches 1 as $k \to 0$. Therefore, since $E\left[\hat{c}_{IDR}\right] = \frac{k}{\frac{c+k}{c}-1}$, when $k \to 0$, $\hat{c}_{IDR}$ converges to the undetermined form $\frac{0}{0}$. Therefore, very little variation from 1 in the denominator of $\hat{c}_{IDR}$ can have a large impact on the variability of the whole ratio.

In order to resolve the choice of k, Petris and Tardella (2007) suggested to look for the perturbation k that minimizes the estimation error: in practice, they suggest calculating the values of the Inflation Density Ratio estimator for a grid of different perturbation values, $\hat{c}_{IDR}(k)$, $k = 1, ..., K$, and of choosing the optimal k^* as the k for which $\widehat{\text{RMSE}}_{\hat{c}_{IDR}(k)}$ is minimum. Moreover, this procedure for the calibration of k requires an extra burden of computations alleviated by the fact that one can use the same sample from $\tilde{g}$: the only quantity one has to evaluate K times is the inflated density. Once the ratio of the perturbed and the original density is obtained, the computation of the empirical harmonic mean and of the $\hat{c}_{IDR}(k)$ is straightforward. For practical purposes, the computation of the inflated density can be easily implemented using the R package HI (Petris and Tardella (2003), R Development Core Team (2008)).

3.4.2.1 IDR: numerical examples

In order to better understand how the choice of k is carried out in practical situations, consider the following toy example: let $X \sim N(0,1)$ and simulate a sample $\theta_1, ..., \theta_{10000}$ from it. The true value of the normalizing constant is, clearly, $c = 1$. Table 3.2 shows the different values of the inflated density ratio estimator $\hat{c}_{IDR}$ for 10 different choices of the perturbation k. Table 3.2 also contains the asymptotic confidence interval (CI) for $\log(\hat{c}_{IDR})$ defined as:

$$[\log(\hat{c}_{IDR}) - \log(1 + 3\text{RMSE}_{\hat{c}_{IDR}}), \log(\hat{c}_{IDR}) + \log(1 + 3\text{RMSE}_{\hat{c}_{IDR}})]$$

and the asymptotic standard deviation ($Asy.sd$) defined as

$$Asy.sd(\hat{c}_{IDR}) = \hat{c}_{IDR} \frac{\sqrt{\text{Var}\left[\frac{g_{P_k}}{g}\right]}}{k}.$$

Since $\text{RMSE}_{\hat{c}_{IDR}} = \frac{Asy.sd}{\sqrt{n}}$, we choose k that minimizes $\text{RMSE}_{\hat{c}_{IDR}}$ or, equivalently, minimizes $Asy.sd$.

Plots in Figure 3.2 show the $\text{RMSE}_{\hat{c}_{IDR}}$ and $\hat{c}_{IDR}$ for different perturbation values, k: as we can see from Table 3.2 and Figure 3.2, the minimum error is reached for $k^\star = 1$, with some decreasing pattern. In order to verify if $k^* = 1$

Table 3.2: An application of the inflated density ratio method: $\tilde{g}$ is a standard normal distribution and $c = 1$. The table shows the $\hat{c}_{IDR}$ values for different choice of k, $k = 1, ..., 10$: $\log(\hat{c}_{IDR})$ is the logarithmic transformation of the estimator, CI the asymptotic confidence interval, *Asy.sd* the asymptotic standard deviation, and n the sample size.

k	$\hat{c}_{IDR}$	$\log(\hat{c}_{IDR})$	RMSE	CI	*Asy.sd*	n
1	1.0035	0.0035	0.0121	$[-0.0322, 0.0405]$	1.2112	10^4
2	1.0074	0.0074	0.0190	$[-0.0480, 0.0660]$	1.8968	10^4
3	1.0081	0.0081	0.0311	$[-0.0812, 0.1061]$	3.1129	10^4
4	1.0100	0.0099	0.0536	$[-0.1392, 0.1853]$	5.3606	10^4
5	1.0188	0.0186	0.0901	$[-0.2206, 0.3338]$	9.0097	10^4
6	1.0467	0.0456	0.1371	$[-0.2988, 0.5752]$	13.7051	10^4
7	1.1086	0.1031	0.1808	$[-0.3303, 0.8851]$	18.0836	10^4
8	1.2202	0.1990	0.2036	$[-0.2778, 1.1430]$	20.3644	10^4
9	1.3707	0.3153	0.2047	$[-0.1634, 1.2672]$	20.4668	10^4
10	1.5230	0.4207	0.2047	$[-0.0580, 1.3726]$	20.4668	10^4

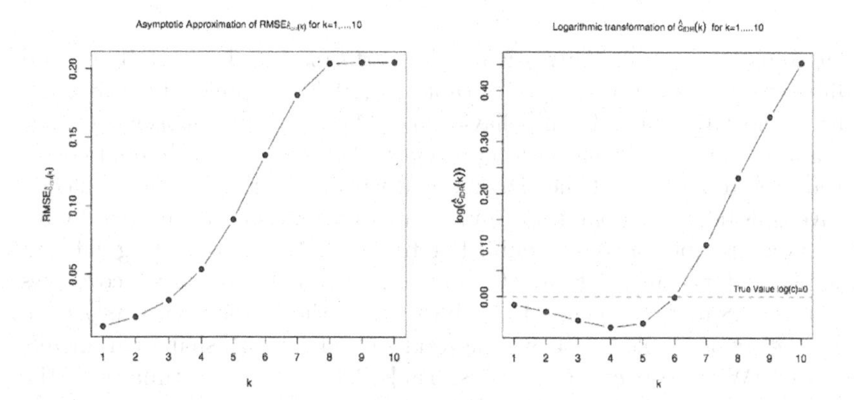

Figure 3.2: Left: $\text{RMSE}_{\hat{c}_{IDR}}$ for perturbation values $k = 1, ..., 10$. Right: $\log(\hat{c}_{IDR})$ for $k = 1, ..., 10$. As it can be noted that the minimum value of the relative mean square error corresponds to $k = 1$. It is then necessary to consider smaller values of k.

is a local minimum for the $\text{RMSE}_{\hat{c}_{IDR}}$, smaller k values should be investigated: we then estimate $\hat{c}_{IDR}$ for perturbation values smaller than 1 and compare the $\text{RMSE}_{\hat{c}_{IDR}}$ with those reported in Table 3.2.

As shown in Table 3.3, the minimum Relative Mean Square Error value is reached for $k = 10^{-9}$. The asymptotic confidence interval of the logarithmic transformation of c, $\log(\hat{c}_{IDR})$, contains the logarithmic transformation of the true value $\log(c) = 0$. Therefore, the method estimates the normalizing constant in a reliable way.

Table 3.3: Normal example where $\tilde{g} \sim N(0,1)$ and $c = 1$. The sample size is 10^4. The table reports $\log(\hat{c}_{IDR})$ values for different k, $k = 10^{-10}, ..., 0$: CI the asymptotic confidence interval, *Asy.sd* the asymptotic standard deviation, and n the sample size. The minimum value is reached for $k = 10^{-9}$.

$-\log_{10}(k)$	$\log(\hat{c}_{IDR})$	RMSE	CI	*Asy.sd*
10	$3.521\ 10^{-3}$	$7.4974\ 10^{-2}$	$[-1.893\ 10^{-2}, 2.648\ 10^{-2}]$	0.757
9	$3.518\ 10^{-3}$	$7.4973\ 10^{-2}$	$[-1.893\ 10^{-2}, 2.648\ 10^{-2}]$	0.756
8	$3.519\ 10^{-3}$	$7.4974\ 10^{-2}$	$[-1.893\ 10^{-2}, 2.648\ 10^{-2}]$	0.757
7	$3.519\ 10^{-3}$	$7.4974\ 10^{-2}$	$[-1.893\ 10^{-2}, 2.648\ 10^{-2}]$	0.757
6	$3.519\ 10^{-3}$	$7.4974\ 10^{-2}$	$[-1.893\ 10^{-2}, 2.648\ 10^{-2}]$	0.757
5	$3.519\ 10^{-3}$	$7.4975\ 10^{-2}$	$[-1.893\ 10^{-2}, 2.648\ 10^{-2}]$	0.757
4	$3.519\ 10^{-3}$	$7.4978\ 10^{-2}$	$[-1.893\ 10^{-2}, 2.648\ 10^{-2}]$	0.757
3	$3.521\ 10^{-3}$	$7.5015\ 10^{-2}$	$[-1.894\ 10^{-2}, 2.650\ 10^{-2}]$	0.757
2	$3.537\ 10^{-3}$	$7.5386\ 10^{-2}$	$[-1.9032\ 10^{-2}, 2.6626\ 10^{-2}]$	0.761
1	$3.707\ 10^{-3}$	$7.9121\ 10^{-2}$	$[-1.9955\ 10^{-2}, 2.7943\ 10^{-2}]$	0.798
0	$5.582\ 10^{-3}$	$1.2112\ 10^{-1}$	$[-3.0038\ 10^{-2}, 4.2519\ 10^{-2}]$	1.2087

Of course, one has to investigate the performance of the IDR method in different controlled settings. In Table 3.4, IDR is applied to univariate distributions with different tail behavior and different degree of skewness. The true value of the normalizing constant is reported in the second column. In order to study the robustness of the IDR method with respect to the tail behavior, we have considered two particular univariate distributions: a centered double exponential distribution and a double Pareto distribution. Considering different shape parameters, ranging from 0.01 to 100, IDR confidence interval coverages range from 0.8 to 1 confirming the robustness of the method with respect to the tail behavior. Table 3.4 shows the results obtained with shape parameters equal to 1. With respect to skewness, Table 3.4 shows the estimates of the normalizing constant of three skewed normal distributions with increasing

Table 3.4: IDR method applied to univariate distributions with different tail behavior and different degrees of skewness.

	c	$\hat{c}_{IDR}$	$\widehat{\text{RMSE}}_{\hat{c}}$	$\widehat{\text{CI}}$
Cauchy	1	1.000	0.001	$[0.997, 1.003]$
Skew Normal (skew=10)	1	1.000	0.006	$[0.982, 1.018]$
Skew Normal (skew=50)	1	0.979	0.015	$[0.934, 1.024]$
Skew Normal (skew=100)	2	2.272	0.105	$[1.998, 2.546]$
Double Exp	2	2.000	$2.477 \cdot 10^{-4}$	$[1.999, 2.001]$
Double Pareto	2	2.005	0.0028	$[1.996, 2.014]$

Table 3.5: IDR for multivariate distributions. The dimension of the distribution is defined in the second column of the table.

	Dim	c	$\hat{c}_{IDR}$	$\widehat{\text{RMSE}}_{\hat{c}}$	$\widehat{\text{CI}}$
Normal	2	1	1.000	10^{-5}	$[0.999, 1.000]$
Normal	10	1	1.003	$4.658 \cdot 10^{-3}$	$[0.989, 1.016]$
Normal	100	1	1.009	0.01	$[0.979, 1.039]$
Skew Normal (skew=10)	10	1	0.970	0.062	$[0.783, 1.156]$
Skew Normal (skew=10)	30	10	11.060	0.0470	$[10.919, 11.201]$
Student T (df=10)	50	1	0.986	0.010	$[0.956, 1.016]$
Log Gamma	10	1	1.045	0.025	$[0.970, 1.119]$

degree of skewness. In all cases, $\hat{c}_{IDR}$ is very close to the true value and the asymptotic confidence interval always contains the true value c.

The estimation of the normalizing constant in multivariate settings is a more complex issue from a computational point of view. Table 3.5 shows the estimates obtained with the IDR method for different multivariate distributions. Also in these cases, the estimation of normalizing constants is accurate even for distributions living in very high dimensional spaces (e.g., 100-dimensional). The robustness with respect to some degree of skewness and tail behavior is confirmed also for multivariate distributions.

We can conclude that IDR estimates are quite robust with respect to the tail behavior of the target distribution. It also performs well with some degree of skewness. IDR seems to be robust with respect to tail behavior. The performance of IDR is still satisfactory in some multivariate cases.

Other examples of the inflated density ratio method for simulated and real data can be found in the original paper (Petris and Tardella, 2007). Using real data, they also compare the method with the generalized harmonic mean approach (Raftery et al. (2007a)) showing that two estimators yield to comparable results. In Arima (2009) simple antithetic variate tricks allow the IDR estimator to perform well even for those distributions with severe variations from the symmetric Gaussian case such as asymmetric and even some multimodal distributions. In the next section, we extend the IDR method in order to use $\hat{c}_{IDR}$ in more complex settings such as phylogenetic models.

3.4.2.2 IDR for substitution models

We show how one can implement the inflated density ratio approach in order to compute the marginal likelihood of phylogenetic models. In this section, we show how to compute the marginal likelihood when it involves integration of substitution model parameters θ and the branch lengths ν_τ, which are both defined in continuous subspaces. Indeed, the approach can be used as a building block to integrate also over the tree topology τ.

For a fixed topology τ and a sequence alignment X, the parameters of a phylogenetic model M_τ are denoted as $\omega = (\theta, \nu_\tau) \in \Omega_\tau = \Theta \times N_\tau$. The joint posterior distribution on ω is given by

$$p(\theta, \nu_\tau | X, M_\tau) = \frac{p(X|\theta, \nu_\tau, M_\tau)\pi_\Theta(\theta)\pi_\mathcal{V}(\nu_\tau)}{m(X|M_\tau)}, \tag{3.22}$$

where

$$m(X|M_\tau) = \int_\Theta \int_{N_\tau} p(X|\theta, \nu_\tau, M_\tau)\pi_\Theta(\theta)\pi_\mathcal{V}(\nu_\tau) d\theta d\nu_\tau \tag{3.23}$$

is the marginal likelihood we aim at estimating.

When the topology τ is fixed, the parameter space Ω_τ is continuous. Hence, in order to apply the IDR method, we only need the following three ingredients:

(i) a sample $(\theta^{(1)}, \nu_\tau^{(1)}), ..., (\theta^{(T)}, \nu_\tau^{(T)})$ from the posterior distribution, $p(\theta, \nu_\tau | X, M_\tau)$;

(ii) the likelihood and the prior distribution evaluated at each posterior sampled value $(\theta^{(k)}, \nu_\tau^{(k)})$, that is, $p(X|\theta^{(k)}, \nu_\tau^{(k)}, M_\tau)$ and $\pi(\theta^{(k)}, \nu_\tau^{(k)}) = \pi_\Theta(\theta^{(k)})\pi_\mathcal{V}(\nu_\tau^{(k)})$; and

(iii) the likelihood and prior evaluated at the perturbed values $\phi_n(\theta^{(k)}; r_k)$.

The first ingredient is just the usual output of the Markov chain Monte Carlo simulations derived from model M and data X. The computation of the likelihood and the joint prior is usually already coded within available software. The first one is accomplished through the pruning algorithm, while computing the prior is straightforward. Indeed, a necessary condition for the inflation idea to be implemented as prescribed in Petris and Tardella (2007) is that the posterior density must have full support on the whole real k-dimensional space. In our phylogenetic models, this is not always the case; hence we explain simple and fully automatic remedies to overcome this kind of obstacle.

We start with branch length parameters which are constrained to lie in the positive half-line. In that case, the remedy is straightforward. One can reparameterize with a simple logarithmic transformation

$$\nu_\tau^{'} = \log(\nu_\tau) \tag{3.24}$$

so that the support corresponding to the reparameterized density becomes unconstrained. Obviously, the $\log(\nu_\tau)$ reparameterization calls for the appropriate Jacobian when evaluating the corresponding transformed density. For model parameters with linear constraints like the substitution $\theta = \{\rho, \pi\}$, a little less obvious transformation is needed. In this case, $\theta = \{\rho, \pi\}$ are subject to the following set of constraints:

$$\sum_{i \in \{A,T,C,G\}} \pi_i = 1,$$

$$\sum_{j \in \{A,T,C,G\}} \rho_{ij}\pi_j = 0 \qquad \forall \; i \in \{A,T,C,G\}.$$

Similar to the first simplex constraint, the last set of constraints together with the reversibility can be rephrased (Gowri-Shankar, 2006) in terms of another simplex constraint concerning only the extra-diagonal entries of the substitution rate matrix (3.2), namely

$$\rho_{AC} + \rho_{AG} + \rho_{AT} + \rho_{CG} + \rho_{CT} + \rho_{GT} = 1.$$

In order to bypass the constrained nature of the parameter space, we have relied on the so-called *additive logistic transformation* (Tiao and Cuttman, 1965; Aitichinson, 1984), which is a one-to-one transformation from $\mathbb{R}^{D-1}$ to the $(D-1)$-dimensional simplex

$$S^D = \{(x_1, ..., x_D) : x_1 > 0, ..., x_D > 0; x_1 + ...x_D = 1\}.$$

Hence, we can use its inverse, called *additive log-ratio transformation*, which is defined as follows

$$y_i = \log\left(\frac{x_i}{x_D}\right), \qquad i = 1, ..., D-1$$

for any $x = (x_1, ..., x_D) \in S^D$. Here, the x_i's are the ρ's and $D = 6$. Applying these transformations to nucleotide frequencies π_i and to exchangeability parameters ρ's, the transformed parameters assume values in the entire real support $\Re^{D-1}$ and the IDR estimator can be applied. Again, the reparameterization calls for the appropriate change-of-measure Jacobian when evaluating the corresponding transformed density (see Aitichinson (1984) for details).

3.4.3 Thermodynamic integration estimators

Starting with the pioneering work of Lartillot and Philippe (2006a), the thermodynamic integration technique has been fruitfully employed for marginal likelihood approximation in phylogenetic substitution models. Indeed, the thermodynamic integration has a longer history that dates back at least to the work of Ogata (1989). The interested reader can refer to Gelman and Meng (1998) for a historic overview and also to Chen and Shao (1997) for a more technical insight.

We will briefly recall the core ideas, which start from the so-called path sampling identity for estimating the ratio of two normalizing constants corresponding to two different densities say $c_f = \int_\Omega f(\theta)d\theta$ and $c_g = \int_\Omega g(\theta)d\theta$

$$\log \frac{c_g}{c_f} = \int_{[0,1]} Q(\beta)d\beta = \int_{[0,1]} \left[\int_\Omega \frac{\partial}{\partial\beta} \log q(\theta,\beta) \frac{q(\theta,\beta)}{c(\beta)} d\theta\right] d\beta, \qquad (3.25)$$

where $q(\theta, \beta) \geq 0$ is a so-called path function which has to guarantee the following conditions

$$f(\theta) = q(\theta, 0) \text{ and } g(\theta) = q(\theta, 1),$$

providing a smooth link (path) between f and g parameterized by β and where $c(\beta) = c_{q(\cdot,\beta)}$ is the normalizing constant of the path function $q(\theta, \beta)$ regarded as a density function in θ. An alternative expression of the path sampling identity is

$$\int_0^1 E_{\tilde{q}(\theta,\beta)} \left[\frac{\partial}{\partial \beta} \log q(\theta, \beta) \right] d\beta, \tag{3.26}$$

where $\tilde{q}(\theta, \beta) = q(\theta, \beta)/c(\beta)$. This identity is generally used to build up approximations of the log ratio of normalizing constants using simulations for approximating the inner expectations and discretizing over $\beta \in [0, 1]$ for approximating the outer integral. One of the most favorite path functions is the geometric path

$$q(\theta, \beta) = g(\theta)^\beta f(\theta)^{1-\beta},$$

which yields the following simplified expression for the inner integral

$$Q(\beta) = \int_\Omega \log \frac{g(\theta)}{f(\theta)} \frac{g(\theta)^\beta f(\theta)^{1-\beta}}{c(\beta)} d\theta.$$

Indeed, the starting density f of the path can be chosen arbitrarily. In the next section we will consider a general path sampling identity as in (3.26) in which $g(\theta) = L(\theta)\pi(\theta)$ and $f(\theta) = \pi_0(\theta)$ where $\pi_0(\theta)$ is an auxiliary distribution which is chosen as close as possible to the target g. Potentially setting $\pi_0 \propto L(\theta)\pi(\theta)$ would yield a perfect approximation. In practice, the resulting generalized path sampling estimator (GPS) is computed as

$$\begin{aligned} &\log \hat{c}_{GPS} \\ &= \frac{1}{2} \sum_{t=1}^{T-1} (\beta_{t+1} - \beta_t) \left[\frac{1}{n} \sum_{i=1}^{n} \left(\log \pi(\theta_{i;\beta_{t+1}}) + \log L(\theta_{i;\beta_{t+1}}) - \log \pi_0(\theta_{i;\beta_{t+1}}) \right) \right. \\ &\quad \left. + \frac{1}{n} \sum_{i=1}^{n} \left(\log \pi(\theta_{i;\beta_t}) + \log L(\theta_{i;\beta_t}) - \log \pi_0(\theta_{i;\beta_t}) \right) \right], \end{aligned} \tag{3.27}$$

where θ_{i,β_t} are sampled from $\tilde{q}(\theta; \beta_t) = \frac{(\pi(\theta)L(\theta))^{\beta_t} \pi_0(\theta)^{1-\beta_t}}{c(\beta_t)}$ and $\beta_1, \beta_2, ..., \beta_T$ are a suitably chosen ordered grid of points in $[0, 1]$ with $\beta_1 = 0$ and $\beta_T = 1$. A simplified version of the GPS estimator has been considered in Lartillot and Philippe (2006a). They used a geometric path for estimating model evidence using $g(\theta) = \pi(\theta)L(\theta)$ and $f(\theta) = \pi(\theta)$, so that $c_f = 1$. The corresponding path sampling estimator of the marginal likelihood on the logarithmic scale

$\log c_g$ can be estimated as

$$\log \hat{c}_{PS} = \frac{1}{2}\sum_{t=1}^{T-1}(\beta_{t+1} - \beta_t)\left[\frac{1}{n}\sum_{i=1}^{n} L(\theta_{i;\beta_{t+1}}) + \frac{1}{n}\sum_{i=1}^{n} L(\theta_{i;\beta_t})\right], \quad (3.28)$$

where θ_{i,β_t} are sampled from the so-called power posterior $\tilde{q}(\theta;\beta_t) = \frac{\pi(\theta)L(\theta)^{\beta_t}}{c(\beta_t)}$.

However, it can be easily argued that the choice of $\pi(\theta)$ as starting density f is hardly a good choice for something which ought to be as close as possible to the target (Lefebvre et al., 2010). Moreover, it has been shown in Calderhead and Girolami (2009) and in Fan et al. (2011) that the choice of a uniform grid is not optimal, and using a grid which is denser in the lower part of the unit interval may substantially improve the accuracy and the efficiency of the resulting estimator. The values $\beta_1, \beta_2, ..., \beta_T$ are set equal to the quantiles of an auxiliary density w on $[0,1]$. Numerical examples in Subsection 3.4.3.2 and Subsection 3.5 show that the choice of the reference auxiliary density w and the grid size play a relevant role in the GPS and PS estimates. In order to appropriately calibrate the starting reference distribution f, one could rely on the simple idea of the *reference density* π_0 proposed in Fan et al. (2011) which has led to remarkable improvements of the so-called stepping-stone method. The reference distribution, rather than being the prior density, can be chosen closer to the posterior by taking a product density with independent marginal distributions within the family of the marginal prior distributions. The parameters of the reference marginals are set possibly close to the posterior counterpart by means of the marginal posterior simulations. In the next sections, we have used this reference distribution π_0 according to Fan et al. (2011).

3.4.3.1 Generalized stepping-stone estimator (GSS)

The generalized stepping-stone method (GSS) presented in Fan et al. (2011), is a modification of the stepping-stone method (SS) introduced in Xie et al. (2011). This method is closely related to thermodynamic integration and shares with it the need to sample from a discretized grid of intermediate distributions in addition to the posterior.

Using the same notation of the previous section, the generalized stepping-stone estimator is defined from the identity

$$c_{GSS} = \prod_{t=1}^{T}\frac{c(\beta_t)}{c(\beta_{t-1})}, \quad (3.29)$$

where

$$r_t = \frac{c(\beta_t)}{c(\beta_{t-1})} = E_{\tilde{q}(;\beta_{t-1})}\left[\left(\frac{g(\theta)}{f(\theta)}\right)^{\beta_t - \beta_{t-1}}\right] \quad (3.30)$$

and $\tilde{q}(;\beta_t) = \frac{q(;\beta_t)}{c(\beta_t)} = \frac{g(\theta)^{\beta_t} f(\theta)^{1-\beta_t}}{c(\beta_t)}$. In Fan et al. (2011), $g(\theta)$ is taken as the product of likelihood and prior, and $f(\theta)$ is a reference distribution $\pi_0(\theta)$, which yields

$$r_t = E_{\tilde{q}(;\beta_{t-1})}\left[\left(\frac{L(\theta)\pi(\theta)}{\pi_0(\theta)}\right)^{\beta_t-\beta_{t-1}}\right], \tag{3.31}$$

where $\tilde{q}(;\beta_t) = \frac{(L(\theta)\pi(\theta))^{\beta_t}\pi_0(\theta)^{1-\beta_t}}{c(\beta_t)}$. The density $\tilde{q}(;\beta_t)$ is equivalent to the posterior distribution when $\beta = 1$ and it is equivalent to the reference distribution when $\beta = 0$. The expression in (3.31) can be estimated using

$$\hat{r}_t = \frac{1}{n}\sum_{i=1}^{n}\left(\frac{L(\theta_{i,\beta_{t-1}})\pi(\theta_{i,\beta_{t-1}})}{\pi_0(\theta_{i,\beta_{t-1}})}\right)^{\beta_t-\beta_{t-1}},$$

where $\theta_{1,\beta_t}, \theta_{2,\beta_t}, ..., \theta_{n,\beta_t} \sim \tilde{q}(\theta;\beta_t)$. Hence the GSS estimator is defined as

$$\widehat{c}_{GSS} = \prod_{t=1}^{T}\hat{r}_t.$$

In Fan et al. (2011), the reference distribution $\pi_0(.)$ is constructed using the same parameterized functional form of the prior distribution and calibrated using samples from the posterior distribution $\tilde{q}(\cdot;\beta = 1)$: parameters of the reference distribution are set equal to the empirical means of the same parameters simulated from $\tilde{q}(\cdot;\beta = 1)$. The relative mean square error of $\hat{c}_{GSS}$ can be approximated as follows

$$\widehat{\text{RMSE}}_{\widehat{c}_{GSS}} \approx \frac{1}{n^2}\sum_{t=1}^{T}\sum_{i=1}^{n}\left(\frac{\widehat{c}_{GSS,i,t}}{\widehat{c}_{GSS,t}} - 1\right)^2. \tag{3.32}$$

The GSS reduces to the original stepping-stone (SS) method proposed in Xie et al. (2011) if the reference distribution is equal to the actual prior:

$$\hat{r}_t^{(SS)} = \frac{1}{n}\sum_{i=1}^{n}(L(\theta_{i,\beta_{t-1}}))^{\beta_t-\beta_{t-1}}, \tag{3.33}$$

where $\theta_{i,\beta_t} \sim \tilde{q}(;\beta_t) = \frac{L(\theta)^{\beta_t}\pi(\theta)}{c(\beta_t)}$ and the stepping-stone estimator is defined as

$$\hat{c}_{SS} = \prod_{t=1}^{T}\hat{r}_t^{(SS)}. \tag{3.34}$$

In Fan et al. (2011), it is argued that GSS is more reliable and efficient than the other methods, such as the harmonic mean, the stepping-stone method, and the path sampling estimator.

3.4.3.2 Comparative performance: the linear model

We have considered a controlled simulation setting as in Lartillot et al. (2007) and Calderhead and Girolami (2009). A simple linear model is considered where the marginal likelihood can be computed exactly. Data X are generated from the following

$$X_i = Z_i^T B + u_i, \qquad i = 1, 2, ..., n,$$

where the $n \times k$ matrix Z of covariates are fixed uniformly, $B = (b_1, ..., b_k)$ is a vector of regression coefficients which have been randomly fixed, and $u_i \sim N(0, \sigma^2)$. The exact closed form formula for the marginal likelihood when the prior distribution on B is a multivariate normal with null mean vector and diagonal covariance matrix η^2 is the following:

$$p(X|\sigma^2, \eta^2) = (2\pi)^{k/2} |\Omega|^{-1/2} \exp\left\{ -\frac{1}{2} X^T \Omega^{-1} X \right\},$$

where $\Omega = \sigma^2 I + \eta^2 Z Z^T$.

We first verified the comparative performance of two alternative classes of marginal likelihood estimators: the first one relying on the GHM formula and using only a single simulation from the posterior, and the second one relying on thermodynamic integration and requiring a sufficiently dense grid of simulations from the auxiliary distributions $\tilde{q}(\cdot, \beta_t)$. As representative of the first class, we have considered the classical harmonic mean estimator, the generalized harmonic mean estimator, and the IDR estimators. The generalized harmonic mean estimator has been implemented as in (3.11), using as instrumental function f the reference distribution π_0 as suggested in the implementation of GSS. For the second class we have considered the generalized path sampling again with $f = \pi_0$ and the choice of $\beta_1, \beta_2, ..., \beta_d$ set equal to the quantiles of an auxiliary density w chosen as a Beta distribution with parameters $\alpha_1 = 0.3$ and $\alpha_2 = 1.0$.

We have simulated 10 datasets from the above model in 4 simulation settings (A, B, C, D). They differ in the way the design matrix Z and the true regressors B are fixed so that they yield different posterior correlation structures. They are ordered according to the increasing presence of correlation ranging from $(-0.13; 0.13)$ in setting A to $(-0.33; 0.58)$ in setting D. The 4 settings have been replicated for different choices of n ($n = 200$; $n = 1000$) and of k ($k = 20$; $k = 100$). Thermodynamic integral estimators are based on 10000 draws from $\tilde{q}(\cdot, \beta_t)$ for each β_t in the grid, while GHM estimators are based only on 10000 from the posterior $\tilde{q}(\cdot, \beta_t = 1)$. In Table 3.6 we can see that overall the comparative performance of the alternative methods is certainly in favor of the thermodynamic integration side. However, one can notice that in most simulation settings in which the correlation structure is mild and hence the reference π_0 is reasonably close to the target posterior, GHM can be competitive. Notice that the reference distribution has been chosen as a multivariate normal so that only the posterior means and variances are

Table 3.6: Linear model simulated data: actual relative mean square error of alternative marginal likelihood estimators resulting from the simulation of ten different datasets in four simulation settings (A, B, C, D) with different correlation structures. The best performing estimators are highlighted in bold.

	n=200				
	HM	GHM	IDR	GSS	GPS
Sim. A (k=20)	0.24184	0.00002	0.00323	**0.00001**	**0.00001**
Sim. B (k=20)	0.27126	0.00002	0.00360	**0.00001**	**0.00001**
Sim. C (k=20)	0.29329	0.00695	0.00281	**0.00005**	0.00007
Sim. D (k=20)	0.28573	0.00653	0.00163	0.00126	**0.00041**
Sim. A (k=100)	0.60359	0.00657	0.00193	**0.00002**	**0.00002**
Sim. B (k=100)	0.63408	0.00613	0.00179	**0.00002**	**0.00002**
Sim. C (k=100)	0.65727	0.06714	0.00149	0.00314	**0.00008**
Sim. D (k=100)	0.64734	0.04089	**0.00161**	0.00268	0.00405
	n=1000				
	HM	GHM	IDR	GSS	GPS
Sim. A (k=20)	0.07016	$\mathbf{< 10^{-5}}$	0.00063	$\mathbf{< 10^{-5}}$	$\mathbf{< 10^{-5}}$
Sim. B (k=20)	0.07903	$\mathbf{< 10^{-5}}$	0.00073	$\mathbf{< 10^{-5}}$	$\mathbf{< 10^{-5}}$
Sim. C (k=20)	0.08789	0.00124	0.00062	**0.00001**	0.00002
Sim. D (k=20)	0.08271	0.00158	0.00089	**0.00010**	0.00012
Sim. A (k=100)	0.27634	0.00006	0.00073	$\mathbf{< 10^{-5}}$	$\mathbf{< 10^{-5}}$
Sim. B (k=100)	0.29862	0.00006	0.00073	$< 10^{-5}$	$< 10^{-5}$
Sim. C (k=100)	0.32460	0.02243	0.00062	0.00028	**0.00022**
Sim. D (k=100)	0.30450	0.01108	**0.00076**	0.00098	0.00109

matched in the same way the reference distribution is calibrated in Fan et al. (2011). This means that there is a correlation structure that is overlooked in the starting density. As evident in setting D, a stronger correlation structure may cause a loss of performance of both of GHM and thermodynamic integration methods up to making IDR more competitive. Indeed, GSS and GPS can outperform the other methods also in simulation D provided that the grid size is increased from 20 to 100 (results not reported here). The main message of this example is that although the thermodynamic integration approach will eventually outperform the simpler GHM and IDR estimators, the latter can still be reliable in all simulation settings, and to some extent competitive when a reference distribution is not too close to the target posterior and/or a very dense grid cannot be used.

3.5 Marginal likelihood for phylogenetic data

In this section, we will investigate to what extent simpler estimators based solely on recycling posterior simulations can be effectively used in the phylogenetic context. As a benchmark, we will use the most powerful and tunable tools based on the thermodynamic integration which require simulations from a grid of distributions. In particular, the successful implementation of the IDR estimator is illustrated with simulated and real data.

All estimates have been computed by means of the MCMC output of the simulations from the posterior distribution obtained using Phycas software (Lewis et al., 2008); the likelihood has been computed using the R package PML, while the reparameterization on $\Re^k$ and IDR perturbation $g_{P_k}(\theta)$ have called for specifically developed R functions. GSS has been implemented using the open-source software Phycas (Lewis et al., 2008). We consider a grid of 20 β values and, following an initial burn-in phase consisting of 50000 cycles at $\beta = 1$ and then 10000 MCMC cycles for each β including the cold chain at $\beta = 1$. A thinning rate of 2 has been used for all chains. In these examples, a grid of 20 β values resulted sufficiently large since estimates obtained with a larger grid size do not appreciably change.

In order to have a fair comparison of the estimation methods, all the other estimators have been computed using 200000 random draws from the posterior only.

3.5.1 Hadamard data: marginal likelihood computation

We use as a first benchmark the synthetic dataset Hadamard already employed in (Felsenstein, 2004, p. 280). It consists of 200 nucleotide sites and four species, A, B, C, D. This dataset was simulated from the Jukes-Cantor model (JC69) and the true tree is shown in the left panel of Figure 3.3.

For the true topology, we compute the marginal likelihood for the JC69 model and for the GTR+Γ model: parameters lie in a 5-dimensional space for the JC69 model and in 14-dimensional space for the GTR+Γ model. The simulated values from the Metropolis-Coupled algorithm have been rearranged to evaluate the model evidence of both models using different variants of the generalized harmonic mean and thermodynamic integration approaches.

GTR+Γ and JC69 models have been implemented in Phycas which provides as output the simulated Markov chains. Simulated values from $\tilde{q}(\cdot, \beta_t = 1)$ can be used as a sample from the posterior distribution in order to make inference on the model parameters.

In Table 3.7, we list the corresponding values of the IDR estimator on the log scale ($\log \hat{c}_{IDR}$), the Relative Mean Square Error estimate ($\widehat{\text{RMSE}}_{\hat{c}_{IDR}}$) as in (3.20), and the confidence interval $\widehat{\text{CI}}$ for different perturbation masses k. In order to take into account the autocorrelation of the posterior simulated values, a correction has been applied to $\widehat{\text{RMSE}}_{\hat{c}_{IDR}}$ replacing n in (3.20) with the effective sample size given by

$$n_{ESS} = n \times \frac{1}{1 + 2\sum_{i=1}^{I} \hat{\rho}_i} \tag{3.35}$$

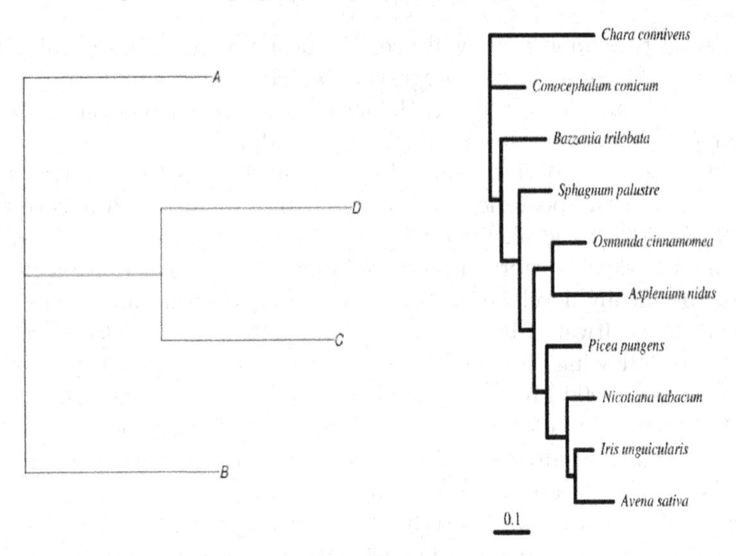

Figure 3.3: True phylogenetic trees of the two datasets used as benchmarks: Hadamard (left) and Green (right).

Table 3.7: Inflated density ratio method applied to Hadamard data with a JC69 model with a 5-dimensional parameter space. IDR estimates on the log scale for a small regular grid of perturbation values. The relative mean square errors $\widehat{\text{RMSE}}_{\hat{c}_{IDR}}$ are computed as in (3.20) without accounting for autocorrelation. $\widehat{\text{RMSE}^*}_{\hat{c}_{IDR}}$ are the relative mean square errors corrected for the autocorrelation. $\widehat{\text{CI}}$ are confidence intervals on a log scale. Since the smallest error in the grid corresponds to a perturbation value $k^{opt} = 10^{-6}$, the IDR estimator for the JC69 model is taken to be $\log \hat{c}_{IDR} = -592.2874$.

k	$\log \hat{c}_{IDR}$	$\widehat{\text{RMSE}}_{\hat{c}_{IDR}}$	$\widehat{\text{RMSE}^*}_{\hat{c}_{IDR}}$	$\widehat{\text{CI}}$
10^{-7}	-592.2877	0.0217	0.0644	$[-592.3504, -592.2202]$
10^{-6}	-592.2874	0.0033	0.0099	$[-592.2986, -592.2761]$
10^{-5}	-592.2934	0.0037	0.0101	$[-592.3035, -592.2832]$
10^{-4}	-592.5918	0.0041	0.0122	$[-592.6039, -592.5794]$

with $I = 10$. Since the optimal corrected error $\widehat{\text{RMSE}^*}_{\hat{c}_{IDR}}$ corresponds to a perturbation value $k^{opt} = 10^{-6}$, the IDR estimator (on a logarithmic scale) for the JC model is $\log \hat{c}_{IDR} = -592.2874$.

For comparative purposes, we considered the results of the IDR method with those obtained with the original harmonic mean (HM) and a generalized version of the harmonic mean based on the same reference distribution π_0, which is used to implement the generalized stepping stone as in Fan et al. (2011). Indeed, to verify the impact of the choice of the initial distribution, we also evaluate the estimates resulting from GSS and SS. The path sampling estimator and its generalized version have also been computed. The evaluation of all these thermodynamic alternatives was simple to implement since the reference distribution as well as the prior together with their evaluation on the sampled parameters are available from the Phycas output.

RMSEs of the harmonic mean estimators have been computed by adapting the formula (3.20), while RMSEs of the thermodynamic estimators have been computed by adapting the formula (3.32). For the adjusted RMSEs, the effective sample size in (3.35) has been used. For each method, the Monte Carlo relative error of the estimate has been computed by re-estimating the model 10 times ($\widehat{\text{RMSE}}_{\hat{c},MC}$) and recording the corresponding different values of $\hat{c}$. Although it is known that under critical conditions such MC error is not sufficient to guarantee its precision, it still remains a necessary premise for an accurate estimate.

Table 3.8 shows the estimates for the JC69 model and for GTR+Γ. The methods produce different results which were overall closer in the JC model while sometimes very distant in the GTR+Γ case. The closest marginal likelihood estimates are those from GSS, GPS, and GHM, all methods relying on the same reference distribution. Also, SS estimates are quite close to those obtained with GSS. IDR and PS produce estimates which are very similar in both models although only slightly different from the previous ones. On the other hand, HM estimates can be very far apart. This is particularly evident in the GTR+Γ case, confirming that HM can be badly biased and unstable resulting in a seriously doubtful reliability. In the JC model, all methods, with the exception of HM, produce similar results: the corresponding estimates of the relative error, once adjusted for the autocorrelation, show that GPS and GSS are the most stable methods followed by IDR and PS. On the other hand, for the GTR+Γ model, all estimates, with the exception of HM, are comparable, although IDR results in the largest estimated error. This is also confirmed by the Monte Carlo error. Indeed, we notice that the Monte Carlo estimates of the relative error do not always reflect well the theoretical estimates even when not accounting for the autocorrelation. This could be explained by the presence of some bias in the corresponding estimates possibly due to choice of a not sufficiently dense grid or a not appropriate starting density such as the prior in the path sampling. This is particularly evident always for HM and for GHM, and SS especially in the GRT+Γ model. For GSS, GPS, PS, and IDR, the Monte Carlo errors always fall in between the one estimated with an independent sample and the one accounting for the presence of autocorrelation.

In order to choose among the two models, Bayes factors have been computed. Considering the reference values for the Bayes factor defined in Kass and Raftery (1995), all methods, with the exception of HM, strongly and consistently give support to the Jukes-Cantor model, which is known in this case to be the true model. We also point out that the estimates of Bayes factors are all very similar.

Table 3.8: Hadamard data: marginal likelihood estimates of the GTR+Γ model ($\omega = \Re^{14}$) obtained with the HM, GHM, IDR, GSS, SS, PS, and GPS approaches. For GSS and SS estimators we show the estimates obtained when the β_k are evenly spaced quantiles from a Uniform distribution ($U(0,1)$) and a Beta distribution with parameters $\beta_1 = 0.3$ and $\beta_2 = 1.0$. For GPS and PS, the β_k are evenly spaced quantiles from a Beta distribution with parameters $\beta_1 = 0.3$ and $\beta_2 = 1.0$. Three different RMSE estimates are provided: for the harmonic mean estimators, $\widehat{\text{RMSE}}_{\hat{c}}$ has been computed adapting the expression (3.20), while for thermodynamic estimators, it has been computed by adapting the expression (3.32). $\widehat{\text{RMSE}^*}_{\hat{c}}$ is the RMSE corrected for the autocorrelation. $\widehat{\text{RMSE}}_{\hat{c},MC}$ comes from 10 Monte Carlo independent replicates of the estimation.

JC69				
Method	$\log(\widehat{c})$	$\widehat{\text{RMSE}}_{\hat{c}}$	$\widehat{\text{RMSE}^*}_{\hat{c}}$	$\widehat{\text{RMSE}}_{\hat{c},MC}$
HM	−590.3502	111.5118	361.6873	0.5500
GHM	−591.9880	0.1949	0.5825	0.0038
IDR	−592.2883	0.0068	0.0223	0.0102
$\text{SS}_{U(0,1)}$	−591.9896	0.0421	0.1141	0.0109
$\text{SS}_{Beta(0.3,1)}$	−591.9923	0.0353	0.1041	0.0045
$\text{GSS}_{U(0,1)}$	−592.0057	0.0006	0.0021	0.0008
$\text{GSS}_{Beta(0.3,1)}$	−591.9885	0.0008	0.0027	0.0005
GPS	−591.9882	0.0008	0.0016	0.0005
PS	−592.0336	0.0105	0.0343	0.0060
GTR+Γ				
Method	$\log(\widehat{c})$	$\widehat{\text{RMSE}}_{\hat{c}}$	$\widehat{\text{RMSE}^*}_{\hat{c}}$	$\widehat{\text{RMSE}}_{\hat{c},MC}$
HM	−596.4504	178.2420	494.4335	0.8641
GHM	−603.8314	0.9089	1.7845	0.0026
IDR	−604.6316	0.1429	0.2479	0.1664
$\text{SS}_{U(0,1)}$	−603.8979	0.1960	0.2458	0.0591
$\text{SS}_{Beta(0.3,1)}$	−603.8865	0.0996	0.1400	0.0349
$\text{GSS}_{U(0,1)}$	−603.9066	0.0035	0.0250	0.0030
$\text{GSS}_{Beta(0.3,1)}$	−603.8921	0.0040	0.0674	0.0056
GPS	−603.8283	0.0033	0.0132	0.0048
PS	−604.0438	0.0182	0.0929	0.0389

We briefly mention the fact that the IDR method can also be extended for dealing with selecting competing trees when the topology is not fixed in advance. For a fixed substitution model, competing trees can be compared by considering the evidence of the data for a fixed tree topology. The evidence in support of each tree topology $\tau_i \in \mathcal{T}$ can be evaluated in terms of its posterior probability $p(\tau_i|X)$ derived from the Bayes theorem as

$$p(\tau_i|X) = \frac{p(X|\tau_i)\pi_{\mathcal{T}}(\tau_i)}{\sum_{\tau \in \mathcal{T}} p(X|\tau)\pi_{\mathcal{T}}(\tau)},$$

where the experimental evidence in favor of the model M_{τ_i} with fixed tree topology τ_i is contained in the marginal likelihood

$$m(X|M_{\tau_i}) = p(X|\tau_i) = \int_{\Omega_{\tau_i}} p(X|\omega_i, M_{\tau_i})\pi(\omega_i|\tau_i)d\omega_i,$$

where the continuous parameters $\omega_i \in \Omega_{\tau_i}$ of the evolutionary process corresponding to M_{τ_i} are integrated out as nuisance parameters. Indeed, when prior beliefs on trees are set equal $\pi(\tau_i) = \pi(\tau_j)$ comparative evidence discriminating τ_i against τ_j, is summarized in the Bayes factor

$$BF_{ij} = \frac{m(X|M_{\tau_i})}{m(X|M_{\tau_j})}. \tag{3.36}$$

Numerical examples and more detailed discussion can be found in Arima and Tardella (2012).

3.5.2 Green plant *rbc*L example

We analyze the same Green Plant data used in Xie et al. (2011). Green Plant data consist of 10 taxa and DNA sequences of the chloroplast-encoded large subunit of the RuBisCO gene (*rbc*L) (1296 nucleotide sites). Following Xie et al. (2011), we use the topology shown in the right panel of Figure 3.3.

Again, we fit JC69 and GTR+Γ models and compare HM, GHM, IDR, SS, GSS, GPS, and PS marginal likelihood estimates. The original HM method appears to be unstable and remarkably different from all the other methods. GHM, GSS, and GPS give the most similar estimates. On the other hand, IDR, although showing larger error estimates, agrees with all the other methods and is very similar to PS.

Similar arguments apply when the GTR+Γ model is considered: all methods, with the exception of HM, produce very close estimates and, when accounting for autocorrelation, error estimates are also comparable. However, the SS method produces different results when applied to GTR+Γ model according to different choices of grid points: it appears that the auxiliary distribution whose quantiles are used to set the β grid plays a key role here in better estimating the marginal likelihood.

As far as model comparison is concerned, the $GTR + \Gamma$ model is strongly supported by all methods. Also in this case, HM reveals itself as unreliable and not conclusive because of the larger error. On the other hand, the other methods give similar results. Despite some difference in the estimates produced by IDR, those differences are mitigated when BF is considered with IDR providing slightly stronger evidence in favor of GTR+Γ model.

Table 3.9: Green data: marginal likelihood estimates of the GTR+Γ model ($\omega = \Re^{26}$) and JC69 model ($\omega = \Re^{17}$) obtained with the HM, GHM, IDR, GSS, and SS methods. For GSS and SS estimators we show the estimates obtained when the β_k are evenly spaced quantiles from a Uniform distribution ($U(0,1)$) and a Beta distribution with parameters $\beta_1 = 0.3$ and $\beta_2 = 1.0$. For GPS and PS, the β_k are evenly spaced quantiles from a Beta distribution with parameters $\beta_1 = 0.3$ and $\beta_2 = 1.0$. Three different RMSE estimates are provided: for the harmonic mean estimators, $\widehat{\text{RMSE}}_{\hat{c}}$ has been computed adapting the expression (3.20), while for thermodynamic estimators, it has been computed adapting the expression (3.32). $\widehat{\text{RMSE}^*}_{\hat{c}}$ is the RMSE corrected for the autocorrelation. $\widehat{\text{RMSE}}_{\hat{c},MC}$ comes from 10 Monte Carlo independent replicates of the estimation.

JC69				
Method	$\log(\widehat{c})$	$\widehat{\text{RMSE}}_{\hat{c}}$	$\widehat{\text{RMSE}^*}_{\hat{c}}$	$\widehat{\text{RMSE}}_{\hat{c},MC}$
HM	−7374.8690	139.3546	453.5142	0.4764
GHM	−7398.5300	0.5945	1.7858	0.0012
IDR	−7399.8450	0.1723	0.3014	0.1959
$\text{SS}_{U(0,1)}$	−7399.6500	0.6824	1.5566	1.8476
$\text{SS}_{Beta(0.3,1)}$	−7398.5360	0.1168	0.1884	0.0606
$\text{GSS}_{U(0,1)}$	−7398.6410	0.0026	0.0084	0.0015
$\text{GSS}_{Beta(0.3,1)}$	−7398.529	0.0019	0.0061	0.0015
GPS	−7398.5280	0.0036	0.0065	0.0017
PS	−7399.1250	0.0293	0.0964	0.0514
GTR+Γ				
Method	$\log(\widehat{c})$	$\widehat{\text{RMSE}}_{\hat{c}}$	$\widehat{\text{RMSE}^*}_{\hat{c}}$	$\widehat{\text{RMSE}}_{\hat{c},MC}$
HM	−6637.366	219.9642	619.4541	1.5106
GHM	−6679.766	0.0110	0.1325	0.0279
IDR	−6679.679	0.2600	0.3501	0.3492
$\text{SS}_{U(0,1)}$	−6700.856	0.9540	6.8098	$> 10^3$
$\text{SS}_{Beta(0.3,1)}$	−6679.802	0.2489	0.4344	0.0891
$\text{GSS}_{U(0,1)}$	−6679.653	0.0110	0.1326	0.0280
$\text{GSS}_{Beta(0.3,1)}$	−6679.859	0.0094	0.2503	0.0336
GPS	−6679.856	0.0163	0.0824	0.0305
PS	−6681.270	0.0396	0.2777	0.1220

In these examples, GSS and GPS resulted in the most stable methods with respect to estimation error and somewhat robust with respect to the specification of the auxiliary distribution used to set the grid of β values. The stability of the results is possibly due to the fact that the reference distribution π_0 is well calibrated and closely resembles the posterior distribution. In Table 3.10, we investigate the robustness of the GSS estimates with respect to

Table 3.10: Sensitivity of the GSS method with respect to perturbation of the reference distribution. Values in brackets correspond to the estimated RMSE.

λ	ν_τ	ρ	Γ
0.5	-6676.399 (1.6×10^{-2})	-6678.473 (1.3×10^{-4})	-6679.538 (1.2×10^{-4})
1.1	-6679.616 (1.1×10^{-4})	$-6679.745(1.2 \times 10^{-4})$	-6679.681 (1.3×10^{-4})
1.2	-6679.485 (1.1×10^{-4})	-6679.771 (1.2×10^{-4})	-6679.673 (1.1×10^{-4})
1.5	-6678.894 (1.1×10^{-4})	-6679.682 (1.2×10^{-4})	-6679.641 (1.2×10^{-4})
2.0	-6677.695 (1.1×10^{-4})	-6679.160 (1.3×10^{-4})	-6679.575 (0.9×10^{-4})

change of scale of the reference distribution. In particular, for the branch length parameters ν_τ, the reference distribution is specified as the product of independent and identically distributed gamma distributions with shape α_1 and rate α_2, which are estimated using simulations from the posterior distribution. We modify the scale of the reference distribution considering as reference a gamma distribution with shape $\frac{\alpha_1}{\lambda}$ and rate $\alpha_2\lambda$ where λ is the perturbation factor. The same is done for the reference distribution component of the Γ parameter. For the substitution rates, we perturbed the original reference distribution multiplying all Dirichlet parameters by λ. Table 3.10 shows that GSS can be affected by some variation in the scale of the reference distribution.

3.6 Discussion

In this chapter, we have investigated the possibility of using alternative simple effective recipes for evaluating model evidence of competing phylogenetic substitution models. In a Bayesian framework, several methods have been proposed in order to approximate the marginal likelihood of a single model and then eventually estimate the Bayes factor of two competing models. None of the methods proposed so far can be considered to outperform the others simultaneously in terms of simplicity of implementation, computational burden, and precision of the estimates. One of the most widely used methods until now has been the harmonic mean: the simplicity of implementation combined with a relatively light computational burden are two appealing features which explain why the HM is still one of the most favorite options for routine implementation (see von Reumont et al. (2009)). However, the simplicity of HM is often not matched with its accuracy, and very recent literature has highlighted the unreliability of HM estimators in phylogenetic models (Lartillot and Philippe, 2006a; Xie et al., 2011; Fan et al., 2011) as well as in more general biological applications (Calderhead and Girolami, 2009).

In this chapter, we have provided evidence of improved effectiveness of a simple alternative marginal likelihood estimator belonging to the class of generalized harmonic mean estimators. In particular, we have introduced a new marginal likelihood estimator, the inflated density ratio (IDR) estimator and we have also borrowed from Fan et al. (2011) the use of a suitable auxiliary distribution in the GHM formula. They share the original simplicity and computation feasibility of the HM estimator but, unlike it, they enjoy important theoretical properties, such as the finiteness of the variance. Moreover, they allow one to recycle posterior simulations, which is particularly appealing in those contexts — such as phylogenetic models — where the computational burden of the simulation is heavier than the evaluation of the likelihood, posterior densities and the like. We have verified the reliability of the IDR estimator as well as the improved GHM in some of the most common phylogenetic substitution models under different model complexities. In this study, the topology was fixed when estimating the marginal likelihood, even though there is often interest in estimating marginal likelihood when the topology is unknown. The most widely used methods for estimating the marginal likelihood are applied to a fixed topology (Lartillot and Philippe, 2006a; Xie et al., 2011; Fan et al., 2011); however, Chapter 5 introduces a tree topology reference distribution for use with GSS and Chapter 6 introduces a variable-topology version of the IDR method. More work is needed in order to determine the efficiency of these methods and their applicability in real contexts.

Overall, we have compared two alternative classes of marginal likelihood estimators: the first one relying on the GHM formula and using only a single simulation from the posterior and the second one relying on thermodynamic integration and requiring simulations from the auxiliary distribution. As representative of the first class, we have considered the classical harmonic mean estimator, the generalized harmonic mean estimator, and the IDR estimator. For the second class we have considered classical path sampling based on the prior as starting auxiliary density as well as the generalized path sampling with an appropriately chosen reference distribution. We have also considered the stepping-stone method and the generalized stepping stone, which can be considered, so far, as one of the most reliable methods for ML approximation.

Focusing only on applications of the first class of estimators, for phylogenetic examples we can conclude that GHM, relying on the same reference distribution of GSS and GPS, was the best-performing estimator. On the other hand, IDR still turned out to be a reliable tool for comparing model evidence of substitution models in agreement with all the other methods, while in simulated examples, IDR was competitive and outperformed GHM when the reference distribution could not be calibrated sufficiently close to the target posterior for the presence of unmodeled correlation in the class of candidate reference distributions. With respect to the second class, although GSS and GPS turned out to be very accurate and performed very similarly in the phylogenetic examples, in the simulated examples GPS was more robust to possible miscalibration of the reference distribution. Moreover, we have shown that when implementing GPS and GSS, the first important feature is the suitable choice of the reference distribution: we have investigated the degree of sensitivity of the final estimates to the change of scale of the reference as well as of the correlation patterns, and GPS was found to be more robust than GSS. The final important feature to be addressed is the choice of the grid size and of the grid points: in fact, one can consider the marginal likelihood estimator to be reliable only when, increasing the grid size, it does not change appreciably. However, once a reference distribution is chosen very close to the target distribution, one can obtain accurate estimators when the grid size is as low as 20 grid points. Whether or not one has calibrated a suitable reference, IDR and GHM can still be considered reliable alternatives that can be useful when a simpler solution is sought or when a sophisticated calibration of the reference cannot be successfully obtained.

We believe that since no method can be considered as an intrinsically infallible benchmark for estimating the marginal likelihood, a consensus of multiple methods on the plausible order of magnitude of the marginal likelihood may well result in a safer estimate and a better-grounded scientific conclusion.

4

Bayesian model selection in phylogenetics and genealogy-based population genetics

Guy Baele
Department of Microbiology and Immunology, Rega Institute, KU Leuven, Leuven, Belgium

Philippe Lemey
Department of Microbiology and Immunology, Rega Institute, KU Leuven, Leuven, Belgium

CONTENTS

In this chapter, we discuss recent advances in the field of Bayesian model testing and focus on methods that aim at either estimating (log) marginal likelihoods or at directly estimating (log) Bayes factors. We start by introducing several of the most popular (log) marginal likelihood estimators, which are attractive from a computational perspective. Because these estimators have recently been shown to perform poorly, we discuss computationally more demanding, but also more accurate path sampling approaches that can be used to either estimate (log) marginal likelihoods for different models, but also to directly estimate (log) Bayes factors between two competing models. For a specific class of evolutionary models, i.e., the relaxed molecular clock models, we also discuss how such methods compare to specific Bayesian model averaging approaches that allow constructing a classifier to approximate (log) Bayes factors between the models in the candidate model set. To demonstrate their practical use, we apply the presented techniques in a simulation study on relaxed molecular clocks and in a demographic model selection study that focuses on an HIV-1 dataset.

4.1 Introduction

Bayesian inference has become increasingly popular in molecular phylogenetics over the past decades, with Markov chain Monte Carlo (MCMC) integration revolutionizing the field (Yang and Rannala, 1997a). While MCMC has provided the opportunity to infer posterior distributions under complex phylogenetic models, the computational demands associated with increasing model complexity and large quantities of data have considerably hampered the ability to assess model performance. Comparing alternative models according to objective criteria in a formal model selection procedure is, however, an essential approach to phylogenetic hypothesis testing (Suchard et al., 2001; Huelsenbeck et al., 2001b). The aim of model selection is not necessarily to find the true model that generated the data, but to select a model that best balances simplicity with flexibility and biological realism in capturing the key features of the data (Steel, 2005).

A standard approach to perform model selection in a Bayesian phylogenetic framework operates through the evaluation of Bayes factors (Sinshheimer et al., 1996; Suchard et al., 2001). The Bayes factor is a ratio of two marginal likelihoods (i.e., two normalizing constants of the form $p(Y \mid M)$, with Y the observed data and M an evolutionary model under evaluation) obtained for the two models, M_0 and M_1, under comparison (Jeffreys, 1935):

$$B_{10} = \frac{p(Y \mid M_1)}{p(Y \mid M_0)}.$$

In order to evaluate model fit and calculate Bayes factors, the normalization constant or marginal likelihood $p(Y \mid M)$, which measures the average fit of a model to the data, is of primary importance. Calculation of the marginal likelihood of model M requires integration of its likelihood across parameter values, weighted by the model's prior distribution

$$p(Y \mid M) = \int_{\theta \in \Theta} p(Y \mid \theta, M)\, p(\theta \mid M)\, \mathrm{d}\theta.$$

Among several models, one is led to choose the model yielding the largest marginal likelihood. Bayes factors offer advantages over likelihood-ratio tests that compare nested models and that only garner evidence in favor of rejecting less-complex models. Instead, the Bayes factor evaluates the relative merits of both competing models. Bayes factors do not require models to be nested and the marginal likelihood naturally penalizes for model complexity. Given that modeling assumptions may impact model fit with several orders of magnitude, the log Bayes factor is often calculated. Kass and Raftery (1995) introduce different gradations to assess the log Bayes factor as evidence against M_0. A value between 0 and 1 is not worth more than a bare mention, whereas a value between 1 and 3 is considered to give positive evidence against M_0. Values larger than 3 and 5 are considered to, respectively, provide strong and very strong evidence against M_0.

Although researchers have proposed several useful methods to evaluate Bayes factors in phylogenetics, they are often limited to specific model selection situations. A detailed description of such estimators is beyond the scope of this chapter and we refer the reader to Lartillot and Philippe (2006a) for more information. Among the few methods of potentially general applicability, phylogenetics has readily adopted (i) importance sampling (IS) estimators (Newton and Raftery, 1994c) and (ii) path sampling (PS) estimators (Ogata, 1989; Gelman and Meng, 1998) to compute marginal likelihoods for competing models. Occasionally, phylogeneticists refer to path sampling as "thermodynamic integration" (Lartillot and Philippe, 2006a) in deference to its application in the physics literature. The path sampling approach represents a very general estimator because it can be applied to any model for which MCMC samples can be obtained.

In this chapter, we first discuss several importance sampling (IS) estimators, which are relatively simple in nature and do not require additional calculations, because they rely on the regular MCMC analysis that is performed to estimate parameters. We then go on to discuss the more recently developed path sampling approaches, which do require additional programming and calculations, but they offer increased accuracy over the simpler importance sampling methods.

4.2 Prior and posterior-based estimators

4.2.1 Integrating the likelihood against the model prior

The simple Monte Carlo estimator of integrating the likelihood against the model prior is the simplest estimator of the (log) marginal likelihood. Often confusingly called the "arithmetic mean estimator" (Xie et al., 2011), or the "prior arithmetic mean estimator" (see next section) (Lartillot and Philippe, 2006a), it uses the prior as the importance distribution. The marginal likelihood estimate is simply the arithmetic mean of the likelihoods of parameter values sampled from the prior distribution:

$$\hat{p}(Y \mid M) = \frac{1}{n} \sum_{i=1}^{n} p(Y \mid \theta_i, M),$$

where $\{\theta_i,\ i = 1, 2, \ldots, n\}$ is a sample from the prior distribution $p(\theta \mid M)$. This approach is unbiased; however, if the likelihood is sharp compared with the prior, the marginal likelihood estimate can have an unacceptably high variance. This is because the magnitude of the estimate often depends critically on a few points sampled in the area of high likelihood (Xie et al., 2011). Most of the parameters will correspond to small likelihood values, making the simulation process quite inefficient. Being dominated by a few large values of the likelihood, this estimator inevitably yields large variance, which makes its convergence to a Gaussian distribution slow because it requires a tremendous number of samples (Newton and Raftery, 1994c; Lartillot and Philippe, 2006a).

4.2.2 The arithmetic mean estimator

Instead of computing the prior mean, Aitkin (1991) proposed to use the posterior mean under each model and this has been used in the past (see, e.g., Aris-Brosou and Yang, 2002) because it can be readily calculated by sampling from the MCMC:

$$\hat{p}(Y \mid M) = \frac{1}{n} \sum_{i=1}^{n} p(Y \mid \theta_i, M),$$

where $\{\theta_i,\ i = 1, 2, \ldots, n\}$ is an MCMC sample from the posterior distribution. This method, also known as the arithmetic mean estimator to calculate posterior Bayes factors (Aitkin, 1991), has been controversial since its inception. Use of the posterior Bayes factor can lead to paradoxes in inference, whereas the method does not correspond to any sensible prior, nor is it a coherent Bayesian procedure (Kadane and Lazar, 2004). Although proposed as an estimator that reduces sensitivity to variations in the prior, its use should hence be avoided at all cost.

4.2.3 The harmonic mean estimator(s)

The harmonic mean estimator (HME), or posterior harmonic mean estimator (Lartillot and Philippe, 2006a), is less sensitive to the occasional occurrence of high-likelihood values and more sensitive to low values than the arithmetic mean estimator of Aitkin (1991). Because high extremes are more likely to be a problem than low extremes, the harmonic mean estimator should perform better than the arithmetic mean estimator. Although the general stability of the harmonic mean estimator has been questioned, it should be sufficiently accurate for comparison of models with large differences in model likelihoods provided the sample from the posterior distribution is large (Nylander et al., 2004).

The harmonic mean estimate of the marginal likelihood only requires samples from the posterior and can therefore be calculated from an MCMC sample that is obtained by standard Bayesian phylogenetic analyses under a particular model. If one collects a sample, $\{\theta_i,\ i = 1, 2, \ldots, n\}$, from the posterior distribution, the HME is estimated as follows:

$$\hat{p}(Y \mid M) = \frac{n}{\sum_{i=1}^{n} \frac{1}{p(Y|\theta_i, M)}}. \tag{4.1}$$

The HME is almost guaranteed to converge to the marginal likelihood but it does not satisfy a Gaussian central limit theorem (Newton and Raftery, 1994c). The HME is sensitive to the occasional occurrence of small likelihood values that have a large effect on the estimator value and hence has an infinite variance in many practical situations. To circumvent this problem, Newton and Raftery (1994c) proposed the stabilized harmonic mean estimator (sHME), based on a mixture of the prior and the posterior:

$$\hat{p}(Y \mid M) = \frac{\sum_{i=1}^{n} p(Y \mid \theta_i, M)/\{\delta p(Y \mid M) + (1-\delta)p(Y \mid \theta_i, M)\}}{\sum_{i=1}^{n}\{\delta p(Y \mid M) + (1-\delta)p(Y \mid \theta_i, M)\}^{-1}}. \tag{4.2}$$

Unlike the HME, the sHME estimator does satisfy a Gaussian central limit theorem. A drawback of the sHME estimator is that it requires simulation from the prior as well as from the posterior (Newton and Raftery, 1994c). However, simulation from the prior as well as the posterior may be avoided, without sacrificing the appealing aspects of the estimator shown in (4.2), by instead "simulating all n values from the posterior distribution and *imagining* that an additional $n\delta/(1-\delta)$ values are drawn from the prior, all with likelihoods equal to their expected value" (Newton and Raftery, 1994c, p. 22). Often, δ values of 0.01 (Newton and Raftery, 1994c) or 0.1 are used (Lartillot and Philippe, 2006a). This yields the following approximation to (4.2):

$$\hat{p}(Y \mid M) = \frac{\delta n/(1-\delta) + \sum_{i=1}^{n} p(Y \mid \theta_i, M)/\{\delta p(Y \mid M) + (1-\delta)p(Y \mid \theta_i, M)\}}{\delta n/(1-\delta)p(Y \mid M) + \sum_{i=1}^{n}\{\delta p(Y \mid M) + (1-\delta)p(Y \mid \theta_i, M)\}^{-1}}. \tag{4.3}$$

This estimator may be evaluated by using a simple iterative scheme and should converge quickly. Newton and Raftery (1994c) state that in their experience, only a single step in the iteration is required. Since their inception, the HME and sHME have received much criticism in the statistical literature. In the discussion of the paper by Newton and Raftery (1994c), Radford Neal clarifies the problem with the HME and sHME estimators shown in (4.1) and (4.3), respectively, which are based solely on the likelihoods of values sampled from the posterior. As Neal states, in many cases the posterior is determined largely by the likelihood, with little or small contribution of the prior. Even replacing the prior with an improper distribution will often have little effect on what a typical sample drawn from the posterior looks like. The value of the marginal likelihood $p(Y \mid M)$ depends strongly on the prior, however; with an improper prior, it is reduced to 0. In such situations, the HME and sHME estimators cannot achieve good performance.

4.2.4 Akaike and Bayesian Information Criterion through Markov chain Monte Carlo: AICM and BICM

In the maximum-likelihood framework, approaches like the Akaike Information Criterion (AIC; Akaike, 1973) and Bayesian Information Criterion (BIC; Schwarz, 1978) penalize the addition of extra parameters unless they provide a sufficient improvement in fit of the model to the data. Unlike traditional maximum likelihood ratio tests (LRTs), these approaches enable the comparison of any two models, even if they are not nested (Posada and Buckley, 2004). Indeed, allowing progressively more parameters always leads to an improvement of fit between the data and the model, but the extensive addition of parameters comes at a price — the predictive power of the model (the information that the data can reveal about the underlying generative process) may be washed out by parameter abundance.

Raftery et al. (2007b) introduce the AICM as a posterior simulation-based analog of the AIC model selection criterion. AICM has the advantage that, like the harmonic mean estimator of marginal likelihood, one may estimate the AICM directly from posterior samples generated by MCMC with little additional work. Raftery et al. (2007b) show that, asymptotically, the posterior distribution of a model's log likelihood ℓ follows

$$\ell_{\max} - \ell \sim \text{Gamma}(\gamma, 1), \tag{4.4}$$

where $\ell_{\max}$ represents the maximum possible log likelihood, $\gamma = k/2$, and k represents the effective number of parameters in the model. The density function of a $\text{Gamma}(\gamma, 1)$ distribution is

$$f(x) = \frac{x^{\gamma-1}\, e^{-x}}{\Gamma(\gamma)},$$

and thus the density function of the log likelihood becomes

$$f(\ell) = \frac{e^{\ell-\ell_{\max}} (\ell_{\max} - \ell)^{\gamma-1}}{\Gamma(\gamma)}. \tag{4.5}$$

Alternatively, the posterior distribution of log likelihoods may be described in terms of a deviance $\mathcal{D} = -2\ell$, such that the posterior deviance is distributed according to a shifted chi-squared distribution

$$\mathcal{D} - \mathcal{D}_{\min} \sim \chi^2(2\gamma),$$

with density function

$$f(\mathcal{D}) = \frac{2^{-\gamma} e^{(\mathcal{D}_{\min}-\mathcal{D})/2} (\mathcal{D} - \mathcal{D}_{\min})^{\gamma-1}}{\Gamma(\gamma)}.$$

Equation (4.4) suggests a method-of-moments estimate of γ as $\hat{\gamma} = s_\ell^2$ and $\hat{\ell}_{\max} = \bar{\ell} + s_\ell^2$, where $\bar{\ell}$ and s_ℓ^2 are the sample mean and variance of the posterior log likelihoods (Raftery et al., 2007b). Thus, an estimate of the effective number of parameters k equals $2s_\ell^2$.

The AIC (Akaike, 1973), used for model comparison in the maximum-likelihood framework, is defined as

$$\text{AIC} = 2k - 2\ell_{\max}.$$

Models with lower values of AIC are preferred over models with higher values. An increase in the number of parameters k penalizes more complex models. Here, we follow Raftery et al. (2007b) in estimating AICM as

$$\begin{aligned} \text{AICM} &= 2\hat{k} - 2\hat{\ell}_{\max} \\ &= 2(2s_\ell^2) - 2(\bar{\ell} + s_\ell^2) \\ &= 2s_\ell^2 - 2\bar{\ell}, \end{aligned} \tag{4.6}$$

a function of the posterior sample mean and variance of the log likelihood. The AICM is similar in spirit to the DIC, the deviance information criterion (Gelman et al., 2004).

We note that a posterior simulation–based analog of the Bayesian information criterion (BIC) through Markov chain Monte Carlo (BICM) has also been proposed (Raftery et al., 2007b), an approach that has a more direct Bayesian justification but requires specification of a sample size for each parameter, which may be problematical in some applications.

4.2.5 Generalized harmonic mean estimator (GHME) and inflated density ratio (IDR)

4.2.5.1 GHME

As stated earlier in this chapter, the HME is one of the most popular approaches for estimating marginal likelihoods in phylogenetics. The HME can be regarded as an easy-to-apply instance of a more general class of estimators called generalized harmonic mean estimators (GHME) (Arima and Tardella, 2012). Because the marginal likelihood is essentially the normalizing constant of the unnormalized posterior density, Arima and Tardella (2012) illustrate that the GHME estimator provides a general solution for estimating the normalizing constant of a non-negative, integrable density g defined as

$$c = \int_{\Omega} g(\theta) d\theta, \tag{4.7}$$

where $\theta \in \Omega \subset \Re^k$ and $g(\theta)$ is the unnormalized version of the probability distribution $\tilde{g}(\theta)$. The GHME estimator, denoted as $\hat{c}_{\text{GHME}}$, takes the form of

$$\hat{c}_{\text{GHME}} = \frac{1}{\frac{1}{n}\sum_{i=1}^{n} \frac{f(\theta_i)}{g(\theta_i)}}, \tag{4.8}$$

where $\theta_1, \theta_2, \ldots, \theta_n$ are sampled from $\tilde{g}$ and where f is a convenient instrumental Lebesgue-integrable function, which is only required to have a support that is contained in that of g and to satisfy

$$\int_{\Omega} f(\theta) d\theta = 1. \tag{4.9}$$

To estimate the marginal likelihood, considered as the normalizing constant of the unnormalized posterior density $g(\theta) = \pi_{\Theta}(\theta)L(\theta)$, and taking $f(\theta) = \pi_{\Theta}(\theta)$, the GHME estimator reduces to the HME. Arima and Tardella (2012) consider the performance (in terms of the relative mean square error) of such a GHME method, where g is taken as the posterior density up to the unknown proportionality constant, while f is taken as an auxiliary probability density which is possibly a convenient fully available probability density approximating the posterior. In principle, setting this auxiliary density as close as possible to the posterior yields a very efficient estimator. Arima and Tardella (2012) propose using the reference distribution $\pi_{(0)}$, as suggested in the implementation of generalized stepping-stone sampling (GSS; see later in this chapter), as the instrumental function f. While progress is being made on developing reference distributions for tree topologies that provide a good approximation of the posterior (see Chapter 5), topological reference distributions have not yet been incorporated into widely used software and are still lacking for rooted trees, thus limiting the applicability of methods such as GHME and GSS.

4.2.5.2 Inflated density ratio (IDR)

The inflated density ratio (IDR) estimator is also discussed in the work of Arima and Tardella (2012) (see also Chapters 3 and 6) and emerges from a different formulation of the GHME estimator, based on a particular choice of the instrumental density $f(\theta)$ as originally proposed in Petris and Tardella (2007). The instrumental $f(\theta)$ is obtained through a perturbation of the original target function g. The perturbed density, denoted with g_{Pk}, is defined so that its total mass has some known functional relation to the total mass c of the target density g as in (4.7). In particular, g_{Pk} is obtained as a parametric inflation of g so that

$$\int_{\Omega} g_{Pk}(\theta) = c + k, \tag{4.10}$$

where k is a known inflation constant which can be arbitrarily fixed. The perturbed density allows one to define an instrumental density $f_k(\theta) = \frac{g_{Pk}(\theta) - g(\theta)}{k}$, which satisfies the requirement in Equation (4.9) needed to define the GHME estimator as in (4.8). The IDR estimator $\hat{c}_{\text{IDR}}$ for c is then equal to

$$\hat{c}_{\text{IDR}} = \frac{k}{\frac{1}{n}\sum_{i=1}^{n} \frac{g_{Pk}(\theta_i)}{g(\theta_i)} - 1},$$

where $\theta_1, \theta_2, \ldots, \theta_n$ is a sample from the normalized target density $\tilde{g}$. We refer to the relevant description in this book for a more in-depth explanation and application of the IDR estimator. The perturbation approach, like GSS, is only now starting to accommodate state spaces such as the space of all tree topologies (see Chapter 6).

4.3 Path sampling approaches

Bayesian phylogenetics requires a sensible balance between parameter-richness and biological realism. A good model captures the essential features of the underlying process that generates the data without introducing unnecessary error, bias, and variance. Accurate model comparisons are therefore a crucial part of any phylogenetic study, even though the model will always be misspecified in the sense that all evolutionary models are severe simplifications of reality. Path sampling methods provide for model comparisons while accommodating uncertainty in tree topology and without the inaccuracy of some of the most frequently used importance sampling methods.

4.3.1 Path sampling

Lartillot and Philippe (2006a) introduced path sampling (PS) in the field of phylogenetics and showed that it outperforms the importance sampling methods across different scenarios. Interestingly, they remain well behaved in cases with high dimensions where importance sampling methods fail, even when the latter are based on extensive sampling from the posterior. Here, we present an overview of the two approaches they introduced, one to estimate the marginal likelihood for a specific model and one to directly estimate (log) Bayes factors between two competing models.

Following the description by Lartillot and Philippe (2006a), we consider two unnormalized densities $q_0(\theta)$ and $q_1(\theta)$, defined on the same parameter space Θ. The corresponding true probability densities are denoted by

$$p_i(\theta) = \frac{1}{Z_i} q_i(\theta), i = 0, 1, \tag{4.11}$$

where

$$Z_i = \int_\Theta q_i(\theta) d\theta, i = 0, 1, \tag{4.12}$$

are the normalization constants. Typically, in a Bayesian phylogenetics context, $q_i(\theta) = p(Y \mid \theta, M_i)p(\theta \mid M_i)$, $Z_i = p(Y \mid M_i)$ and $p_i(\theta) = p(\theta \mid Y, M_i)$. The goal is to perform a numerical evaluation of the log ratio

$$\mu = \log\left(\frac{Z_1}{Z_0}\right) = \log Z_1 - \log Z_0. \tag{4.13}$$

Lartillot and Philippe (2006a) propose to do this by defining a continuous and differentiable path $(q_\beta)_{0 \le \beta \le 1}$ in the space of unnormalized densities, joining q_0 and q_1. By extension, for any $\beta, 0 \le \beta \le 1$, p_β and Z_β are defined as

$$p_\beta(\theta) = \frac{1}{Z_\beta} q_\beta(\theta)$$

and

$$Z_\beta = \int_\Theta q_\beta(\theta) d\theta.$$

When β tends to 0 (resp. 1), p_β converges pointwise to p_0 (resp. p_1) and Z_β to Z_0 (resp. Z_1).

Further derivation work by Lartillot and Philippe (2006a) results in the following expression for the log ratio of interest

$$\mu = \log Z_1 - \log Z_0 = \int_0^1 E_\beta[U(\theta, \beta)] d\beta, \tag{4.14}$$

where $U(\theta, \beta) = \frac{\partial}{\partial \beta} \log q_\beta(\theta)$ and the expectation E_β is taken with respect to the density $p_\beta(\theta)$. The authors present a continuous (or quasistatic) estimator for μ, improving upon a previously introduced discrete estimator (Lartillot and Philippe, 2004). The quasistatic method most often consists of equilibrating an MCMC under $\beta = 1$, then smoothly decreasing the value of β, and continuing the MCMC sampling until $\beta = 0$ is reached. During this procedure, points θ_k are saved before each update of β. If we denote $(\beta_k, \theta_k)_{k=0 \ldots K}$ as the series of points obtained, the quasistatic estimate of $\log Z_1 - \log Z_0$ is given by

$$\log p(Y \mid M) = \frac{1}{2K} \sum_{k=0}^{K-1} (\log p(Y \mid \theta_k, M) + \log p(Y \mid \theta_{k+1}, M)). \tag{4.15}$$

Lartillot and Philippe (2006a) consider two implementation schemes of the path sampling approach, one to estimate marginal likelihoods and one to directly estimate Bayes factors. Both are discussed here in the next two subsections.

4.3.1.1 Marginal likelihood estimation

Most implementations of PS rely on drawing MCMC samples from a series of distributions, each representing a power posterior that differs only in its power, along the path going from the unnormalized posterior defined by the model M to the prior. Lartillot and Philippe (2006a) define this path to be:

$$q_\beta(\theta) = p(Y \mid \theta, M)^\beta p(\theta \mid M), \tag{4.16}$$

where $p(Y \mid \theta, M)$ is the likelihood function and $p(\theta \mid M)$ is the prior. Hence, the power posterior is equivalent to the posterior distribution when $\beta = 1.0$, i.e., $q_1(\theta) = p(Y \mid \theta, M)p(\theta \mid M)$, and is equivalent to the prior distribution when $\beta = 0.0$, i.e., $q_0(\theta) = p(\theta \mid M)$. The corresponding normalization constants are $Z_0 = 1$ and $Z_1 = p(Y \mid M)$. Lartillot and Philippe (2006a) show that for U in (4.14) it then holds that

$$U(\theta, \beta) = \frac{\partial}{\partial \beta} \log q_\beta(\theta) = \log p(Y \mid \theta, M), \tag{4.17}$$

and therefore that

$$\log p(Y \mid M) = \log Z_1 - \log Z_0 = \int_0^1 E_\beta[\log p(Y \mid \theta, M)] d\beta. \tag{4.18}$$

Lartillot and Philippe (2006a) propose to evenly spread the different values of that power β between 0.0 to 1.0 and use Simpson's triangulation method to derive an expression for the marginal likelihood. The authors propose to collect one sample from each power posterior, before β is updated. Assuming $K + 1$ path steps, this yields a collection of samples $(\beta_k, \theta_k)_{k=0...K}$, with $\beta_0 = 0$ and $\beta_K = 1$, which are used to calculate the estimate for the marginal likelihood using (4.15). In our implementation of path sampling in BEAST (Drummond et al., 2012) (Baele et al., 2012a), we have, however, chosen to use multiple samples per β, requiring a small adaptation of (4.15) in that each log likelihood is replaced by the mean log likelihood of the samples taken at each β.

Lepage et al. (2007) advocate the use of a sigmoidal function that places most power values near the extremes of the unit interval in their model-switch PS analysis, and Friel and Pettitt (2008) use equally spaced points in the interval [0,1] elevated to the fourth or fifth power. The rationale for these approaches is to place most of the power values at points where the power posterior is changing rapidly. Xie et al. (2011) find that the efficiency of PS could dramatically improve by choosing β values according to evenly spaced quantiles of a Beta$(\alpha, 1.0)$ distribution rather than spacing β values evenly from 0.0 to 1.0; this is a generalization of the approach by Friel and Pettitt (2008).

4.3.1.2 Direct Bayes factor estimation

Lartillot and Philippe (2006a) note that the difference between the logarithm of the marginal likelihoods of two phylogenetic models can be small compared to the values of the two log marginal likelihoods estimates; this can lead to a poor estimate of the Bayes factor, unless the precision on each marginal likelihood estimate is very high. To circumvent this issue, a single path can be constructed to connect the two competing models in the space of unnormalized densities, which allows directly calculating the Bayes factor along this single path (Gelman and Meng, 1998).

Lartillot and Philippe (2006a) propose the following approach for direct (log) Bayes factor estimation using path sampling. Suppose that one wants to compare two models M_0 and M_1 that are defined on the same parameter space Θ. The direct (log) Bayes factor or model-switch scheme involves a path that goes directly from model M_0 to model M_1:

$$q_\beta(\theta) = [p(Y \mid \theta, M_0)p(\theta \mid M_0)]^{1-\beta}[p(Y \mid \theta, M_1)p(\theta \mid M_1)]^{\beta}.$$

This leads to the following, for $\beta = 0$ or 1:

$$q_0(\theta) = p(Y \mid \theta, M_0)p(\theta \mid M_0), \tag{4.19}$$

$$Z_0 = p(Y \mid M_0), \tag{4.20}$$

$$q_1(\theta) = p(Y \mid \theta, M_1)p(\theta \mid M_1), \tag{4.21}$$

$$Z_1 = p(Y \mid M_1).$$

Lartillot and Philippe (2006a) show that for U in (4.14) it then holds that

$$U(\theta, \beta) = \log p(Y \mid \theta, M_1) + \log p(\theta \mid M_1) - \log p(Y \mid \theta, M_0) - \log p(\theta \mid M_0).$$

By construction, this approach often results in a lower estimation error for the Bayes factor in phylogenetics (Rodrigue et al., 2006). In line with this, Baele et al. (2013a) have demonstrated that the direct (log) Bayes factor estimator generally has lower variance than the (log) Bayes factor estimator obtained through the ratio of marginal likelihoods estimated in the context of stepping-stone sampling (see 4.3.2). Estimator efficiency also depends on the path construction and hence, other paths (i.e., other functions used to connect the posteriors) between two arbitrary models may be devised. For highly structured models, such as those applied in phylogenetics, constructing an efficient path between two arbitrary models is not a generic exercise and requires expert knowledge, in particular when it comes to identifying mismatches between the parameterization of different models.

4.3.2 Stepping-stone sampling

Recently, Xie et al. (2011) introduced a new method, called stepping-stone sampling (SS), that employs ideas from both IS and PS to estimate the marginal likelihood in a series (the "stepping stones") that bridges the posterior and prior distribution of a model. The authors show that SS yields a substantially less-biased estimator than PS, and that SS requires significantly fewer path steps than PS to accurately estimate the marginal likelihood with a relatively small discretization bias. Whereas Xie et al. (2011) essentially only introduced a marginal likelihood estimator using SS, this approach can be extended to also directly estimate (log) Bayes factors between two competing models, which was shown recently (Baele et al., 2013a).

4.3.2.1 Marginal likelihood estimation

Analogous to PS (Lartillot and Philippe, 2006a), the implementation of SS relies on drawing MCMC samples from a series of power posteriors along the path going from the prior to the unnormalized posterior defined by the model M (Xie et al., 2011). Consider the unnormalized power posterior density function $q_\beta(\theta)$, which has normalizing constant c_β, yielding the normalized power posterior density p_β:

$$\begin{aligned} q_\beta(\theta) &= p(Y \mid \theta, M)^\beta p(\theta \mid M), \\ p_\beta &= q_\beta / c_\beta, \end{aligned} \tag{4.22}$$

where $p(Y \mid \theta, M)$ is the likelihood function and $p(\theta \mid M)$ the prior. The power posterior is again equivalent to the prior and posterior distribution when $\beta = 0.0$ and $\beta = 1.0$, respectively. The goal is to estimate the ratio $c_{1.0}/c_{0.0}$, which is equivalent to the marginal likelihood because $c_{0.0} = 1.0$ if the prior is proper (Xie et al., 2011). This ratio can be expressed as a product of K ratios:

$$r_{\text{SS}} = \frac{c_{1.0}}{c_{0.0}} = \prod_{k=1}^{K} r_{\text{SS},k} = \prod_{k=1}^{K} \frac{c_{\beta_k}}{c_{\beta_{k-1}}},$$

where $\beta_k = k/K$ for $k = 1, 2, \ldots, K$. The basic idea of SS is to estimate each ratio $c_{\beta_k}/c_{\beta_{k-1}}$ in the product by importance sampling, using $p_{\beta_{k-1}}$ as the importance-sampling density. Using this approach, Xie et al. (2011) show that

$$\hat{r}_{\text{SS},k} = \frac{1}{n} \sum_{i=1}^{n} p(Y \mid \theta_i, M)^{\beta_k - \beta_{k-1}}, \tag{4.23}$$

where $\theta_{k-1,i}$ is an MCMC sample from $p_{\beta_{k-1}}$ and $p(Y \mid \theta_i, M)$ is the likelihood of that sampled parameter vector. By factoring out the largest sampled likelihood term to improve numerical stability, i.e., $L_{\max,k} = \max_{1 \le i \le n} p(Y \mid \theta_{k-1,i}, M)$ and by combining all K ratios, the SS estimate of the marginal likelihood is

$$\hat{r}_{\text{SS}} = \prod_{k=1}^{K} \hat{r}_{\text{SS},k} = \prod_{k=1}^{K} \frac{1}{n} (L_{\max,k})^{\beta_k - \beta_{k-1}} \sum_{i=1}^{n} \left(\frac{p(Y \mid \theta_{k-1,i}, M)}{L_{\max,k}} \right)^{\beta_k - \beta_{k-1}}.$$

Being a product of independent unbiased estimators, $\hat{r}_{SS}$ is itself unbiased. On the log scale, we arrive at

$$\log \hat{r}_{\mathrm{SS}} = \sum_{k=1}^{K} [(\beta_k - \beta_{k-1}) \log L_{\max,k}] + \sum_{k=1}^{K} \log \left(\frac{1}{n} \sum_{i=1}^{n} \exp \left\{ (\beta_k - \beta_{k-1}) \times [\log p(Y \mid \theta_{k-1,i}, M) - \log L_{\max,k}] \right\} \right). \tag{4.24}$$

Xie et al. (2011) report that, although $\hat{r}_{SS}$ is unbiased, changing to the log scale introduces a bias. This bias appears to be directly proportional to the variance of $\log \hat{r}_{SS}$, which can be alleviated by increasing K.

As stated before, Xie et al. (2011) show that the efficiency of PS can dramatically improve by choosing β values according to evenly spaced quantiles of a Beta$(\alpha, 1.0)$ distribution rather than spacing β values evenly from 0.0 to 1.0. A value of $\alpha = 0.3$ is close to optimal for their Gaussian model example and may perhaps be generalized to other problems. The choice $\alpha = 0.3$ implies that half of the β values being evaluated are less than 0.1. The authors state that the positive skewness of this distribution is useful because (with sufficient and informative data) the likelihood only begins losing control over the power posterior for β values near 0, and at that point, the target distribution changes rapidly from something resembling the posterior to something resembling the prior. Conditioning on the total number of β values evaluated, placing most of the computational effort on β values near zero, results in increased accuracy. In BEAST, we provide flexibility to the user in specifying the spread of power values. However, the paper presenting results of the BEAST implementation (Baele et al., 2012a) follows the recommendation by Xie et al. (2011) .

4.3.2.2 Direct Bayes factor estimation

Recently an extension of SS to directly calculate (log) Bayes factors was presented (Baele et al., 2013a). This approach represents the stepping-stone version of model-switch thermodynamic integration, with the term "model-switch" referring to a single path that directly connects the two models in the space of unnormalized densities. Whereas such a general approach to directly estimate (log) Bayes factors is relatively new in the field of phylogenetics (Lartillot and Philippe, 2006a), the idea was first introduced in the statistics literature by Meng and Wong (1996), who propose a number of approaches to calculate the ratio of two normalizing constants.

Consider again the unnormalized density function q_β, which constitutes a direct path between the two competing models M_0 and M_1 (Lartillot and Philippe, 2006a) and has normalizing constant c_β yielding the normalized density p_β:

$$q_\beta(\theta) = [p(Y \mid \theta, M_0)p(\theta \mid M_0)]^{1-\beta}[p(Y \mid \theta, M_1)p(\theta \mid M_1)]^\beta,$$
$$p_\beta(\theta) = q_\beta(\theta)/c_\beta(\theta),$$
$$c_\beta = \int_\Theta q_\beta(\theta)d\theta,$$

where again Y represents the data, θ is the vector of model parameters, $M_i, i = 0, 1$ are the two models under consideration, $p(Y \mid \theta, M_i), i = 0, 1$ are the likelihood functions and $p(\theta \mid M_i), i = 0, 1$ the priors. Similar to the original stepping-stone method, the goal is to estimate the ratio $c_{1.0}/c_{0.0}$ and this ratio can be expressed as a product of K ratios:

$$r = \frac{c_{1.0}}{c_{0.0}} = \prod_{k=1}^{K} \frac{c_{\beta_k}}{c_{\beta_{k-1}}},$$

where $0 = \beta_0 < \ldots < \beta_{k-1} < \beta_k < \ldots < \beta_K = 1$. Each ratio $c_{\beta_k}/c_{\beta_{k-1}}$ can be estimated by importance sampling, using $p_{\beta_{k-1}}$ as the importance sampling density. Because $p_{\beta_{k-1}}$ is only slightly different from p_{β_k}, it serves as a useful importance distribution.

An estimator $\hat{r}_k$ is constructed using samples $\theta_{k-1,i} (i = 1, 2, \ldots, n)$ from $p_{\beta_{k-1}}$:

$$\hat{r}_k = \frac{1}{n} \sum_{i=1}^{n} \left[\frac{p(Y \mid \theta_{k-1,i}, M_1)p(\theta_{k-1,i} \mid M_1)}{p(Y \mid \theta_{k-1,i}, M_0)p(\theta_{k-1,i} \mid M_0)}\right]^{\beta_k - \beta_{k-1}}.$$

Numerical stability can be improved by factoring out the largest sampled term $\eta_k = \max_{1 \le i \le n}\{p(Y \mid \theta_{k-1,i}, M_1)p(\theta_{k-1,i} \mid M_1)/p(Y \mid \theta_{k-1,i}, M_0)p(\theta_{k-1,i} \mid M_0)\}$. Combining all K ratios, the SS estimate of the Bayes factor is

$$\hat{r} = \prod_{k=1}^{K} \hat{r}_k = \prod_{k=1}^{K} \frac{1}{n}(\eta_k)^{\beta_k - \beta_{k-1}} \sum_{i=1}^{n} \left[\frac{p(Y \mid \theta_{k-1,i}, M_1)p(\theta_{k-1,i} \mid M_1)}{\eta_k p(Y \mid \theta_{k-1,i}, M_0)p(\theta_{k-1,i} \mid M_0)}\right]^{\beta_k - \beta_{k-1}}.$$

As for the marginal likelihood estimator based on stepping-stone sampling (Xie et al., 2011), $\hat{r}$ is unbiased because it results from a product of independent unbiased estimators. On the log scale, the necessary calculations and the summation over all K ratios yields the following overall estimator for $\log \hat{r}_k$:

$$\begin{aligned}
\log \hat{r} &= \sum_{k=1}^{K} \log \hat{r}_k \\
&= \sum_{k=1}^{K} [(\beta_k - \beta_{k-1})\log \eta_k] \\
&+ \sum_{k=1}^{K} \log \left(\frac{1}{n} \sum_{i=1}^{n} \left[\frac{p(Y \mid \theta_{k-1,i}, M_1)p(\theta_{k-1,i} \mid M_1)}{\eta_k p(Y \mid \theta_{k-1,i}, M_0)p(\theta_{k-1,i} \mid M_0)}\right]^{\beta_k - \beta_{k-1}}\right).
\end{aligned}$$

4.3.3 Generalized stepping-stone sampling

Although associated with lower estimation error, constructing a path between two arbitrary models is technically challenging and difficult to implement in a generic fashion. The question that naturally arises from this is whether marginal likelihoods can be estimated for each model separately but with shorter paths from posterior to prior. Successfully shortening this path would reduce the computational effort associated with marginal likelihood estimation while offering comparable accuracy. Based on the geometric path approach by Lefebvre et al. (2010), Fan et al. (2011) propose a more general version of SS (i.e., generalized stepping-stone sampling; GSS) that is considerably more efficient and does not require sampling from distributions close to the prior (which can be problematic for vague priors). This generalized SS (or GSS) introduces a reference distribution, which in practice is a product of independent probability densities parameterized using samples from the posterior distribution. While SS does not require samples from the posterior distribution, in practice, the posterior is typically explored as a first step in the chain. GSS on the other hand uses these samples from the posterior to parameterize its reference distribution or alternatively, allows for samples from a previous extensive MCMC analysis of the posterior to be used to parameterize the reference distribution.

To derive the log marginal likelihood estimator using GSS, consider the unnormalized density function q_β, which has normalizing constant c_β yielding the normalized density p_β:

$$q_\beta = [p(Y \mid \theta, M)p(\theta \mid M)]^\beta [p_0(\theta \mid M)]^{1-\beta},$$

$$p_\beta = q_\beta / c_\beta,$$

and

$$c_\beta = \int_\Theta q_\beta d\theta,$$

where $p(Y \mid \theta, M)$ is again the likelihood function, $p(\theta \mid M)$ the actual model prior, and $p_0(\theta \mid M)$ the reference distribution. GSS proposes to collect samples from a different series of power posteriors than PS and SS, since the latter methods construct power posteriors that can be difficult to sample if β is close to 0.0 and the prior is diffuse (which is frequently the case). In the case of GSS, the power posterior is equivalent to the posterior distribution when $\beta = 1.0$ but equivalent to the reference distribution when $\beta = 0$, which is different from the PS and SS approaches, where the actual prior is sampled when $\beta = 0$. Using a reference distribution facilitates sampling from p_β, regardless of the value of β. GSS also has a practical advantage. Software packages often employ rescaling techniques to avoid under- or over-flows (i.e., rounding errors) when calculating (partial) likelihoods, especially on larger phylogenies (Yang, 2000b; Suchard and Rambaut, 2009), which seem to occur more frequently for β values close to 0.0 when performing PS and SS. Given that GSS constructs a path from the posterior to the reference distribution, and hence avoids collecting samples close to the actual prior, it is expected that such rounding errors will occur less frequently than with PS/SS, leading to overall faster calculations.

The goal of GSS is to estimate the ratio $c_{1.0}/c_{0.0}$, which is equivalent to the marginal likelihood as $c_{0.0} = 1.0$ if the reference distribution is proper. This ratio is estimated as a product of K ratios:

$$r = \frac{c_{1.0}}{c_{0.0}} = \prod_{k=1}^{K} \frac{c_{\beta_k}}{c_{\beta_{k-1}}},$$

where $0 = \beta_0 < \ldots < \beta_{k-1} < \beta_k < \ldots < \beta_K = 1$. Each ratio $r_k = c_{\beta_k}/c_{\beta_{k-1}}$ can be estimated by importance sampling, using $p_{\beta_{k-1}}$ again as the appropriate importance sampling density. An estimator $\hat{r}_k$ for the ratio $c_{\beta_k}/c_{\beta_{k-1}}$ is constructed using samples $\theta_{k-1,i} (i = 1, 2, \ldots, n)$ from $p_{\beta_{k-1}}$ and numerical stability is improved by factoring out the largest sampled term, $\eta_k = \max_{1 \leq i \leq n}\{p(Y \mid \theta_{k-1,i}, M)p(\theta_{k-1,i} \mid M)/p_0(\theta_{k-1,i} \mid M)\}$:

$$\hat{r}_k = \frac{1}{n}(\eta_k)^{\beta_k - \beta_{k-1}} \sum_{i=1}^{n} \left[\frac{p(Y \mid \theta_{k-1,i}, M)p(\theta_{k-1,i} \mid M)}{\eta_k p_0(\theta_{k-1,i} \mid M)}\right]^{\beta_k - \beta_{k-1}}.$$

Transformation to the log scale and summing over all K ratios yields the overall estimator:

$$\begin{aligned}
\log \hat{r} &= \sum_{k=1}^{K} \log \hat{r}_k \\
&= \sum_{k=1}^{K} [(\beta_k - \beta_{k-1}) \log \eta_k] \\
&\quad + \sum_{k=1}^{K} \log \left\{ \frac{1}{n} \sum_{i=1}^{n} \left[\frac{p(Y \mid \theta_{k-1,i}, M)p(\theta_{k-1,i} \mid M)}{\eta_k p_0(\theta_{k-1,i} \mid M)}\right]^{\beta_k - \beta_{k-1}} \right\}.
\end{aligned}$$

This approach reduces to the SS method if the reference distribution is equal to the actual prior (Fan et al., 2011). However, Fan et al. (2011) suggest choosing a reference distribution closer to the posterior than the actual prior and, in practice, using samples from the posterior distribution ($\beta_k = 1$) to parameterize the joint reference distribution $p_0(\theta \mid M)$. Note also that, if the reference distribution equals the posterior distribution, the marginal likelihood can be estimated exactly (i.e., in this case, MCMC sampling error does not affect the estimate).

A simulation study by Fan et al. (2011) demonstrated that the HME is far less repeatable than GSS while SS has intermediate in repeatability. This is of practical importance, as it means that less computational effort needs to be spent on GSS compared to SS to yield estimates with the same amount of variance. The increased stability and efficiency of the GSS estimator of (log) marginal likelihood is a direct consequence of choosing a reference distribution that approximates the posterior rather than the prior. Fan et al. (2011) indicate that the estimator is more stable because there is more similarity across the series of power posterior distributions being explored, and sampling does not become problematic when the power β is close to zero. The authors state that sampling actually becomes more straightforward at the reference distribution end of the path. Bayesian phylogenetics can, however, not fully exploit the advantages of GSS as the method has heretofore been restricted to evaluation on a fixed phylogenetic tree topology (but see Chapters 5 and 6). Integrating over plausible tree topologies poses a considerable challenge to the application of generalized SS because it requires a reference distribution for topologies that provides a good approximation of the posterior tree distribution.

4.4 Simulation study: uncorrelated relaxed clocks

In this section, we conduct a simulation study to assess the performance of a series of model selection approaches in the context of relaxed molecular clock models, and compare this to a *maximum a posteriori (MAP)* classifier that is produced by a model averaging approach. We present these clock models, discuss the differences between model selection and model averaging, and highlight the importance of using proper priors on all the parameters of our full probabilistic model. We extend a previously analyzed simulation study to further enhance our understanding of various model selection approaches and conclude that frequently used estimators such as the HME perform poorly for the models we examine.

4.4.1 Relaxed molecular clock models

Until about a decade ago, the phylogenetic community generally considered branch lengths on a tree to represent genetic distance in units of substitutions per site. The introduction of maximum-likelihood estimation to evaluate evolutionary trees (Felsenstein, 1981) in the early 80's made it possible to infer unrooted tree topologies in a model-based framework. At the basis of this approach lies the assumption that every branch has an independent rate of molecular evolution, allowing users to infer unrooted phylogenies but not to estimate molecular rates or divergence times, because the individual contributions of rate and time to molecular evolution cannot be separated. If the rate and time along each branch can only be estimated as their product, then the position of the root of the tree cannot be estimated without additional assumptions such as an outgroup or a non-reversible substitution process (Drummond et al., 2006).

To obtain a phylogeny that is rooted and has the notion of time, an assumption is needed that describes the relationship between the accumulation of genetic divergence and time. Such an assumption leads to biologically more relevant estimates of branch lengths and allows reconstructing the temporal aspect of evolutionary history and dissecting the processes involved (Drummond et al., 2006). The classical approach assumes that rates conform to a (strict) "molecular clock" which dictates that the rates are equal across all branches on a tree (Zuckerkandl and Pauling, 1965). In reality, however, rates of substitution differ across species as a result of variation in mutation rates, metabolic rates, generation times, population size and structure, selection, etc. Unfortunately, rate variation among lineages can seriously mislead not only divergence date estimation (Yoder and Yang, 2000b) but also phylogenetic inference (Felsenstein, 1978a). Despite the fact that statistical testing showed departures from clock-like evolution in many datasets and the consequences this may have, evolutionary biologists had to resort to the strict clock model to estimate divergence times and rates of evolution.

To relax the strict clock assumption, local molecular clock models have been proposed that allow rates to vary on particular branches (Rambaut and Bromham, 1998), but this is only useful when a prior hypothesis can be specified and the approach may suffer from identifiability issues when specifying too many separate parameters. Models have also been developed that describe a process of change in evolutionary rates throughout the tree, such as autocorrelated clocks, where the rate on a branch depends on the rate of its parent branch (Aris-Brosou and Yang, 2002). Relaxed molecular clock models have, in recent years, been accepted into the broader field of phylogenetics because of their biological relevance. They improve the accuracy of phylogenetic estimation and hence also divergence time estimation (Felsenstein, 1978a). For a more thorough overview on molecular clock model development, we refer the reader to Drummond et al. (2006) (also see Chapter 13).

An alternative to autocorrelated relaxed clocks is the use of uncorrelated relaxed clocks, which do not assume *a priori* correlation of rates on adjacent branches in a tree. Instead, branch-specific rates are drawn independently and identically from an underlying rate distribution (Drummond et al., 2006). Originally, two candidate distributions were proposed to draw rates on branches: the uncorrelated exponential distribution (often denoted UCED) and the uncorrelated lognormal distribution (often denoted UCLD):

$$r \sim \text{Exp}(\lambda) \tag{4.25}$$

and

$$r \sim \text{LogNormal}(\mu, \sigma^2). \tag{4.26}$$

More recently, the use of an inverse Gaussian (IG) distribution as a model for the distribution of rates across branches was also investigated (Li and Drummond, 2012). While the probability distribution function of the IG is similar to the lognormal when the coefficient of variation is low for the latter (i.e., less than 1), the IG distribution is more liberal in allowing for relatively faster rates of evolution within the tree, which may be suitable for datasets where there are taxa with exceptionally fast rates (such as mammalian datasets where rodent lineages have accelerated rates of substitution (Drummond et al., 2006)). While we do not use the IG distribution in this chapter, we discuss it here for the sake of completeness:

$$r \sim \text{InverseGaussian}(\mu, \lambda), \tag{4.27}$$

where μ is the mean and λ is the shape parameter. We refer to Li and Drummond (2012) for implementation details of the IG distribution. While the Bayesian model averaging (BMA) approach is applied to the UCED and UCLD models in this chapter, an example of BMA with the three uncorrelated relaxed molecular clock models can be found in the work of Li and Drummond (2012).

As shown by Drummond et al. (2006), (4.25), (4.26), and (4.27) can be reformulated as a full likelihood model, in which case the branch rates are constrained so as to fit one of the proposed distributions exactly. The parameters of the rate distribution are then parameters of the likelihood model, analogous to the way rate heterogeneity among sites is modeled (Yang, 1996). In a Bayesian framework such as BEAST, which we use throughout this chapter, these parameters are co-estimated with the substitution parameters, the tree topology, and other parameters in the full probabilistic model. It is important to note that in BEAST, the branch rates are not integrated analytically according to the Felsenstein approach, but they are instead integrated using MCMC (Drummond et al., 2006). In other words, a data augmentation approach is used to estimate the branch rates across the tree, which is achieved by assigning a unique rate category $c \in \{1, 2, \ldots, 2n-2\}$ to each branch of the tree (with n the number of taxa in the alignment). The rate category c is converted to a rate as follows:

$$r_c = D^{-1}\left(\frac{c - 1/2}{2n - 2}\right),$$

with the function $D^{-1}(x)$ the inverse function of the probability distribution function of the relaxed-clock model specified by (4.25), (4.26), and (4.27), which discretizes the underlying rate distribution in $2n - 2$ rate categories. The assignment of rate categories c to branches is sampled via MCMC to integrate the branch rates out by using various so-called moves (see Drummond et al. (2006)). This is a computationally interesting alternative to other data augmentation approaches, that calculate a multinomial distribution for the branch rate at each of the branches of the underlying tree, and involves recalculating the likelihood for each possible rate category at that branch.

4.4.2 Comparing model selection to model averaging

Like any other statistical modeling technique, relaxed molecular clock methods in Bayesian phylogenetics suffer from problems of model misspecification and uncertainty. Model misspecification is a deep-rooted problem that plagues a range of applications of mathematics across the sciences and can cause errors and bias in the resulting analysis (Li and Drummond, 2012). Complex processes in the field of molecular evolution are no exceptions to this, and evolutionary models will always be misspecified in the sense that they are all severe simplifications of reality (Li and Drummond, 2012). The aim of model selection in phylogenetics is thus not to find the true model that generated the data, but to select a model that best balances simplicity with flexibility and biological realism in capturing the key features of the data.

In this chapter, we describe several procedures to compare a wide range of evolutionary hypotheses and select the most likely model given the data at hand. Model selection, however, and the search for a single best-fit model typically ignore uncertainty about the "correct" model specification and can lead to over-confident inferences (Hoeting et al., 1999). Bayesian model averaging (BMA) approaches have been proposed to explicitly address model uncertainty by forming a posterior distribution over a set of candidate models (Madigan and Raftery, 1994). In general, BMA does not provide a method for identifying the most likely model within the candidate set of models, as the aim of BMA is often orthogonal to model selection efforts. However, specific BMA constructions can yield estimates of the posterior probabilities of each of the candidate models (Carlin and Chib, 1995). Carlin and Chib (1995) consider the case of choosing between K (possibly non-nested) models for an observed data vector Y, with a distinct parameter vector $\theta_j, j = 1, \ldots, K$ corresponding to each model and M be an integer-valued parameter that indexes the model. The goal is to estimate the posterior probabilities of each of the K models: $p(M = j \mid Y), j = 1, \ldots, K$, with the prior distributions $p(\theta_j), j = 1, \ldots, K$ as part of the Bayesian model specification and allowed to depend on the model indicator M. In other words, corresponding to model j are the likelihood $p(Y \mid \theta_j, M = j)$ and the prior $p(\theta_j \mid M = j)$. Since the primary goal of the approach of Carlin and Chib (1995) is the computation of (log) Bayes factors, each prior is assumed to be proper (though possibly quite vague). Complete independence among the various θ_j given the model indicator M is assumed, allowing the Bayesian model specification to be completed by choosing proper "pseudopriors" $p(\theta_j \mid M \neq j)$. Given prior model probabilities $\pi_j \equiv P(M = j)$ such that $\sum_{j=1}^{K} \pi_j = 1$, the full conditional distribution of each θ_j is:

$$p(\theta_j \mid \theta_{i \neq j}, M, Y) \propto \begin{cases} p(Y \mid \theta_j, M = j) p(\theta_j \mid M = j), & M = j, \\ p(\theta_j \mid M \neq j), & M \neq j, \end{cases} \tag{4.28}$$

i.e., when $M = j$, a new parameter value is generated from the usual model j full conditional; when $M \neq j$, a new parameter value is generated from the "pseudoprior". While it would be tempting to skip the generation of actual "pseudoprior" values and instead to simply keep $\theta_j^{(g)}$ at its current value when $M^{(g)} \neq j$, such an approach is not a Gibbs sampler in the strict sense (Carlin and Chib, 1995). The full conditional distribution for M is

$$p(M = j \mid \theta, Y) = \frac{p(Y \mid \theta_j, M = j) \left\{ \prod_{i=1}^{K} p(\theta_i \mid M = j) \right\} \pi_j}{\sum_{k=1}^{K} p(\theta_k \mid M = k) \left\{ \prod_{i=1}^{K} p(\theta_i \mid M = k) \right\} \pi_k},$$

with $\theta = \{\theta_1, \ldots, \theta_K\}$. The ratio

$$\hat{p}(M = j \mid Y) = \frac{\text{number of } M^{(g)} = j}{\text{total number of } M^{(g)}}, \quad j = 1, \ldots, K,$$

with $\{M^{(g)}, g = 1, \ldots, G\}$ a sample from the marginal posterior distribution for M, provides simple estimates that may be used to compute the Bayes factor between any two of the models (Carlin and Chib, 1995). In other words, dividing the posterior odds of two models by their prior odds returns their Bayes factor, or ratio of marginal likelihoods. This construction offers a way to quantify posterior probability support for the different models within the candidate set of models. A potential limitation to this approach centers on the accuracy to which one can estimate each model's posterior probability, since these estimates often fall very close to 0 or 1 without adjusting the prior probabilities (Suchard et al., 2005).

In recent work, Li and Drummond (2012) developed a BMA approach that considers a candidate set of two relaxed molecular clock models in Bayesian phylogenetics, the UCED and UCLD. Li and Drummond (2012) show that their BMA method accurately recovers the true underlying distribution of rates. The authors use the IG distribution in a consistency check of Bayes factor computations, i.e., the Bayes factor is computed using their BMA approach with or without a third model (i.e., the IG distribution) in the mixture. Li and Drummond (2012) state that, if the MCMC is run sufficiently long, the Bayes factor computed for a pair of models should be the same whether a third model is present in the set of models or not. Using a large number of genes, the authors found a very strong correlation of the log Bayes factors between the UCED and UCLD models, estimated with and without the IG distribution in the mixture. Since their construction returns estimates of the posterior probability of each model, Li and Drummond (2012) also examine the performance of identifying the maximum *a posteriori* (MAP) model under BMA as a model selection criterion. The authors found that across a large set of alignments taken from a dataset of 12 mammalian species, a model with log-normally distributed rates is more likely than exponentially distributed rates in most of the alignments. This is not surprising because the exponential distribution has a mode at 0 whereas the log-normal distribution allows for a more flexible modeling of the rates.

Given the concurrent development and application of state-of-the-art approaches to estimate (log) marginal likelihoods to perform model selection, such as PS and SS (Baele et al., 2012a), and the development of a Bayesian model averaging approach that avoids conditioning on a single model but averages over a set of relaxed-clock models and allows one to estimate the Bayes factor in favor of the maximum *a posteriori* (MAP) clock model, a comparison between model selection and model averaging for relaxed molecular clocks seemed to be in order (Baele et al., 2013b). Both model selection and model averaging have reported dramatic improvements over the frequently used HME, but they had not been compared to each other yet. Baele et al. (2013b) therefore compared these two recent developments with the HME, stabilized/smoothed HME (sHME), and AICM, using both synthetic and empirical data. The comparison shows reassuringly that MAP identification and its Bayes factor provide similar performance to PS and SS, and that these approaches considerably outperform HME, sHME, and AICM in selecting the correct underlying clock model. Further, the authors illustrate the importance of using proper priors on a large set of empirical datasets. While the log Bayes factors calculated by PS, SS, and MAP were strongly correlated when analyzing empirical datasets, and the classification outcomes of the performed simulations were highly similar, the performance of approaches seemed somewhat unsatisfactory for simulated datasets using Yule trees (Baele et al., 2013b). We here report and expand on these simulations to obtain a better understanding of the behavior of Bayesian model selection and averaging tools.

4.4.3 Simulation settings

In previous work (Baele et al., 2013b), we simulated data using different assumptions concerning the underlying tree. We first assumed a balanced tree with 32 taxa and an additional outgroup, with branch lengths set to 5 time units, except the outgroup branch which had a length of 30 to make the tree ultrametric. For each of the branches on a tree, we assigned a rate of substitution drawn from either a UCED with a mean of 0.005 or a UCLD with a mean of 0.005 and variance of 0.004. One hundred realizations of rates were simulated under each of the two relaxed-clock models and alignments of 1000 bp in length were subsequently generated using Seq-Gen (Rambaut and Grassly, 1997) under a Hasegawa-Kishino-Yano (Hasegawa et al., 1985, HKY) model with gamma-distributed rate heterogeneity across sites (Yang, 1996) with a transition–transversion ratio of 3.0 and a gamma shape parameter of 0.5. We used the simulation settings described in Li and Drummond (2012), but specified proper prior distributions in our analyses. A second series of simulations were generated using a collection of 100 trees with 32 taxa generated using a Yule birth process with a birth rate of 0.2, leading to an average branch length of ~ 2.5 time units. We applied the same uncorrelated molecular clock parameterizations as for the balanced trees. One hundred realizations of rates were simulated under each of the two relaxed-clock models and alignments of 1000 bp, 2500 bp, and 5000 bp in length were subsequently generated using Seq-Gen (Rambaut and Grassly, 1997) under the same model settings. A third and final set of simulations were generated in a similar fashion as the second set, but with 64 taxa instead of 32.

In BEAST, default model specifications do not always include proper prior probability distributions (integrating to 1). The frequently used constant function on an infinite interval is often inaccurately called a uniform distribution, although it is actually an example of an improper prior. In general, the use of such priors can lead to posterior distributions that do not exist (i.e., are not probability distributions). In practice, since standard floating-point representations of numbers have a maximum attainable value, the implementation of this prior in a computer can actually be regarded as proper, but regardless, such priors are effectively improper and provide a great challenge to MCMC sampling. This is evident in the results presented in previous work (Baele et al., 2013b). Here, we follow the prior specifications from this study: a Yule pure birth process was used as a prior on the speciation process to analyze the simulations (Yule, 1924), with a diffuse normally distributed prior on the log birth rate ($\log\mu = 1.0$, $\log\sigma = 1.25$); a diffuse normally distributed prior on the log transition–transversion parameter of the HKY model ($\log\mu = 1.0$, $\log\sigma = 1.25$) for the simulations; an exponential prior (with mean 0.5) on the rate heterogeneity parameter (Yang, 1996); an exponential prior (with mean 1/3) on the standard deviation of the UCLD clock model; and a Dirichlet(1,1,1,1) distribution on the base frequencies.

4.4.4 Results

Marginal likelihoods are calculated using PS (Lartillot and Philippe, 2006a), SS (Xie et al., 2011), HME (Newton and Raftery, 1994c), and sHME (Redelings and Suchard, 2005). We use the recommendation of Xie et al. (2011) in selecting β values between the path from prior to posterior according to evenly spaced quantiles of a Beta(α, 1.0) distribution and therefore choose $\alpha = 0.3$, which places most of the computational effort on β values near zero. In addition, we employ the AICM to perform model selection (Raftery et al., 2007b). For more detail on these methods and their implementation in BEAST (Drummond et al., 2012), we refer interested readers to Baele et al. (2012a). BMA between the UCED and UCLD clock models was performed using the approach of Li and Drummond (2012) that assumes equal prior probability for each model, i.e., there is no prior knowledge as to which model is preferred. As a consequence, the ratio of the posterior probabilities of the two models equals the Bayes factor as the prior odds equals 1. We use a cutoff of 0 for the log Bayes factor in judging which of the two uncorrelated clock models is preferred. The results for the different simulation scenarios are shown in Tables 4.1 and 4.2.

As discussed in previous work (Baele et al., 2013b), and as shown in Table 4.1, for the data simulated under the UCED clock model and a balanced tree, the HME, PS, SS, and MAP recover the true model in high frequencies (between 90% and 94%) and the sHME and AICM always recover the true model. However, when the data are simulated under the UCLD clock model assuming a balanced tree, the HME, sHME, and AICM recover the true model only in very low frequencies, whereas PS, SS, and MAP almost always classify

Table 4.1: Model selection performance for 100 simulated datasets, consisting of 32 taxa, under either a balanced or Yule tree and two relaxed molecular clock models using HME, sHME, AICM, PS, SS, and the maximum *a posteriori model* estimated under BMA. The columns report the number of correct classifications obtained out of 100 simulations.

Tree	Clock	Length	HME	sHME	AICM	PS	SS	MAP
Balanced	UCED	1.000	92	100	100	94	94	90
Balanced	UCLD	1.000	28	5	1	99	99	99
Yule	UCED	1.000	92	100	100	99	99	97
Yule	UCLD	1.000	11	1	1	61	61	65
Yule	UCED	2.500	89	100	100	98	98	99
Yule	UCLD	2.500	26	5	9	83	82	81
Yule	UCED	5.000	78	99	99	98	98	98
Yule	UCLD	5.000	38	11	12	82	82	82

Table 4.2: Model selection performance for 100 simulated datasets, consisting of 64 taxa, under a Yule tree and two relaxed molecular clock models using HME, sHME, AICM, PS, SS, and the maximum *a posteriori model* estimated under BMA. The columns report the number of correct classifications obtained out of 100 simulations.

Tree	Clock	Length	HME	sHME	AICM	PS	SS	MAP
Yule	UCED	1.000	91	100	100	99	99	98
Yule	UCLD	1.000	9	0	1	89	88	92

the relaxed-clock model correctly for the simulations under a balanced tree. The results for MAP are consistent with those reported by Li and Drummond (2012) (where 83 and 100 correct classifications were obtained using the MAP under the UCED and UCLD clock model, respectively). The increase in number of correct classifications we observe may be attributed to the use of proper priors in the analyses performed in this study.

For the datasets simulated under 100 different Yule trees and the UCED clock model, the performance of the various methods remains high. However, when sequence data is simulated under the UCLD clock model using the same collection of Yule trees, all approaches take a performance hit. While HME, sHME, and AICM were already performing poorly, they now almost never correctly recover the clock model under which the data was generated. Also, the state-of-the-art PS, SS, and MAP methods suffer from a considerable decrease in performance, recovering the true clock model in between 61% to 65% of the datasets. PS, SS, and MAP still clearly outperform HME, sHME, and AICM when simulating under a Yule birth process, but no longer achieve the near-ideal performance of the simulations performed under a balanced tree.

Even though PS, SS, and MAP offer improved performance over the HME, sHME, and AICM estimators for the datasets generated under the Yule trees, we here perform additional simulations to gain a better understanding of the performance of the various methods. Sequence length will determine our ability to estimate branch lengths and rate variability across branches, and this may therefore impact the performance of our model selection procedures. Therefore, we increase the sequence length to 2500 bp and 5000 bp, using the same set of underlying Yule trees, and analyze these datasets using the same proper priors and model selection approaches.

Using simulated sequences of 2500 bp, all methods still achieve near-ideal performance for data simulated under the UCED clock, with only the HME showing a small drop in the number of correct classifications. When the longer sequence data is simulated under a UCLD clock model, the performance of PS, SS, and MAP increases to between 81% and 83% compared to the simulated datasets of 1000 bp, which we consider to be much more acceptable, given the additional complexity and computational demands of these methods. The performance of the HME, sHME, and AICM also increases compared to the simulated datasets of 1000 bp, although their overall performance remains poor. Using simulated sequences of 5000 bp, the performance of PS, SS, and MAP is nearly identical to that for the sequences of 2500 bp, both for the data simulated under the UCED and the UCLD clock models. The performance of the HME drops marginally for the data sets simulated under the UCED model, but remains acceptable. The HME, sHME, and AICM increases slightly compared to the simulated datasets of 2500 bp, but remains poor.

While the results in Table 4.1 show that for sequences of over 2500 bp in length, the performance of PS, SS, and MAP is within acceptable ranges, there appears to be limit in performance increase that can be achieved by longer sequences. We subsequently examine the impact of the number of taxa by simulating additional data using Yule trees, keeping the sequence length at 1000 bp, but increasing the number of taxa from 32 to 64. Currently, this may still represent moderate size dataset, but such a simulation scenario may allow extrapolating to even larger datasets. The results for this simulation scenario are shown in Table 4.2.

For simulated datasets consisting of 64 taxa, the HME performs really well and the sHME and AICM yield perfect performance when the data are simulated under the UCED clock model. In this case, the performance of PS, SS, and MAP is also near perfect and again consistent among estimators. When the data is simulated under the UCLD model, the posterior-based estimators HME, sHME, and AICM yield extremely poor performance. However, PS, SS, and MAP achieve the best performance seen so far for the Yule trees, outperforming the previous scenarios that contain much longer sequences. This bodes really well for the extrapolation of the performance of PS, SS, and MAP towards real-world datasets, given that enough sequences are present.

In conclusion, Tables 4.1 and 4.2 show that PS, SS, and MAP offer similar performance in assessing the correct relaxed molecular clock model and clearly outperform the HME, sHME and AICM. These results again stress the poor performance of frequently used posterior-based estimators such as the HME. We have shown here that PS, SS, and MAP can reliably be used for real-world datasets, with performance increasing with the number of sequences and the number of sites. This may not be surprising as with more taxa, many more branches will be present on which the rate variation process can be realized. Sequence length determines how well we can estimate those branches but there is a limit to which they contribute in distinguishing between different processes of rate variation. In the next section, we perform a thorough model comparison of demographic models for an HIV-1 dataset for which selection of the best demographic model has proven difficult.

4.5 Practical example on demographic models

In this section, we focus our attention on the analysis of a real-world HIV-1 dataset. After an initial discussion of this dataset and the demographic models that are typically compared to obtain an estimate of the change in effective population size through time, we examine the computational requirements needed to accurately estimate marginal likelihoods for this specific dataset. Compared to our initial analysis of this dataset (Baele et al., 2012a), we here opt for using the Bayesian skyride model as the multiple change-point (MCP) model of choice, rather than the Bayesian skyline model, a decision we motivate in this section. We again stress the importance of using proper priors, here specifically for the Bayesian skyride model, and try to determine the best-fitting demographic model to the data.

4.5.1 HIV-1 group M dataset

We revisit a Bayesian evolutionary reconstruction of the HIV-1 group M epidemic history originally performed by Worobey et al. (2008). This study examines sequence data from a 1960 specimen from Leopoldville in the Belgian Congo (now Kinshasa, Democratic Republic of the Congo) that shows considerable divergence from the 1959 (ZR59) sequence (Zhu et al., 1998), the oldest and only known sequence sampled before 1976 at that time. Because sequences pre-dating the recognition of AIDS are critical to defining the time of origin and the timescale of virus evolution, the authors include these in a relaxed molecular clock analysis and estimated an origin of group M near the beginning of the twentieth century (Worobey et al., 2008).

By estimating the demographic history using several coalescent models (or coalescent priors; see next section), for a dataset both including and excluding the 1959 and 1960 sequences, the authors show that the inclusion of the 1959 and 1960 sequences seemed to improve estimation of the most recent common ancestor (TMRCA) of the M group, limiting the influence of the coalescent tree prior on the posterior TMRCA distributions compared with the dataset that excluded these earliest cases of HIV-1. This approach robustly incorporates phylogenetic uncertainty and accounts for the possibility of variable substitution rates among lineages and differences in the demographic history of the virus (Worobey et al., 2008). Because coalescent priors may impact TMRCA estimation, and the viral population history may itself be of considerable interest, model comparison of different coalescent models seems in order. In the original analysis (Worobey et al., 2008), the HME suggests that a constant population size model provided the best fit to the data. This appears to be at odds with the presumed viral population expansion over the last century, which is also evident from the Bayesian skyline plot reconstruction the authors presented. Worobey et al. (2008) state that the inability to reject the constant population size model is counterintuitive because it is clear that the HIV-1 population size has increased notably and speculate that this finding might be attributed to the fact that the simplest model provides a good fit to a relatively short, information-poor alignment, and therefore is preferred over more parameter-rich, but more realistic models.

4.5.2 Demographic/coalescent models

Coalescent theory refers to a group of mathematical models that describe the statistical properties of intra-population phylogenies, also referred to as genealogies (Kingman, 1982b). This theory is used to estimate population genetic parameters, such as the effective population size, from the genealogical coalescent time distribution, which can in turn be estimated from a sample of gene sequences. The shared history of these sampled sequences creates a genealogy, the lineages of which "coalesce" as time progresses into the past, until the most recent common ancestor of the sample is reached (Pybus and Rambaut, 2002). One particular coalescent model, the variable population size model, describes how the shape of the genealogy depends on the demographic history of the sampled population (Griffiths and Tavaré, 1994). It provides a probability distribution for the waiting times between coalescent events (i.e., when two lineages merge) in a genealogy (Pybus and Rambaut, 2002). This distribution depends on the demographic function which represents the effective population sizes at a given time before the present. The demographic function constitutes a demographic hypothesis and is represented using demographic models, i.e., simple mathematical functions that characterize biologically plausible population dynamic histories (Pybus and Rambaut, 2002). Each demographic model has one or more demographic parameters.

Commonly used demographic models include the constant population size (one parameter), exponential growth (assuming a constant growth rate through time; two parameters), expansion growth (assuming an increasing growth rate through time; three parameters), logistic growth (assuming a decreasing growth rate through time; three parameters). However, it is not usually known in advance which scenario is the most plausible historical scenario for a particular population. Hence, often the fit of different demographic models is compared using standard model selection techniques, which is often considered to be time-consuming and does not guarantee that any of the compared models will fit the data adequately (Drummond et al., 2006). The use of an incorrect demographic model will lead to biased and invalid estimates of demographic history. Furthermore, an incorrect demographic model could lead to bias in other estimated parameters of the evolutionary model, such as the time to the TMRCA, when the demographic history is being treated as a nuisance parameter to be averaged over (Worobey et al., 2008; Gill et al., 2012).

To address these concerns, a flexible model called the skyline plot was developed (Pybus et al., 2000a), which is a piecewise-constant model of population size that can fit a wide range of demographic scenarios. It has proved very useful as a model selection tool that visually indicates the most appropriate demographic model for any given dataset (Pybus and Rambaut, 2002). The Bayesian skyline model represents an extension of the "classical" skyline model because it enables inferring demographic history from the sampled gene sequences, rather than from an estimated genealogy, and therefore incorporates the error associated with phylogenetic reconstruction (Drummond et al., 2006). The Bayesian skyline plot model uses standard Markov chain Monte Carlo (MCMC) sampling procedures to estimate a posterior distribution of effective population size through time directly from a sample of gene sequences, given any specified sequence substitution model. The Bayesian skyline plot provides credible intervals for the estimated effective population size at every point in time, back to the most recent common ancestor of the gene sequences. These credible intervals represent both phylogenetic and coalescent uncertainty. In summary, the Bayesian skyline plot demographic model provides a general, non-parametric coalescent prior that does not enforce a particular demographic history (Drummond et al., 2006).

4.5.3 Proper priors and the Bayesian skyride model

Worobey et al. (2008) considered several different demographic models that serve as prior distributions for time-measured trees and offer a glimpse into the population dynamics of the epidemic. In previous work (Baele et al., 2012a), we reanalyzed the HIV-1 dataset by performing two independent analyses for each possible coalescent prior model and applied the HME, AICM, PS, and SS to perform model selection. We found that, depending on the analysis on which we focused, the constant population model was either the best or the worst model according to the HME, highlighting the poor reliability of this estimator. Using PS and SS, however, the Bayesian skyline model outperformed all other models considered (constant size, exponential, expansion and logistic growth), whereas the constant population model was shown to perform considerably worse compared to the other demographic models. These results suggested that the constant population model was originally preferred not because of an information-poor alignment (Worobey et al., 2008), but because the HME fails to provide an adequate classification of the demographic models (Baele et al., 2012a).

In a recent large-scale comparison between model selection and model averaging (Baele et al., 2013b), we showed the importance of using proper priors when comparing relaxed-clock models. Unfortunately, not all of the (demographic) parameters in our analysis of the HIV-1 dataset were equipped with such proper priors (Baele et al., 2012a). Proper priors can be readily specified for the standard parametric models (constant population size, exponential, expansion, and logistic growth models). Prior specification is, however, less clear for the Bayesian skyline model, which is mostly used without providing any explicit priors on the population sizes and the group sizes. Whereas the population sizes can easily be fitted with proper priors, the group sizes are correlated and sum to the number of coalescent events. As far as we know, proper prior specification for this model has so far not been used, i.e., the Bayesian skyline model is missing necessary normalizing constants.

Rather than attempting to provide proper priors for the group sizes of the "unnormalized" Bayesian skyline model, we here consider the Bayesian skyride model (Minin et al., 2008) as our MCP model of choice to analyze the HIV-1 dataset. MCP models approximate the effective population size trajectory with a step function, defined by estimable change-point locations and step heights. The Bayesian skyline model (Drummond et al., 2006), often used in coalescent studies, requires prior specification of the total number of change points, a critical parameter in their model that controls the smoothness of the population size trajectory. As a non-parametric alternative, Minin et al. (2008) propose a Bayesian *skyride* model that smooths population size trajectories explicitly by imposing a Gaussian Markov random field (GMRF) process on the parameters of the piecewise constant population size trajectory. This smoothing prior is made "time-aware" in order to penalize effective population size changes between small consecutive intercoalescent intervals more than changes between intervals of larger size, by equipping each consecutive pair of intercoalescent intervals with an appropriate smoothing weight.

The GMRF smoothing prior is specified on the log effective population sizes and it is determined by a precision parameter (Minin et al., 2008), which requires a proper prior distribution. In light of our findings on the importance of using proper priors (Baele et al., 2013b), we also place an additional prior density on the log population sizes by putting a diffuse proper prior centered at zero on the mean of the log population size parameters. Figure 4.1 shows the Bayesian skyride plot of the HIV-1 dataset, using proper priors on all the parameters. The Bayesian skyride plot suggests a smoother population size trajectory through time compared to the original Bayesian skyline plot, which mostly impacts the population history before 1950 whereas the more recent dynamics are quite similar.

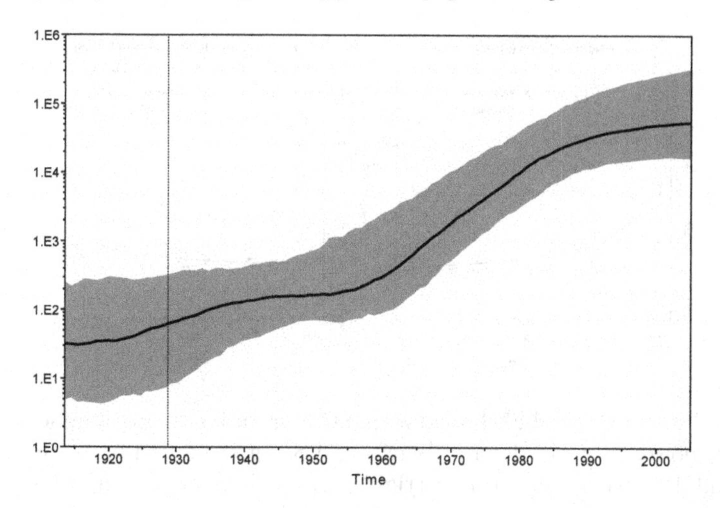

Figure 4.1: Bayesian skyride plot of HIV-1 group M assuming proper priors for all parameters. The plot begins at the median posterior TMRCA (1914). The bold line traces the inferred median effective population size over time with the 95% HPD shaded in blue.

4.5.4 Computational requirements

Xie et al. (2011) analyzed a 10-taxon green plant dataset using DNA sequences of the chloroplast-encoded large subunit of the RuBisCO gene (rbcL), with taxa chosen to be broadly representative of green plant diversity. The authors analyzed the dataset to determine how many ratios are needed to estimate the marginal likelihood accurately and found that eight intervals are sufficient for estimating the marginal likelihood using SS. PS, however, requires a larger number of intervals, about four times as many, compared to SS in order to overcome the additional discretization bias. Given a sufficiently large number of ratios, PS and SS estimate virtually identical log marginal likelihoods. Xie et al. (2011) conclude that the number of ratios required for accurate (log) marginal likelihood estimation remains an open question for each dataset under investigation.

The HIV-1 dataset we analyze here consists of 162 taxa and may therefore require significantly more computational effort. As a starting point to determine adequate settings for model comparison on this dataset, we take 64 path steps/ratios (i.e., the maximum number used by Xie et al. (2011)). Starting with a rather low number of iterations for each power posterior and gradually increasing it will allow us to determine an adequate number of iterations per power posterior to yield accurate estimation of the log marginal likelihoods. Subsequently increasing the number of path steps/ratios to 128 will hopefully confirm that sufficient computational effort has been spent to make a distinction between the model fit of the different demographic priors. In Figure 4.2, we

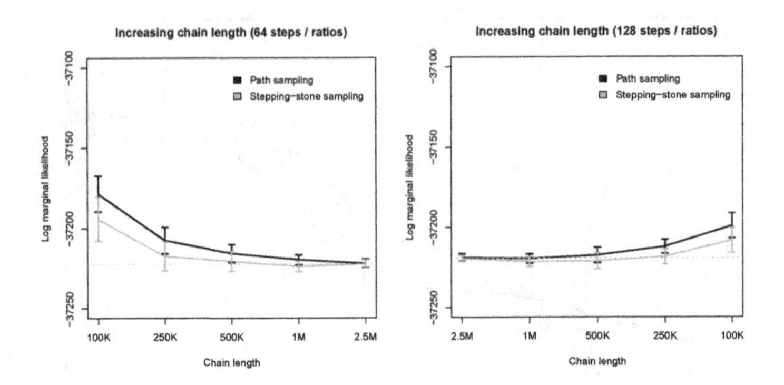

Figure 4.2: Log marginal likelihood estimates for an increasing number of iterations per power posterior for PS (black) and SS (grey) for 64 path steps/ratios (left) and 128 path steps/ratios (right). Error bars represent ± 1 standard deviation based on 10 independent MCMC analyses. Grey dotted lines indicate the mean log marginal likelihood using the most demanding computational settings.

use 64 and 128 path steps/ratios for PS and SS and calculate log marginal likelihoods for the constant population size model by gradually increasing the number of iterations for each power posterior. For each combination of the number of path steps/ratios and the number of iterations, we have performed 10 independent replicates to estimate standard errors.

In terms of the number of iterations required for each power posterior to yield a representative contribution to the (log) marginal likelihood, fixing the sampling frequency to every 1000 iterations, Figure 4.2 suggests that at least 500,000 iterations per power posterior are required. The downward trend observed for both 64 and 128 path steps/ratios stabilizes from 500,000 iterations per power posterior on, but only for SS, while PS requires more samples per power posterior for the (log) marginal likelihood to converge. Setting the number of path steps/ratios to 64 seems sufficient for this dataset, as the increase to 128 path steps/ratios does not alter the (mean log) marginal likelihood once sufficient samples from each power posterior have been obtained. One could also argue that 64 path steps/ratios are too many and this number could be decreased, although this may result in longer chains for each power posterior, as this leads to a decreased number of ratios to approximate the (log) marginal likelihood and hence a more representative sample from each power posterior may be required.

4.5.5 Results

We have performed model selection by calculating log marginal likelihoods for the demographic models shown in Table 4.3 using SS. We used 64 ratios

Table 4.3: HIV-1 group M TMRCA estimates under different coalescent tree priors assuming proper priors. Shown for each coalescent tree prior is the median, with the 95% highest probability distribution of the TMRCA in parentheses. The log Bayes factor differences in estimated marginal likelihood compared with the coalescent model with the strongest support are shown in the last column.

Coalescent tree prior	DRC60 and ZR59 included	log Bayes factor
Constant	1922 (1909 – 1933)	−161.2
Exponential	1913 (1892 – 1929)	0.0
Expansion	1900 (1871 – 1922)	−0.8
Logistic	1914 (1895 – 1929)	−3.4
Bayesian skyride	1914 (1889 – 1930)	−15.4

for SS based on the results in Figure 4.2. In order to generate sufficient samples from the 64 power posteriors, we ran each power posterior for up to 5,000,000 iterations taking a sample every 1000 iterations. Including the burn-in for each power posterior (which is set at 10% by default in BEAST) and running an initial standard MCMC chain to burn in the marginal likelihood estimation process, this results in 357 million BEAST iterations being spent to estimate each model's log marginal likelihood, thereby far exceeding the computational efforts for this dataset in previous analyses (Worobey et al., 2008; Baele et al., 2012a). To study convergence of the results in terms of the ranking of the different demographic models, we also performed log marginal likelihood estimation using less iterations per power posterior, as can be seen in Figure 4.3.

A first conclusion that is directly apparent in Figure 4.3, even when performing a small number of iterations per power posterior, is that the constant population size model yields a much lower model fit than the other demographic models tested. The results from 500,000 iterations indicate that the Bayesian skyride model yields marginally lower marginal likelihoods compared to the three parametric demographic models (i.e., the exponential, expansion, and logistic growth models). In light of the fairly smooth increase in population size (Figure 4.1), there may not be a need to estimate a large number of population sizes as is done for Bayesian skyride model. The top three performing demographic models are only slightly different in their model fit, with the ranking of these three models changing even when increasing the number of iterations per power posterior from 2.5 to 5 million iterations. The expansion growth model ranks close to the exponential and logistic growth models due the "expansion-like" trend seen in Figure 4.1, but estimates the TMRCA to be older than the latter models dictate because of the population size being more constant toward the root in this model. The differences in model fit are, however, small relative to the variance of the marginal likelihood estimator

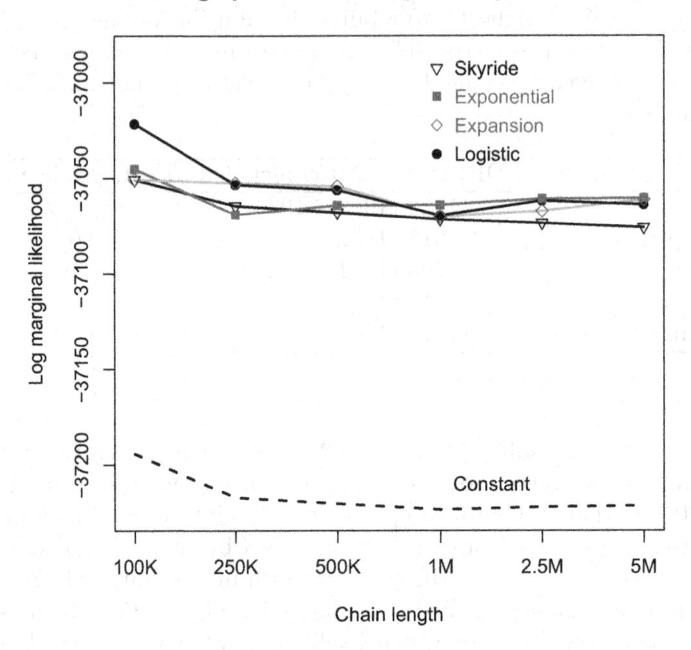

Figure 4.3: Log marginal likelihood estimates for the different demographic models for increasing chain lengths for each power posterior, using SS with 64 ratios.

(i.e., SS), and shows that, even when using a tremendous number of iterations, the difference in performance for certain models remains difficult to assess for this dataset. It may be interesting to examine whether GSS (Fan et al., 2011) could further improve (log) marginal likelihood estimation in this case, but this will require the ability to accommodate phylogenetic uncertainty.

Finally, we list the TMRCA estimates for the HIV-1 group M dataset under different coalescent tree priors, assuming proper priors for all the parameters. Each MCMC analysis was run in BEAST (Drummond and Rambaut, 2007) for 100 million iterations and the ESS values for all the parameters, prior, and likelihood were verified to be above 200. The results are shown in Table 4.3. Compared to the TMRCA estimates published in the original analysis of Worobey et al. (2008), the estimates are largely similar, only differing by 1 or 2 years for some of the demographic models, which is negligible considering their credible intervals.

The constant population size model again yields the youngest TMRCA estimate, while the expansion growth model yields the oldest TMRCA estimate. The Bayesian skyride model yields a TMRCA that is 6 years younger than does the Bayesian skyline model and in doing so corresponds to the TMRCA estimate of the exponential and logistic growth models. The main conclusion of the original paper (Worobey et al., 2008) still holds, i.e., TMRCA dates more recent than the 1930s are strongly rejected by this statistical analyses.

5

Variable tree topology stepping-stone marginal likelihood estimation

Mark T. Holder

Department of Ecology and Evolutionary Biology, University of Kansas, Lawrence, Kansas, USA

Paul O. Lewis

Department of Ecology and Evolutionary Biology, University of Connecticut, Storrs, Connecticut, USA

David L. Swofford

Department of Biology and Institute for Genome Sciences and Policy, Duke University, Durham, North Carolina, USA

David Bryant

Department of Mathematics and Statistics, University of Otago, Dunedin, New Zealand

CONTENTS

5.1 Introduction

The marginal likelihood is central to Bayesian model selection. It is the normalizing constant in Bayes' formula, and the Bayes factor used to compare two models is a ratio of marginal likelihoods. The marginal likelihood is defined as the expected value of the likelihood with respect to the prior. Methods for accurately estimating the marginal likelihood in phylogenetics were developed only recently (Lartillot and Philippe, 2006a; Xie et al., 2011; Fan et al., 2011; Arima and Tardella, 2012). Two of these methods—thermodynamic integration (TI; Lartillot and Philippe, 2006a) and the stepping-stone method (SS; Xie et al., 2011)—allow marginal likelihood estimation when the tree topology varies, but the most efficient methods to date—generalized stepping-stone (GSS; Fan et al., 2011) and the inflated density ratio method (IDR; Arima and Tardella, 2012)—have thus far remained restricted to estimating marginal likelihoods for a fixed tree topology. This chapter is concerned with updating GSS to allow variable tree topology, and the chapter by Wu et al. (Chapter 6) is concerned with updating the IDR method to allow variable tree topology.

5.2 The generalized stepping-stone (GSS) method

The goal of the GSS method (Fan et al., 2011) is to estimate the marginal likelihood,

$$p(\mathbf{y}) = \int_{\Omega} p(\mathbf{y}|\boldsymbol{\theta})p(\boldsymbol{\theta})d\boldsymbol{\theta},$$

where $\boldsymbol{\theta} = (\theta_g : g = 1, \cdots, n_g)$ is a vector of substitution model parameters, $\Omega \in \mathbb{R}^{n_g}$, and $\mathbf{y} = (y_j : j = 1, \cdots, n_j)$ is a vector of site patterns $y_j = (y_{jl} : l = 1, \cdots, n_l)$ where y_{jl} represents the single nucleotide state observed at site j for taxon l. The GSS method works by recognizing that estimating $p(\mathbf{y})$ is equivalent to estimating the ratio c_1/c_0, where c_β is the normalizing constant for a power posterior distribution of the form

$$p_\beta(\boldsymbol{\theta}) = \frac{q_\beta(\boldsymbol{\theta})}{c_\beta} = \frac{[p(\mathbf{y}|\boldsymbol{\theta})p(\boldsymbol{\theta})]^\beta \, p_0(\boldsymbol{\theta})^{1-\beta}}{c_\beta}.$$

Note that c_1 is the marginal likelihood of interest and c_0 is the normalizing constant for the arbitrary reference distribution $p_0(\boldsymbol{\theta})$. In the GSS method, $c_0 = 1$ because $p_0(\boldsymbol{\theta})$ is assumed to be proper.

The ratio $r = c_1/c_0$ is equivalent to following product of n_k ratios,

$$r = \frac{c_1}{c_0} = \left(\frac{c_{\beta_{n_k}}}{c_{\beta_{n_k-1}}}\right)\left(\frac{c_{\beta_{n_k-1}}}{c_{\beta_{n_k-2}}}\right)\cdots\left(\frac{c_{\beta_2}}{c_{\beta_1}}\right)\left(\frac{c_{\beta_1}}{c_{\beta_0}}\right),$$

where $\beta_k = k/n_k$, $k = 0, \cdots, n_k$. Each individual ratio $c_{\beta_k}/c_{\beta_{k-1}}$ composing this product can be estimated accurately using importance sampling, with $c_{\beta_{k-1}}$ serving as the importance distribution. Given MCMC samples $\{\boldsymbol{\theta}^{(i)}_{\beta_{k-1}} : i = 1, \cdots, n\}$ from $p_{\beta_{k-1}}(\boldsymbol{\theta})$, $r_k = c_{\beta_k}/c_{\beta_{k-1}}$ may be estimated as follows (Fan et al., 2011):

$$\hat{r}_k = \frac{1}{n}\sum_{i=1}^{n}\left[\frac{p\left(\mathbf{y}|\boldsymbol{\theta}^{(i)}_{\beta_{k-1}}\right)p\left(\boldsymbol{\theta}^{(i)}_{\beta_{k-1}}\right)}{p_0\left(\boldsymbol{\theta}^{(i)}_{\beta_{k-1}}\right)}\right]^{\beta_k-\beta_{k-1}}.$$

Numerical stability is improved by factoring out the largest term,

$$\eta_k = \max_{1\leq i\leq n}\left\{\frac{p\left(\mathbf{y}|\boldsymbol{\theta}^{(i)}_{\beta_{k-1}}\right)p\left(\boldsymbol{\theta}^{(i)}_{\beta_{k-1}}\right)}{p_0\left(\boldsymbol{\theta}^{(i)}_{\beta_{k-1}}\right)}\right\},$$

yielding the following estimator of the log marginal likelihood:

$$\begin{aligned}\log\hat{r} &= \sum_{k=1}^{n_k}\log\hat{r}_k \\ &= \sum_{k=1}^{n_k}\left[(\beta_k-\beta_{k-1})\log\eta_k\right] \\ &\quad+\sum_{k=1}^{n_k}\log\left[\frac{1}{n}\sum_{i=1}^{n}\left(\frac{p\left(\mathbf{y}|\theta^{(i)}_{\beta_{k-1}}\right)p\left(\theta^{(i)}_{\beta_{k-1}}\right)}{\eta_k p_0\left(\theta^{(i)}_{\beta_{k-1}}\right)}\right)^{\beta_k-\beta_{k-1}}\right].\end{aligned}$$

In Fan et al. (2011), the topology was fixed and $p_0(\boldsymbol{\theta})$ had the same form as the joint prior distribution, with individual components adjusted so that their means and variances matched the corresponding sample means and variances from a pre-existing posterior sample. For example, two parameters may be estimated for the K80 model applied to just 2 sequences: the transition/transversion rate ratio, κ, and the edge length, ν (evolutionary distance between the 2 sequences). Assume that the joint prior density is a product of two Gamma densities (one for ν and the other for κ). If the marginal posterior distribution of ν has sample mean = 0.02 and sample variance = 0.0002, and the marginal posterior distribution of κ has sample mean = 4.0 and sample variance = 0.2, the reference distribution density would be a product of a Gamma(2.00, 0.01) density (for ν) and a Gamma(80.00, 0.05) density (for κ). Here, we generalize GSS further by incorporating the tree topology into the reference distribution.

5.3 Reference distribution for tree topology

In the previous discussion, $\boldsymbol{\theta}$ included all model parameters, including edge lengths. The fact that edge lengths are specific to a particular tree topology T requires modification of notation: $\boldsymbol{\nu}_T$ is now a vector of edge length parameters specific to tree topology T, and $\boldsymbol{\theta}$ now includes all other model parameters. The total parameter space $\Omega \in \mathbb{R}^{n_g + n_t n_b}$, where n_g is now the number of non-tree-specific model parameters and, for unrooted tree topologies with n_l tips,

$$\begin{aligned} n_t &= \frac{(2n_l-5)!}{2^{n_l-3}(n_l-3)!} && \text{(number of tree topologies)}, \\ n_b &= 2n_l - 3 && \text{(edge lengths/tree topology)}. \end{aligned}$$

The GSS method achieves its greater efficiency (over the SS method described by Xie et al., 2011) by using a reference distribution that is closer to the posterior distribution than the prior. As pointed out by Fan et al. (2011), maximum efficiency is obtained if the reference distribution equals the posterior exactly, in which case 1 MCMC sample is sufficient for estimating the marginal likelihood. While this maximum efficiency requires exact knowledge of the very quantity being estimated (the marginal likelihood) and is thus unobtainable, it is desirable to choose a reference distribution that is as similar to the posterior as possible. In addition to being close to the posterior, the reference distribution must be normalized because the GSS method assumes that $c_0 = 1$. Finally, the reference distribution should ideally allow direct sampling (rather than requiring Metropolis-Hastings updates).

In the next section, we describe a reference distribution that possesses all of these desirable properties, and we present an algorithm for sampling trees from this reference distribution. The proposed reference distribution, $p_0(T, \boldsymbol{\nu}_T, \boldsymbol{\theta}) = \pi(T) f(\boldsymbol{\nu}_T|T) f(\boldsymbol{\theta})$, is parameterized using a sample from a preliminary MCMC analysis, called the pilot run, exploring the posterior distribution. The goal is a distribution that samples trees roughly in proportion to the posterior distribution but does not rule out trees not visited in the pilot run. The general approach is:

1. draw a tree topology, T, from a distribution, $\pi(T)$, over all n_t bifurcating topologies (Section 5.3.1);
2. draw $\boldsymbol{\nu}_T$, a vector of n_b edge lengths, from $f(\boldsymbol{\nu}_T|T)$ (Section 5.3.2); and
3. draw substitution model parameters, $\boldsymbol{\theta}$, from $f(\boldsymbol{\theta})$ (Fan et al., 2011).

5.3.1 Tree topology reference distribution

Let a tree topology $T = (V, E)$ comprise a set of vertices, $V = \{v : v = 1, \cdots, L\}$, where $L = 2(n_l - 1)$ for unrooted and $L = 2n_l - 1$ for rooted trees, and a set of edges, $E = \{(i,j) : i,j \in V, i < j\}$, where i and j are the vertices at the ends of the edge (i,j). Vertices are ordered such that for each edge (i,j), j is the parent (i.e., closer to the root) of i. For unrooted trees, the vertex chosen to represent the root is arbitrary and may even be a vertex of degree 1 (i.e., a leaf vertex).

The procedure for generating the topology requires the specification of a focal tree topology, T^*, and split probabilities for every split in T^*. This collection of splits will be denoted $\mathcal{S}(T^*)$, and the split probability for split s will be denoted $p(s)$.

In practice, T^* will be a fully resolved tree topology with high posterior probability based on the pilot run — the MAP (maximum a posteriori) tree, for example. If $0 < p(s) < 1$ for every split in T^*, then the procedure outlined below will specify a probability distribution over all n_t possible tree topologies. To guarantee this, we can base $p(s)$ on the frequency of split s in the pilot run. For example, if split s was sampled n_s times out of a total pilot run length n, and if s partitions the n_l taxa into two subsets of size n_1 and n_2 ($n_1 + n_2 = n_l$), then

$$
\begin{aligned}
n_s^* &= \frac{(2n_1 - 3)!}{2^{n_1-2}(n_1-2)!} + \frac{(2n_2 - 3)!}{2^{n_2-2}(n_2-2)!}, \\
p(s) &= \frac{n_s + (n_s^*/n_t)\epsilon}{n + \epsilon}, \qquad (5.1)
\end{aligned}
$$

where ϵ is a small fraction of the MCMC sample size (e.g., $\epsilon = 0.01n$) and n_s^*/n_t is the induced prior on split s given a prior distribution that places equal weight on all possible unrooted tree topologies.

5.3.1.1 Tree simulation

Algorithm TopoGen describes how to generate a tree topology using the focal tree T^* and split probabilities for every split in $\mathcal{S}(T^*)$. The algorithm chooses which of the splits in $\mathcal{S}(T^*)$ to add. Because the splits are mutually compatible, they can be combined into a (possibly unresolved) tree. Finally, all polytomies are resolved by drawing a topology for their subtree, and rejecting any resolution of that portion of the tree that includes splits found in $\mathcal{S}(T^*)$.

Figure 5.1 illustrates simulation of a 6-taxon tree topology using the TopoGen algorithm. In the example illustrated, $S(T^*)$ comprises 3 splits, 2 of which, by chance, are included in $S(T_u)$, leaving T_u with a single polytomy of degree 4 at node v. The 4-taxon unrooted tree whose tips are vertices in $\mathcal{A}(v)$ has 3 possible resolutions, which are shown along the bottom of Figure 5.1. One of these possible resolutions (left) is invalid because it contains a split in T^*. One of the 2 remaining trees would be returned by the TopoGen algorithm with probability 0.5.

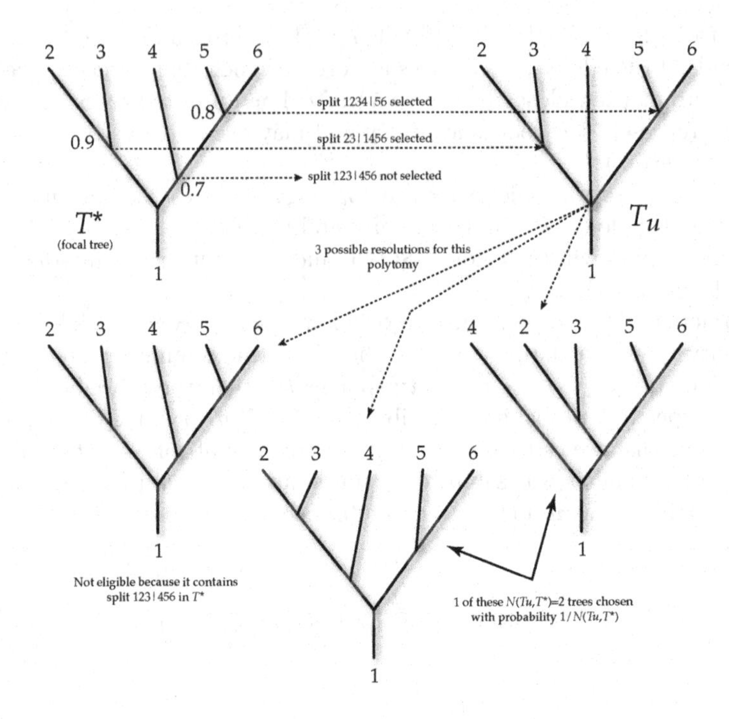

Figure 5.1: An example application of the TopoGen algorithm for a 6-taxon case.

5.3.1.2 Tree topology reference distribution

Let $\mathcal{S}(T_u)$ be the set of splits displayed by T that are members of the set $\mathcal{S}(T^*)$. Let T_u denote the (potentially nonbinary) tree that displays only those splits that are in $\mathcal{S}(T_u)$ (see TopoGen line 5). The probability of the tree topology component of the reference distribution is

$$\pi(T) = \Pr(T|p, \mathcal{S}(T^*)) = \Pr(T_u|p, \mathcal{S}(T^*)) \Pr(T|T_u, \mathcal{S}(T^*)). \tag{5.2}$$

The first term, $\Pr(T_u|p, \mathcal{S}(T^*))$, corresponds to lines 1–5 of the TopoGen algorithm, and equals the product of the probabilities of selecting each of the non-trivial splits found in T_u multiplied by the product of the probabilities of not selecting the other splits. For split s, let $I_{s\in\mathcal{S}(T_u)} = 1$ if $s \in \mathcal{S}(T_u)$ and 0 otherwise. Then, we have

$$\Pr(T_u|p, \mathcal{S}(T^*)) = \prod_{s\in\mathcal{S}(T^*)} p(s)^{I_{s\in\mathcal{S}(T_u)}}(1 - p(s))^{1-I_{s\in\mathcal{S}(T_u)}}. \tag{5.3}$$

From the fact that $\epsilon > 0$ and the definition of $p(s)$ in (5.1), $0 < p(s) < 1$ and therefore $0 < \Pr(T_u|f) < 1$. The independence of the probability terms in (5.3)

Algorithm 1 TopoGen: Generate a tree topology using split selection probabilities.

Require: $\mathcal{S}(T^*)$, a set of compatible splits, and function, p, that maps each split in $\mathcal{S}(T^*)$ to a probability that the split will be included in the tree.

1: $\mathcal{S}(T_u) \leftarrow \emptyset$
2: **for all** $s \in \mathcal{S}(T^*)$ **do**
3: With probability $p(s)$, add s to $\mathcal{S}(T_u)$
4: **end for**
5: Produce the tree T that displays the splits in $\mathcal{S}(T_u)$ (This tree, without the modifications below, will be referred to as T_u.)
6: **while** T is not fully resolved **do**
7: Randomly select a vertex, v, from T that is not fully resolved ($\deg(v) > 3$). Let $\mathcal{A}(v)$ be the set of vertices that are adjacent to v; i.e., $\mathcal{A}(v) = \{a : (i,j) \in E, a \in \{i,j\}, v \in \{i,j\}, a \neq v\}$.
8: Treat the vertices in $\mathcal{A}(v)$ as leaves and randomly choose tree topology, $T_{\mathcal{A}(v)}$, for them from a uniform distribution over all fully resolved unrooted trees with $|\mathcal{A}(v)|$ leaves (thus replacing the polytomy, v, with a set of vertices each of degree 3).
9: If any of the newly added splits (splits that correspond to internal edges in $T_{\mathcal{A}(v)}$) are in $\mathcal{S}(T^*)$, then collapse $T_{\mathcal{A}(v)}$ back into a polytomy (returning the tree to its original state before $T_{\mathcal{A}(v)}$ was added).
10: **end while**
11: **return** T as the simulated tree and $\mathcal{S}(T_u)$ as the set of splits from $\mathcal{S}(T^*)$ which are displayed by T.

---

reflects the fact that split selection in line 3 of the TopoGen algorithm does not depend in any way on splits already chosen to be included in set $\mathcal{S}(T_u)$.

The second term in $\pi(T)$, $\Pr(T|T_u, \mathcal{S}(T^*))$, corresponds to lines 6–10 of the TopoGen algorithm. Let $\mathcal{V}(T_u)$ denote the set of unresolved vertices in T_u:

$$\mathcal{V}(T_u) = \{v : v \in T_u, \deg(v) > 3\}.$$

Let $\mathcal{A}(v)$ denote the vertices that are adjacent to vertex v:

$$\mathcal{A}(v) = \{a : (i,j) \in E, a \in \{i,j\}, v \in \{i,j\}, a \neq v\}.$$

For each vertex v in $\mathcal{V}(T_u)$, we can easily identify the set of vertices $\mathcal{A}(v)$. For $x \in \mathcal{A}(v)$, let x^* be the corresponding vertex in T^*.

Let $T^*_{\mathcal{A}(v)}$ denote the tree that would be obtained from T^* by deleting vertices and edges such that $\mathcal{A}(v)$ is the leaf set of the new tree. The rejection step in TopoGen guarantees that the polytomy-breaking portion of the algorithm is equivalent to drawing from a discrete uniform distribution of all of the trees with leaf-set $\mathcal{A}(v)$ that share no splits with $T^*_{\mathcal{A}(v)}$. Bryant and Steel (2009) provide an algorithm for computing $q_s(T)$, the number of fully resolved unrooted trees that share exactly s splits with tree topology T. Using their notation, the number of trees with leaf-set $\mathcal{A}(v)$ that share zero splits with $T^*_{\mathcal{A}(v)}$ is $q_0(T^*_{\mathcal{A}(v)})$. The algorithms presented below as "Preprocessing For Count Max Diff" and "Count Max Diff" were devised by one of us (DB); they provide a more efficient method of calculating $q_0(T^*_{\mathcal{A}(v)})$.

Algorithm 2 Preprocessing for Count Max Diff: Preprocessing steps for algorithm "Count Max Diff."

Require: T is a binary tree with n leaves rooted at an internal vertex v_0.
  {*Pre-processing*}
1: $b[0] \leftarrow 1$
2: **for** $k = 1, 2, \ldots, (n-3)$ **do**
3:   $b[k] \leftarrow (2k+1)b[k-1]$ {$b[k]$ = number of binary trees on $k+3$ leaves}
4: **end for**
5: **for** v in a post-order traversal of T **do**
6:   **if** v is a leaf **then**
7:     $n[v] \leftarrow 0$
8:   **else**
9:     let v_1, v_2 be the children of v
10:     $n[v] \leftarrow n[v_1] + n[v_2] +$ number of children of v that are internal
11:     {$n[v]$ is the number of internal edges below v}
12:   **end if**
13: **end for**

$N(T_u, T^*)$ is the number of trees that are resolutions of T_u and contain no splits in $\mathcal{S}(T^*)$ other than the splits $\mathcal{S}(T_u)$ that can be found by considering the products of the number of all relevant resolutions around the set of unresolved vertices, $\mathcal{V}(T_u)$:

$$N(T_u, T^*) = \prod_{i \in \mathcal{V}(T_u)} q_0(T^*_{\mathcal{A}(i)}). \tag{5.4}$$

Because the TopoGen algorithm chooses uniformly from these trees,

$$\Pr(T|T_u, \mathcal{S}(T^*)) = \frac{1}{N(T_u, T^*)}. \tag{5.5}$$

Substituting into (5.2) yields:

$$\Pr(T|f, \mathcal{S}(T^*)) = \frac{\prod_{s \in \mathcal{S}(T^*)} p(s)^{I_{s \in \mathcal{S}(T_u)}} (1 - p(s))^{1 - I_{s \in \mathcal{S}(T_u)}}}{N(T_u, T^*)}. \tag{5.6}$$

To evaluate the probability of an arbitrary tree, T, (a tree for which T_u is not available beforehand), let T_u equal the strict consensus of T and T^*.

An example calculation of the probability of a 6-taxon tree T is illustrated in Figure 5.2. Tree T_u represents the strict consensus of T and T^*, and tree $T^*_{\mathcal{A}(v)}$ equals T^* with all vertices pruned except those corresponding to vertices in $\mathcal{A}(v)$ and rooted at internal node v. Results of applying the algorithms

---

Algorithm 3 Count Max Diff: Count the number of binary trees at the maximum RF distance from a focal tree

---

1: **for** vertex v in a post-order traversal of T **do**
2:     **if** v is internal with no children that are internal **then**
3:         $f[v, 0] \leftarrow 1$
4:     **else if** v has one child v_1 that is internal **then**
5:         $f[v, 0] \leftarrow -\sum_{\ell=0}^{n[v_1]} f[v_1, \ell]$
6:         **for** $k = 1, 2, \ldots n[v]$ **do**
7:             $f[v, k] \leftarrow (2k + 1)f[v_1, k - 1]$
8:         **end for**
9:     **else if** v has two children v_1, v_2 that are internal **then**
10:         $F_1 \leftarrow \sum_{\ell=0}^{n[v_1]} f[v_1, \ell]$
11:         $F_2 \leftarrow \sum_{\ell=0}^{n[v_2]} f[v_2, \ell]$
12:         $f[v, 0] \leftarrow F_1 \times F_2$
13:         **for** $k = 1, 2, \ldots n[v]$ **do**
14:             $f[v, k] \leftarrow 0$
15:         **end for**
16:         **for** $k = 1, 2, \ldots (n[v_2] + 1)$ **do**
17:             $f[v, k] \leftarrow f[v, k] - F_1 \times f[v_2, k - 1] \times (2k + 1)$
18:         **end for**
19:         **for** $k = 1, 2, \ldots (n[v_1] + 1)$ **do**
20:             $f[v, k] \leftarrow f[v, k] - F_2 \times f[v_1, k - 1] \times (2k + 1)$
21:         **end for**
22:         **for** $k_1 = 1, 2, \ldots (n[v_1] + 1)$ **do**
23:             **for** $k_2 = 1, 2, \ldots (n[v_2] + 1)$ **do**
24:                 $k \leftarrow k_1 + k_2$
25:                 $f[v, k] \leftarrow f[v, k] + f[v_1, k_1 - 1] \times f[v_2, k_2 - 1] \times \frac{b[k]}{b[k_1 - 1] \times b[k_2 - 1]}$
26:             **end for**
27:         **end for**
28:     **end if**
29: **end for**
30: **return** $\left|\sum_{k=0}^{n[v_0]} f[v_0, k]\right|$

"Preprocessing For Count Max Diff" and "Count Max Diff" are shown beside internal nodes of $T^*_{\mathcal{A}(v)}$, and the calculation of the probability of tree T is given at the bottom.

5.3.2 Edge length reference distribution

An edge length reference distribution can be constructed using a sample from the MCMC pilot run. Let $\mathcal{B} = \{s : n_s > N_{\min}\}$ be the set of splits having sample size at least $N_{\min}$ in the pilot run. The probability density of the edge

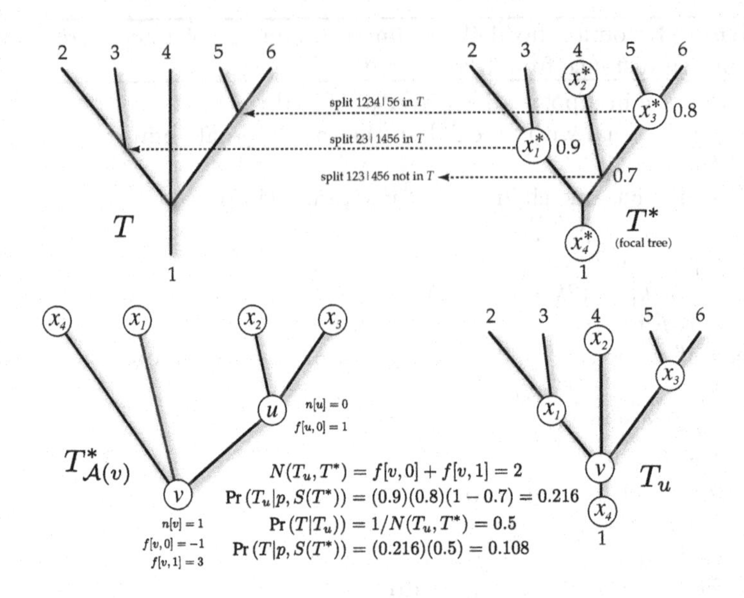

Figure 5.2: Calculation of $\Pr(T|p, S(T^*))$ for a 6-taxon example.

length ν_s corresponding to split s is

$$f(\nu_s) = \begin{cases} \text{Gamma}(a_s, b_s) & \text{if } s \in \mathcal{B}, \\ \text{Gamma}(a_., b_.) & \text{otherwise,} \end{cases}$$

where a_s and b_s are split-specific shape and scale parameters of a Gamma distribution fit to the marginal posterior distribution of ν_s from the pilot run, and $a.$ and $b.$ are the shape and scale parameters of a Gamma distribution fit to the marginal posterior distribution of edge lengths $\nu.$ associated with any split not in $\mathcal{B}$. (The choice of a Gamma distribution here is arbitrary, and could be replaced by a Lognormal distribution, or any other univariate probability distribution with support $(0, \infty)$.) The edge length reference distribution $f(\boldsymbol{\nu}_T|T)$ may now be defined as:

$$f(\boldsymbol{\nu}_T|T) = \prod_{s \in \mathcal{S}(T)} f(\nu_s).$$

The value N_{min} should be chosen large enough to provide reliable estimates of μ_s and σ_s.

5.3.3 Comparison with CCD methods

Larget (2013) and Höhna and Drummond (2011) use conditional clade distributions (CCDs) to provide approximations to marginal posterior distributions of tree topologies. Both approaches use a preliminary sample from the posterior distribution to estimate CCDs and, from those, allow estimation of the marginal posterior probability of an arbitrary tree topology. Larget's method improves upon Höhna and Drummond in being more accurate and not requiring normalization. The reference distribution reported here differs from both of these CCD methods in allowing simulation and calculation of the probability of tree topologies having conditional clade relationships not sampled in the preliminary pilot run. Our method is far less accurate than Larget's approach in general, but reference distributions used for stepping-stone can provide efficiency for marginal likelihood estimation even if not providing the most accurate approximation to the posterior distribution of trees. Nevertheless, development of an approach based on Larget (2013) that allows construction of an irreducible Markov chain is well worthy of future effort.

5.4 Example

Lewis and Trainor (2011) obtained *rbc*L chloroplast DNA sequences of 6 green algae in the genus *Protosiphon* and two closely related genera in order to identify the lone surviving green alga in soil kept dry for 43 years. The phylogeny of these 6 taxa provides a good test of variable-topology marginal likelihood estimation methods because the posterior distribution is not dominated by a single tree topology.

5.4.1 Model details

A general time reversible (GTR) substitution model (Lanave et al., 1984; Tavaré, 1986) allowing invariable sites (Reeves, 1992) and discrete-gamma among-site rate heterogeneity (Yang, 1994b) was used to model evolution of DNA sequences along edges of the tree. The GTR model has 8 free parameters (3 equilibrium relative nucleotide frequencies and 5 exchangeability parameters), with 2 more parameters added to model among-site rate heterogeneity (proportion of invariable sites and discrete gamma shape parameter). Restrictions were placed on these 10 parameters to create a total of 12 different models. An additional 12 models resulted from partitioning the data by codon position. Noninformative Dirichlet distributions were used as priors for relative nucleotide frequencies and exchangeabilities, with a noninformative Beta distribution used for the proportion of invariable sites. An Exponential(1) distribution was used as the prior for the discrete gamma shape, and each edge length was assigned an independent Exponential prior distribution with mean 0.1. The prior probability for each tree topology was 1/105, where applicable (i.e., discrete uniform distribution over all possible unrooted tree topologies). For partitioned models, a noninformative relative rate distribution (eq. 3 in Fan

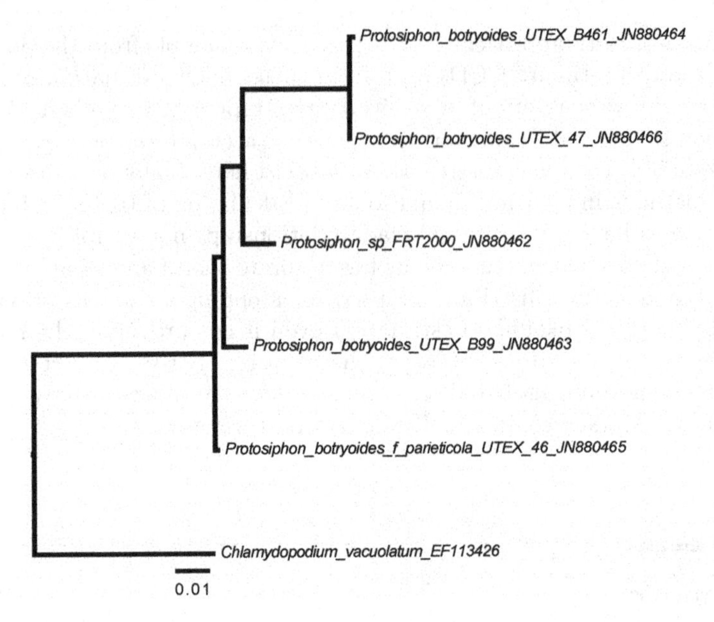

Figure 5.3: Tree topology with maximum marginal likelihood.

et al., 2011) was used as the prior for subset relative rates. In this study, all parameters of the relative rate distribution equaled 1, resulting in the constant relative rate density $2p_1p_2 = 2/9$ (where $p_1 = 1/3$ and $p_2 = 1/3$ are the proportions of sites in the 1st and 2nd position subsets, respectively).

5.4.2 Brute-force approach

It is not possible to simulate sequence data with a known marginal likelihood, and it is not possible to compute the marginal likelihood analytically for phylogenetic datasets of any reasonable size or complexity, so we must resort to other approaches to test the accuracy of particular estimation methods. Because estimating the marginal likelihood with GSS has been demonstrated to be accurate when the tree topology is fixed, one way to test the method proposed here is to estimate the total marginal likelihood from individual fixed tree results. The total marginal likelihood for this 6-taxon example may be written

$$p(\mathbf{y}) = \sum_{i=1}^{105} p(\mathbf{y}|T_i)p(T_i),$$

where T_i is the ith tree topology (out of the 105 tree topologies possible for 6 taxa), $p(T_i) = 1/105$ (discrete uniform prior distribution), and

$$p(\mathbf{y}|T_i) = \int p(\mathbf{y}|T_i, \theta)p(\theta|T_i)d\theta$$

is the conditional marginal likelihood given tree T_i.

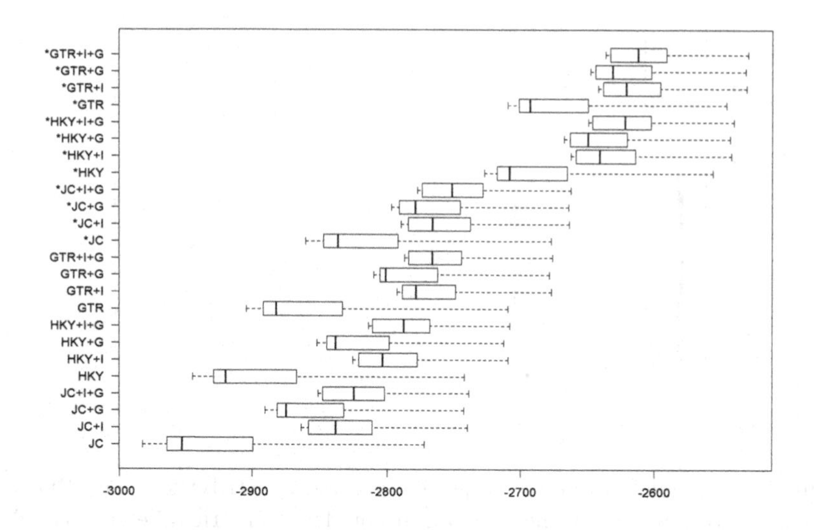

Figure 5.4: Box plots showing mean, 25%, and 75% quantiles and extremes for the 105 log marginal likelihoods estimated for each of the 24 models.

Figure 5.4 provides box plots of $\log p(\mathbf{y}|T_i)$ across the 105 tree topologies for each model. Partitioning sites into three subsets according to codon position provides the most dramatic increase in model fit; however, adding rate heterogeneity (+I, +G, or +I+G) to any unpartitioned model also substantially improves its fit relative to the base model (JC, HKY, or GTR). Despite differences in fit, the tree topology in Figure 5.3 (and Fig. 4 of Lewis and Trainor, 2011) was best according to log marginal likelihood for every one of the 24 models tested.

With estimates of all 105 conditional marginal likelihoods, it is possible to very accurately estimate the tree topology posterior distribution:

$$p(T|\mathbf{y}) = \frac{p(\mathbf{y}|T)p(T)}{p(\mathbf{y})}.$$

Figure 5.5 is a bar plot of the tree topology posterior distribution ordered left to right by Robinson-Foulds symmetric difference (RFSD) distance (Robinson and Foulds, 1981) from the best tree. The best tree has posterior probability 0.584, the 6 tree topologies 1 nearest-neighbor interchange (NNI) away from the best tree (RFSD distance = 2) collectively contributed 0.304 posterior probability, the 24 tree topologies 2 NNI swaps away (RFSD distance = 4) collectively account for 0.112 posterior probability, and none of the 74 tree topologies with the maximum RFSD distance (6) from the best tree topology contributed appreciably to the posterior.

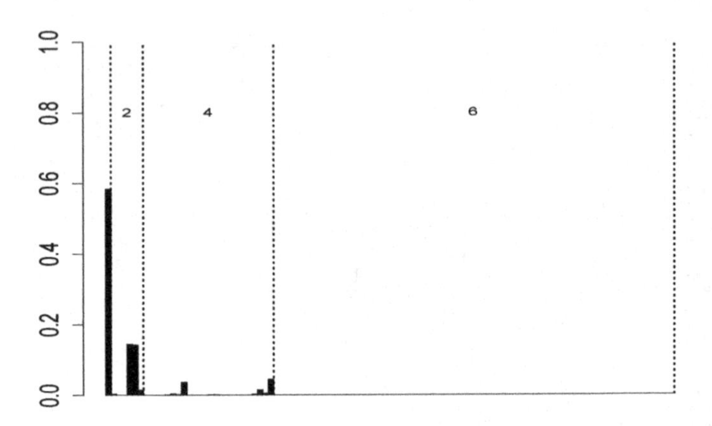

Figure 5.5: Bar plot of the tree topology posterior distribution. The 105-tree topologies are grouped by their Robinson-Foulds symmetric difference (RFSD) distance from the best tree (first bar on left). Groups are separated by vertical dotted lines, and number indicate the RFSD distance for each group.

5.4.3 GSS performance

The variable-topology GSS method performed well over all 24 models tested. Table 5.1 ranks models from best to worst according to their (brute-force) marginal likelihood estimates. For each model, the GSS estimate is given alongside the brute-force estimate, and the column labeled Δ is the difference (GSS estimate minus brute-force estimate). Most absolute differences are less than 0.1 log unit.

Not surprisingly, the accuracy of both the GSS and brute-force approaches depends on the quality of the samples obtained during the stepping-stone MCMC simulation, and both can deviate by several log units from the correct value if care is not taken to ensure that autocorrelation is minimized for all parameters and for all β values. One benefit of using GSS over SS is that in GSS the MCMC simulation crawls between the posterior and the reference distribution, which is similar to the posterior in many respects. The expectation is that an MCMC simulator tuned to the posterior still mixes well when exploring the reference distribution. In contrast, SS crawls from the posterior to the prior. Given that the prior is typically much less informative than the posterior, considerable adjustments must be made to the tuning of the MCMC simulator as the analysis proceeds. Proposals that are bold for the posterior represent tiny steps as the MCMC nears the prior. Despite our expectations, we found that some adjustment to tuning was necessary even for GSS. The reference distribution used is essentially the prior fit to samples drawn from a pilot study of the posterior. This approach matches means and

Table 5.1: Comparison of marginal likelihoods for 24 models estimated using the generalized stepping-stone (GSS) and brute-force approaches. Δ is the difference between brute-force and GSS marginal likelihood estimates. "From best" is the difference in the brute-force estimate from that of the best model (GTR+I+G*). Models indicated by an asterisk (∗) partitioned sites by codon position. All other models used a single model for all sites.

Model	GSS	Brute force	Δ	From best
*GTR+I+G	−2534.57	−2534.66	0.09	0.00
*GTR+I	−2535.59	−2535.57	−0.02	0.91
*GTR+G	−2536.69	−2536.75	0.06	2.09
*HKY+I+G	−2544.49	−2545.04	0.55	10.38
*HKY+I	−2546.78	−2546.75	−0.03	12.09
*HKY+G	−2548.20	−2548.16	−0.04	13.50
*GTR	−2551.09	−2551.10	0.01	16.44
*HKY	−2561.10	−2561.13	0.03	26.47
*JC+I+G	−2667.09	−2667.19	0.10	132.53
*JC+I	−2668.35	−2668.38	0.03	133.72
*JC+G	−2668.94	−2668.99	0.05	134.33
GTR+I+G	−2680.21	−2680.29	0.08	145.63
GTR+I	−2681.01	−2681.00	−0.01	146.34
*JC	−2681.83	−2681.79	−0.04	147.13
GTR+G	−2682.38	−2682.73	0.02	148.07
HKY+I+G	−2712.18	−2712.30	0.12	177.64
HKY+I	−2713.28	−2713.31	0.03	178.65
GTR	−2714.22	−2714.20	−0.02	179.54
HKY+G	−2716.83	−2716.87	0.04	182.21
JC+I+G	−2743.19	−2743.56	0.37	208.90
JC+I	−2744.48	−2744.59	0.11	209.93
HKY	−2747.01	−2746.99	−0.02	212.33
JC+G	−2747.44	−2747.44	0.00	212.78
JC	−2776.58	−2776.52	−0.06	241.86

variances from the posterior in constructing the reference distribution, but fails to capture correlations among parameters. Such correlations are particularly strong between certain edge length parameters, and between edge lengths and rate heterogeneity parameters. As a result, many values sampled from the joint reference distribution may be quite improbable with respect to the posterior, and we found that increasing the boldness of some proposals as a function of β helped keep autocorrelation low.

5.5 Summary

This chapter describes a method for generalizing the generalized stepping-stone (GSS) method to accommodate varying tree topology. The GSS method as originally described (Fan et al., 2011) was designed for estimating the marginal likelihood under a fixed tree topology. Systematists, in particular, use Bayesian methods expressly to estimate the tree topology, and would therefore prefer to base model comparison on $p(\mathbf{y})$ rather than $p(\mathbf{y}|T)$ for some fixed tree topology T. The greater efficiency of the GSS method compared to SS (Xie et al., 2011) is its use of a reference distribution that is close (as measured by Kullback-Leibler divergence) to the posterior distribution. The SS method instead uses the prior distribution as the reference distribution, and the prior is usually quite different than the posterior, usually displaying a much greater variance.

The primary contribution of this chapter is a proposed reference distribution for trees (topology and edge lengths) that assigns high probability to topologies containing splits deemed important by the posterior. In addition to tree topology, edge length distributions are maintained separately for high (posterior)-probability splits to improve the match between reference distribution and posterior distribution. The proposed reference distribution for trees has a known normalizing constant, which is a requirement for any reference distribution used within the context of GSS.

Marginal likelihoods were estimated for 24 models for a 6-taxon example using the variable-topology GSS. The results compared well to marginal likelihoods estimated using a brute-force approach that involved estimating the marginal likelihood separately for each of the 105 possible unrooted tree topologies, then combining these to give the total marginal likelihood. The method proposed here applies to binary unrooted trees. Adapting the method to rooted trees and analyses involving polytomous trees (see Lewis et al., 2005) will require further work.

The GSS method described here is implemented in Phycas version 2.0 (freely available at `phycas.org`). Python and bash scripts for performing all analyses reported here are available in the supplementary materials for the book.

5.6 Funding

This material is based upon work supported by the National Science Foundation under grant numbers DEB-0732920, DEB-1208393 (MTH), and DEB-1036448 (POL). Any opinions, findings, and conclusions or recommendations expressed in this material are those of the authors and do not necessarily reflect the views of the National Science Foundation.

Acknowledgments

This work made extensive use of the computer cluster maintained by the Bioinformatics Facility (Biotechnology*Bioservices Center), University of Connecticut (http://bioinformatics.uconn.edu/).

6

Consistency of marginal likelihood estimation when topology varies

Rui Wu

Department of Statistics, University of Connecticut, Storrs, Connecticut, USA[1]

Ming-Hui Chen

Department of Statistics, University of Connecticut, Storrs, Connecticut, USA

Lynn Kuo

Department of Statistics, University of Connecticut, Storrs, Connecticut, USA

Paul O. Lewis

Department of Ecology and Evolutionary Biology, University of Connecticut, Storrs, Connecticut, USA

CONTENTS

[1]Current affiliation is Novartis Pharmaceuticals Corporation.

6.1 Introduction

Model selection in Bayesian phylogenetics, as in Bayesian statistics in general, often centers around the marginal likelihood, defined as the probability of the data given only the model, marginalized over all model parameters. Models with larger marginal likelihoods fit the data better and thus are preferred over models with a smaller marginal likelihood. Several methods have been used for estimating the marginal likelihood in phylogenetics, including the *harmonic mean* (HM) method (Newton and Raftery, 1994a), *thermodynamic integration* (TI) (Lartillot and Philippe, 2006a), the *stepping-stone* (SS) method (Xie et al., 2011), the *generalized stepping-stone* method (Fan et al., 2011) (see also Chapter 5), and, most recently, the *inflated density ratio* (IDR) method (Arima and Tardella, 2012) (see also Chapter 3). (In this chapter, the abbreviation SS will refer to generalized SS unless otherwise indicated.) Most of these approaches focus on estimating the marginal likelihood when the tree topology is fixed. Specifically, they estimate the marginal likelihood of tree T given model M, $c(T|M) = f(\boldsymbol{y}|T, M)$, where $\boldsymbol{y}$ denotes data. In principle, all of these methods can be used to also estimate the marginal likelihood of the model only (integrating over tree topology), $c(M) = p(\boldsymbol{y}|M)$, although to our knowledge these methods have not been proven to be consistent in this case. Here we prove that HM and SS are statistically consistent estimators of $c(M)$, and generalize the IDR method for use in the variable topology case, proving that it also is a consistent estimator of $c(M)$. The variable tree topology versions of HM, SS, and IDR we denote as VHM, VSS, and VIDR, respectively.

6.2 Notation and definitions

Let $f(\boldsymbol{y}|\boldsymbol{\theta}_T, T, M)$ denote the likelihood, where $\boldsymbol{\theta}_T$ is the collection of all parameters specific to topology T and substitution model M. Also, let $\pi(\boldsymbol{\theta}_T|T, M)$ denote the prior for $\boldsymbol{\theta}_T$ and $p(T|M)$ the prior probability of topology T under model M. Let $\mathcal{T}$ denote the set of all possible topologies.

Using Bayes' theorem, the posterior probability of topology T given data $\boldsymbol{y}$ is

$$p(T|\boldsymbol{y}, M) = \frac{c(T|M)p(T|M)}{\sum_{T' \in \mathcal{T}} c(T'|M)p(T'|M)}, \tag{6.1}$$

where $c(T|M)$ is the marginal likelihood of topology T given model M, and is defined

$$c(T|M) = \int f(\boldsymbol{y}|\boldsymbol{\theta}_T, T, M)\pi(\boldsymbol{\theta}_T|T, M)d\boldsymbol{\theta}_T. \tag{6.2}$$

Marginalizing over tree topology yields the total marginal likelihood, $c(M)$:

$$c(M) = \sum_{T \in \mathcal{T}} c(T|M)p(T|M). \tag{6.3}$$

Treating $\boldsymbol{\theta}_T$ and T as parameters under model M, the joint posterior distribution of $(\boldsymbol{\theta}_T, T)$ is given by

$$\pi(\boldsymbol{\theta}_T, T|M) = \frac{f(\boldsymbol{y}|\boldsymbol{\theta}_T, T, M)\pi(\boldsymbol{\theta}_T|T, M)p(T|M)}{c(M)}. \tag{6.4}$$

6.2.1 Consistency of VHM

The consistency of the *varying topology harmonic mean* (VHM) estimator for $c(M)$, $\hat{c}_{VHM}(M)$, is formally stated in the following theorem.

Theorem 6.2.1. *Suppose $\{(\boldsymbol{\theta}_T^{(i)}, T^{(i)}),\ i = 1, 2, \ldots, n\}$ is a Markov chain Monte Carlo (MCMC) sample from the joint posterior distribution $\pi(\boldsymbol{\theta}_T, T|M, \boldsymbol{y})$. The HM estimator of $c(M)$ is given by*

$$\hat{c}_{VHM}(M) = \left[\frac{1}{n}\sum_{i=1}^{n} \frac{1}{f(\boldsymbol{y}|\boldsymbol{\theta}_T^{(i)}, T^{(i)}, M)}\right]^{-1}.$$

Under certain ergodic conditions,

$$\lim_{n\to\infty} \hat{c}_{VHM}(M) \overset{a.s.}{=} c(M).$$

Proof: Assume that the MCMC sample of topologies $\{T^{(i)},\ i = 1, 2, \ldots, n\}$ has only D distinct values denoted by $\{\tau_j,\ j = 1, 2, \ldots, D\}$. We can write

$$\sum_{i=1}^{n} \frac{1}{f(\boldsymbol{y}|\boldsymbol{\theta}_T^{(i)}, T^{(i)}, M)} = \sum_{j=1}^{D}\sum_{i=1}^{n} 1\{T^{(i)} = \tau_j\}\frac{1}{f(\boldsymbol{y}|\boldsymbol{\theta}_T^{(i)}, T^{(i)}, M)}, \tag{6.5}$$

where the indicator function

$$1\{T^{(i)} = \tau_j\} = \begin{cases} 1 & \text{if } T^{(i)} = \tau_j \\ 0 & \text{if } T^{(i)} \neq \tau_j \end{cases}. \tag{6.6}$$

We further write

$$n(\tau_j) = \sum_{i=1}^{n} 1\{T^{(i)} = \tau_j\}, \tag{6.7}$$

which is the total number of sampled trees having topology τ_j.

Note that when $n(\tau_j) \to \infty$ for some j ($j = 1, \ldots, D$), we have

$$\begin{aligned}
&\frac{1}{n(\tau_j)} \sum_{i=1}^{n} 1\{T^{(i)} = \tau_j\} \frac{1}{f(\boldsymbol{y}|\boldsymbol{\theta}_T^{(i)}, T^{(i)}, M)} \\
&\longrightarrow \int \frac{1}{f(\boldsymbol{y}|\boldsymbol{\theta}_{\tau_j}, \tau_j, M)} \frac{f(\boldsymbol{y}|\boldsymbol{\theta}_{\tau_j}, \tau_j, M)\pi(\boldsymbol{\theta}_{\tau_j}|\tau_j, M)}{c(\tau_j|M)} d\boldsymbol{\theta}_{\tau_j} \\
&= \frac{1}{c(\tau_j|M)} \int \pi(\boldsymbol{\theta}_{\tau_j}|\tau_j, M) d\boldsymbol{\theta}_{\tau_j} \\
&= \frac{1}{c(\tau_j|M)}. \qquad (6.8)
\end{aligned}$$

Furthermore, we have

$$\frac{n(\tau_j)}{n} \longrightarrow p(\tau_j|\boldsymbol{y}, M). \qquad (6.9)$$

Assume that n is sufficiently large and all topologies have been visited. Using (6.1), (6.5), (6.8), and (6.9), we obtain

$$\begin{aligned}
\lim_{n\to\infty} \hat{c}_{VHM}(M) &= \frac{1}{\sum_{j=1}^{D} p(\tau_j|\boldsymbol{y}, M) \times \frac{1}{c(\tau_j|M)}} \\
&= \frac{1}{\sum_{j=1}^{D} \frac{c(\tau_j|M)p(\tau_j|M)}{c(M)} \times \frac{1}{c(\tau_j|M)}} \\
&= \frac{c(M)}{\sum_{j=1}^{D} p(\tau_j|M)} \\
&= c(M). \qquad (6.10)
\end{aligned}$$

Therefore, the VHM estimator $\hat{c}(M)_{VHM}$ is a consistent estimator of $c(M)$.

6.2.2 Consistency of VSS

Define the *power posterior* distribution to be

$$p_\beta(\boldsymbol{\theta}_T, T|M) = \frac{q_\beta(\boldsymbol{\theta}_T|T, M)p_\beta(T|M)}{c_\beta(M)}, \qquad (6.11)$$

where $0.0 \leq \beta \leq 1.0$,

$$\begin{aligned}
p_\beta(T|M) &= p(T|M)^\beta \pi^*(T|M)^{1-\beta}, \qquad (6.12) \\
q_\beta(\boldsymbol{\theta}_T|T, M) &= [f(\boldsymbol{y}|\boldsymbol{\theta}_T, T, M)\pi(\boldsymbol{\theta}_T|T, M)]^\beta [\pi^*(\boldsymbol{\theta}_T|T, M)]^{1-\beta}, \qquad (6.13)
\end{aligned}$$

$\pi^*(\boldsymbol{\theta}_T|T, M)$ is a proper reference distribution under topology T, and $\pi^*(T|M)$ is a proper reference distribution over tree topologies. In generalized SS, the reference distributions $\pi^*(\boldsymbol{\theta}_T|T, M)$ and $\pi^*(T|M)$ need not depend on $\boldsymbol{y}$

and can therefore equal the corresponding prior distributions, $\pi(\boldsymbol{\theta}_T|T, M)$ and $p(T|M)$. The special case of SS when the tree topology is fixed and $\pi^*(\boldsymbol{\theta}_T|T, M) = \pi(\boldsymbol{\theta}_T|T, M)$ was described in Xie et al., (2011). The total marginal likelihood is

$$c_\beta(M) = \sum_{T \in \mathcal{T}} c_\beta(T|M) p_\beta(T|M), \tag{6.14}$$

where

$$c_\beta(T|M) = \int [f(\boldsymbol{y}|\boldsymbol{\theta}_T, T, M)\pi(\boldsymbol{\theta}_T|T, M)]^\beta [\pi^*(\boldsymbol{\theta}_T|T, M)]^{1-\beta} d\boldsymbol{\theta}_T. \tag{6.15}$$

Notice that in (6.11) and (6.13), we view

$$L_\beta(\boldsymbol{\theta}_T|\boldsymbol{y}, T, M) = \left[\frac{f(\boldsymbol{y}|\boldsymbol{\theta}_T, T, M)\pi(\boldsymbol{\theta}_T|T, M)}{\pi^*(\boldsymbol{\theta}_T|T, M)}\right]^\beta \tag{6.16}$$

as the *power likelihood* function. Corresponding to this power likelihood function, the *power posterior probability for tree topology* T is given by

$$p_\beta(T|\boldsymbol{y}, M) = \frac{c_\beta(T|M) p_\beta(T|M)}{c_\beta(M)}. \tag{6.17}$$

Our goal is to estimate the ratio $c_{1.0}(M)/c_{0.0}(M)$, which is equivalent to the marginal likelihood because $c_{0.0}(M) = 1.0$ if the reference distribution is proper (which is assumed in this method).

Note that the ratio $c_{1.0}(M)/c_{0.0}(M)$ can be expressed as a product of K ratios:

$$\begin{aligned} r_{VSS}(M) &= \frac{c_{1.0}(M)}{c_{0.0}(M)} \\ &= \prod_{k=1}^{K} \frac{c_{\beta_k}(M)}{c_{\beta_{k-1}}(M)}, \end{aligned} \tag{6.18}$$

where $\beta_k = k/K, k = 1, 2, \ldots, K$.

Theorem 6.2.2. *Let*

$$r_{SS,k}(M) = \frac{c_{\beta_k}(M)}{c_{\beta_{k-1}}(M)}$$

be the kth ratio. For each ratio, we generate $\{(\boldsymbol{\theta}_T^{(i)}, T^{(i)}),\ i = 1, 2, \ldots, n\}$ from $p_\beta(\boldsymbol{\theta}_T, T|M)$ with $\beta = \beta_{k-1}$. Then, a Monte Carlo (MC) estimate of $r_{SS,k}(M)$ is given by

$$\hat{r}_{VSS,k}(M) = \frac{1}{n}\sum_{i=1}^{n} \left[\frac{f(\boldsymbol{y}|\boldsymbol{\theta}_T^{(i)}, T^{(i)}, M)\pi(\boldsymbol{\theta}_T^{(i)}|T^{(i)}, M)}{\pi^*(\boldsymbol{\theta}_T^{(i)}|T^{(i)}, M)}\right]^{\beta_k - \beta_{k-1}}. \tag{6.19}$$

Under certain ergodic conditions,

$$\lim_{n\to\infty} \hat{r}_{VSS,k}(M) \overset{a.s.}{=} r_{VSS,k}(M) = \frac{c_{\beta_k}(M)}{c_{\beta_{k-1}}(M)}.$$

Proof: Assume that the MCMC sample of topologies $\{T^{(i)},\ i = 1, 2, \ldots, n\}$ has only D distinct values denoted by $\{\tau_j,\ j = 1, 2, \ldots, D\}$. Let $n(\tau) = \sum_{i=1}^{n} 1\{T^{(i)} = \tau\}$. Then, (6.19) can be written

$$\hat{r}_{VSS,k}(M) = \sum_{j=1}^{D} \frac{n(\tau_j)}{n} \frac{1}{n(\tau_j)} \sum_{i=1}^{n} 1\{T^{(i)} = \tau_j\} \cdot \left[\frac{f(\boldsymbol{y}|\boldsymbol{\theta}_T^{(i)}, T^{(i)}, M)\pi(\boldsymbol{\theta}_T^{(i)}|T^{(i)}, M)}{\pi^*(\boldsymbol{\theta}_T^{(i)}|T^{(i)}, M)} \right]^{\beta_k - \beta_{k-1}}. \tag{6.20}$$

In (6.20), we have

$$\frac{1}{n(\tau_j)} \sum_{i=1}^{n} 1\{T^{(i)} = \tau_j\} \left[\frac{f(\boldsymbol{y}|\boldsymbol{\theta}_T^{(i)}, T^{(i)}, M)\pi(\boldsymbol{\theta}_T^{(i)}|T^{(i)}, M)}{\pi^*(\boldsymbol{\theta}_T^{(i)}|T^{(i)}, M)} \right]^{\beta_k - \beta_{k-1}} \longrightarrow \frac{c_{\beta_k}(\tau_j|M)}{c_{\beta_{k-1}}(\tau_j|M)} \tag{6.21}$$

as $n(\tau_j) \to \infty$ and

$$\frac{n(\tau_j)}{n} \longrightarrow \frac{c_{\beta_{k-1}}(\tau_j|M) p_\beta(\tau_j|M)}{c_{\beta_{k-1}}(M)} \tag{6.22}$$

as $n \to \infty$. Assume that $\lim_{n\to\infty} n(\tau_j) = \infty$. Then, (6.21) and (6.22) imply that

$$\begin{aligned}
\lim_{n\to\infty} \hat{r}_{VSS,k}(M) &= \sum_{j=1}^{D} \left[\frac{c_{\beta_{k-1}}(\tau_j|M) p_\beta(\tau_j|M)}{c_{\beta_{k-1}}(M)} \right] \times \frac{c_{\beta_k}(\tau_j|M)}{c_{\beta_{k-1}}(\tau_j|M)} \\
&= \frac{\sum_{\tau_j=1}^{D} c_{\beta_k}(\tau_j|M) p_\beta(\tau_j|M)}{c_{\beta_{k-1}}(M)} \\
&= \frac{c_{\beta_k}(M)}{c_{\beta_{k-1}}(M)} \\
&= r_{VSS,k}(M).
\end{aligned} \tag{6.23}$$

Thus, $\hat{r}_{VSS,k}(M)$ is a consistent estimator of $r_{VSS,k}(M)$. Therefore, applying (6.18), the marginal likelihood $c(M)$ can be estimated.

Chapter 5 suggests one possible definition of $\pi^*(T|M)$ that should yield much greater efficiency for large problems (many taxa) compared to setting $\pi^*(T|M) = p(T|M)$. The commonly used flat topology prior ($p(T|M)$ equals the inverse of the total number of distinct tree topologies) quickly becomes a very poor approximation of the marginal tree topology posterior distribution as the number of taxa increases, and is especially problematic when sampling from the reference distribution itself. Even for a problem involving only $n_{\text{taxa}} = 20$ taxa, the total number of tree topologies is:

$$\begin{aligned} n_{\text{trees}} &= \frac{(2n_{\text{taxa}} - 5)!}{2^{n_{\text{taxa}}-3}(n_{\text{taxa}} - 3)!} \\ &= 2.21643095477 \times 10^{20}. \end{aligned}$$

Even a very large sample from $p(T|M) = n_{\text{trees}}^{-1}$ accounts for a vanishingly small fraction of the total probability and predominantly samples tree topologies that would never appear in a sample from the posterior distribution, complicating the design of tree topology specific reference distributions $\pi^*(\boldsymbol{\theta}_T^{(i)}|T^{(i)}, M)$.

6.2.3 Consistency of VIDR

Let the joint posterior kernel of $\boldsymbol{\theta}_T$ given topology T under model M be:

$$q(\boldsymbol{\theta}_T|T) = f(\boldsymbol{y}|\boldsymbol{\theta}_T, T, M)\pi(\boldsymbol{\theta}_T|T, M). \tag{6.24}$$

The posterior distribution of $\boldsymbol{\theta}_T$ given T is given by

$$\pi(\boldsymbol{\theta}_T|T, M, \boldsymbol{y}) = \frac{q(\boldsymbol{\theta}_T|T)}{c(T|M)}, \tag{6.25}$$

where $c(T|M)$ is the marginal likelihood given T.

Extending the *inflated density ratio* (IDR) method (Arima and Tardella, 2012) to the variable topology case, define

$$q_r(\boldsymbol{\theta}_T|T) = \begin{cases} q(0|T) & \text{if } \|\boldsymbol{\theta}_T\| \le r \\ q(h(\boldsymbol{\theta}_T)|T) & \text{if } \|\boldsymbol{\theta}_T\| > r \end{cases}, \tag{6.26}$$

where $h(\boldsymbol{\theta}_T) = (1 - \frac{r^{p_T}}{\|\boldsymbol{\theta}\|^{p_T}})^{1/p_T}\boldsymbol{\theta}_T$, and p_T is the dimension of $\boldsymbol{\theta}_T$.

Suppose $\{(\boldsymbol{\theta}_T^{(i)}, T^{(i)}),\ i = 1, 2, \ldots, n\}$ is an MCMC sample from the joint posterior distribution given in (6.25). We assume once again that the MCMC sample of topologies $\{T^{(i)},\ i = 1, 2, \ldots, n\}$ has only D distinct values denoted by $\{\tau_j,\ j = 1, 2, \ldots, D\}$.

The varying topology IDR estimate of $c(M)$ is given by the following theorem.

Theorem 6.2.3. *Suppose $\{(\boldsymbol{\theta}_T^{(i)}, T^{(i)}),\ i = 1, 2, \ldots, n\}$ is an MCMC sample from the joint posterior distribution given in $\pi(\boldsymbol{\theta}_T, T|M, \boldsymbol{y})$. Then, an estimator of $c(M)$ is given by*

$$\hat{c}_{VIDR}(M) = \frac{1}{\frac{1}{n}\sum_{i=1}^{n} \frac{1}{q(0|T^{(i)})b_{T^{(i)}}}\left[\frac{q_r(\boldsymbol{\theta}^{(i)}|T^{(i)})}{q(\boldsymbol{\theta}^{(i)}|T^{(i)})} - 1\right]},$$

where $b_T = \{$*Volume of the ball* $: \|\boldsymbol{\theta}_T\| \leq r\}$ *given* T. *Under certain ergodic conditions, we have*

$$\lim_{n\to\infty} \hat{c}_{VIDR}(M) \stackrel{a.s.}{=} c(M).$$

Proof: Write $n(\tau) = \sum_{i=1}^{n} 1\{T^{(i)} = \tau\}$, where the indicator function $1\{T^{(i)} = \tau\} = 1$ if $T^{(i)} = \tau$ and 0 if $T^{(i)} \neq \tau$. Thus, $n(\tau)$ is the total number of sampled trees having topology τ. When $n(\tau) \to \infty$, we have

$$\begin{aligned}
&\frac{1}{n(\tau)} \sum_{i=1}^{n} 1(T^{(i)} = \tau) \frac{1}{q(0|\tau) b_\tau} \left(\frac{q_r(\boldsymbol{\theta}_\tau^{(i)}|\tau)}{q(\boldsymbol{\theta}_\tau^{(i)}|\tau)} - 1 \right) \\
&\to \int \frac{1}{q(0|\tau) b_\tau} \left(\frac{q_r(\boldsymbol{\theta}_\tau|\tau)}{q(\boldsymbol{\theta}_\tau|\tau)} - 1 \right) \frac{f(\boldsymbol{y}|\boldsymbol{\theta}_\tau, \tau, M)\pi(\boldsymbol{\theta}_\tau|\tau, M)}{c(\tau|M)} d\boldsymbol{\theta}_\tau \\
&= \frac{1}{c(\tau|M)},
\end{aligned} \tag{6.27}$$

and the posterior probability of topology T given $\boldsymbol{y}$ and M is

$$p(\tau|\boldsymbol{y}, M) = \lim_{n\to\infty} \frac{n(\tau)}{n}. \tag{6.28}$$

Therefore, we have

$$\begin{aligned}
\lim_{n\to\infty} \hat{c}_{VIDR}(M) &= \lim_{n\to\infty} \frac{1}{\sum_{\tau=1}^{D} \frac{1}{n} \sum_{i=1}^{n} 1(T^{(i)} = \tau) \frac{1}{q(0|\tau) b_\tau} \left[\frac{q_r(\boldsymbol{\theta}_\tau^{(i)}|\tau)}{q(\boldsymbol{\theta}_\tau^{(i)}|\tau)} - 1 \right]} \\
&= \frac{1}{\sum_{\tau=1}^{D} \frac{c(\tau|M)p(\tau|M)}{c(M)} \frac{1}{c(\tau|M)}} \\
&= c(M).
\end{aligned} \tag{6.29}$$

6.3 Empirical example

Yang (2000a) examined a phylogenetic example that is simple enough that maximum likelihood estimates of tree topology and edge lengths may be obtained analytically. Unfortunately, analytical estimates of the marginal likelihood are not available even for this simplest example, yet it is simple enough that numerical estimation of the marginal likelihood is straightforward and can be used to test the accuracy of the marginal likelihood estimation methods discussed in this paper. Yang (2000a) provided maximum likelihood estimates of model parameters and the value of the likelihood function at the optimum. We here extend Yang's example to Bayesian phylogenetics by using three methods (HM, SS, and IDR) to estimate the marginal likelihood for each

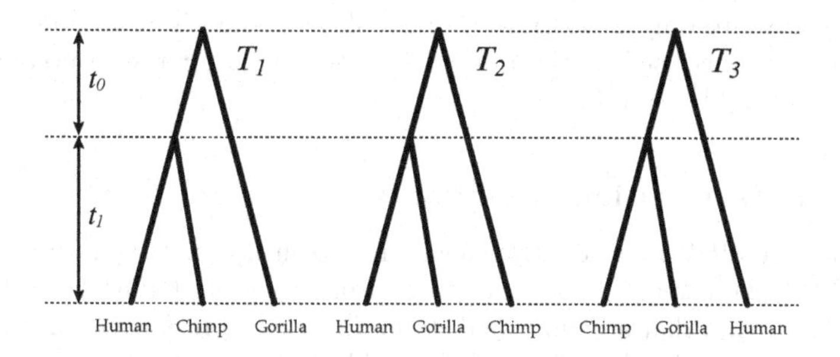

Figure 6.1: Three trees for three species: human (species 1), chimpanzee (species 2), and gorilla (species 3). Parameters t_0 and t_1, each measured in expected number of substitutions per site, together determine all edge lengths.

tree separately, as well as using VHM, VSS, and VIDR to estimate the total (variable tree topology) marginal likelihood.

Yang's example involved a rooted 3-taxon tree, 2-state character data, and the assumption of a strict molecular clock. The data comprise 895 sites from the mitochondrial DNA of 3 primates (human, chimpanzee, and gorilla) published in Brown et al. (1982). Nucleotide data were converted to purines and pyrimidines, reducing the number of states from 4 (A, C, G, T) to 2 (R, Y). The model assumes that the states at the leaves of the tree result from a homogeneous continuous-time Markov process involving only 2 parameters: t_0 and t_1, where t_0+t_1 represents the length (in expected number of substitutions) of the path between the root and any leaf (Figure 6.1). Together with the assumption that trees are ultrametric (due to the strict clock), these two parameters suffice to determine all edge lengths in the tree. The Markov rate matrix for this model is simply

$$R = \begin{bmatrix} -1 & 1 \\ 1 & -1 \end{bmatrix}. \tag{6.30}$$

The transition probability matrix given time t (measured in units of expected number of substitutions/site) is given by

$$P(t) = e^{Rt} = \begin{bmatrix} 0.5(1+e^{-2t}) & 0.5(1-e^{-2t}) \\ 0.5(1-e^{-2t}) & 0.5(1+e^{-2t}) \end{bmatrix}. \tag{6.31}$$

Under this model, there are only 4 distinct data patterns: $n_0 = 762$ sites have the constant pattern, xxx; $n_1 = 54$ sites have the pattern xxy; $n_2 = 41$ sites have the pattern yxx; and $n_3 = 38$ sites have the pattern xyx. Yang (2000a) provides formulas for the likelihood of each tree topology. We assumed an Exponential(1.0) prior for both t_0 and t_1. For stepping-stone analyses, the reference distributions for both t_0 and t_1 were Lognormal distributions with mean and variance equal to the estimated posterior mean and variance. Because there are only 3 possible tree topologies, the discrete uniform tree topology prior was used as the reference distribution for tree topology: i.e., $\pi^*(T_i|M) = p(T_i|M) = 1/3, i = 1,2,3$.

6.3.1 Transformations used for VIDR

Both IDR and VIDR require a transformation to ensure that the posterior density conditional on tree topology has support on the entire k-dimensional space $\mathbb{R}^k$. We have found that an additional standardization transformation improves robustness by removing location and scale differences between parameters. For this example, the transformation used was

$$\boldsymbol{\theta}_T^* = \begin{pmatrix} t_0^* \\ t_1^* \end{pmatrix} = \begin{pmatrix} s_0^2 & s_{01} \\ s_{01} & s_1^2 \end{pmatrix}^{-\frac{1}{2}} \begin{pmatrix} \log(t_0) - u_0 \\ \log(t_1) - u_1 \end{pmatrix} \tag{6.32}$$

$$= \mathbf{S}^{-0.5} \left(\log(\boldsymbol{\theta}_T) - \mathbf{u}\right), \tag{6.33}$$

where $\mathbf{u}$ is the sample mean vector and $\mathbf{S}$ the sample variance–covariance matrix for log-transformed values $\boldsymbol{\theta}_T^{(i)}$ sampled from the posterior distribution $(i = 1, 2, \ldots, n)$. With these transformations, the joint posterior kernel for $\boldsymbol{\theta}_T$ given tree topology T under model M in 6.24 becomes:

$$q(\boldsymbol{\theta}_T^*) = f(\mathbf{y}|\boldsymbol{\theta}_T^*, T, M)\pi(\boldsymbol{\theta}_T^*|T, M) \left|\mathbf{S}^{0.5} \exp\left(\mathbf{S}^{0.5}\boldsymbol{\theta}_T^* + \mathbf{u}\right)\right|. \tag{6.34}$$

6.3.2 Topology-specific marginal likelihood estimation

We used three different methods (HM, SS, IDR) to estimate the marginal likelihood separately for each of the three fully resolved tree topologies. An initial MCMC analysis was performed to estimate marginal likelihoods using HM and IDR, and was used to generate the reference distribution used by SS. This first analysis involved 10100 update cycles, sampling every 10 cycles after the first 100. Each parameter was updated once per cycle by proposing a new value t^* using current value t and Uniform(0,1) random deviate u as follows: $t^* = t + \delta(u - 0.5)$. If $t^* < 0$, t^* is assigned the value $-t^*$. The Hastings ratio for this proposal is 1.0 (symmetric). The values of δ used were 0.08 for updating t_0 and 0.1 for updating t_1. Three different values of the radius of inflation (0.01,0.1,1.0) were used for IDR.

Three further MCMC analyses were carried out for purposes of SS estimation, each using a different number of *stepping-stones* (1, 10, and 100). The powers used for the power posteriors in SS analyses were uniformly spaced between 0.0 and 1.0. For estimating each ratio, 10100 update cycles were used, sampling every 10 cycles after the first 100. Thus, the SS analyses performed for this study required approximately 2, 11, and 101 times more computational effort than either HM or IDR.

Table 6.1: Mean and standard error (SE) of the log topology-specific marginal likelihood $\log \hat{c}(T|M)$ of 1000 MCMC simulations from analysis of primate mitochondrial DNA data. Subscripts for IDR indicate the radius of inflation used. Subscripts for SS indicate the number of stepping-stones (ratios) used.

	Tree T_1		Tree T_2		Tree T_3	
Method	Mean	SE	Mean	SE	Mean	SE
$\text{IDR}_{0.01}$	−529.441	0.049	−532.054	0.039	−532.364	0.041
$\text{IDR}_{0.1}$	−529.444	0.037	−532.058	0.035	−532.361	0.035
$\text{IDR}_{1.0}$	−529.445	0.043	−532.059	0.043	−532.363	0.041
SS_1	−529.467	0.050	−532.087	0.056	−532.388	0.053
SS_{10}	−529.441	0.014	−532.056	0.013	−532.359	0.017
SS_{100}	−529.441	0.003	−532.055	0.003	−532.359	0.003
HM	−522.784	0.478	−525.014	0.711	−525.114	0.626

Table 6.1 shows that IDR and SS achieve nearly the same estimated log marginal likelihoods regardless of the radius of inflation (IDR) or the number of stepping-stones (SS). IDR estimates are nearly identical across the three values (spanning two orders of magnitude) used for the radius of inflation. SS estimates are noticeably worse when only one stepping-stone was used; SS analyses using 10 stepping-stones were essentially identical to the 100 stepping-stone analyses, with the exception of MCMC standard error, which was (not surprisingly) much lower for the analysis involving a 10-fold greater sampling effort. Log marginal likelihoods estimated using the HM method, however, were greater by 6.7, 7.0, and 7.2 log units (for trees 1, 2, and 3, respectively) than the highest SS or IDR estimates. The MCMC standard error for HM was also an order of magnitude greater than either of the other two methods. Considering the computational effort involved, the IDR method is the clear winner for estimating marginal likelihoods in the case of a fixed tree topology for this example. It achieves accuracy comparable to SS while requiring a computational effort only slightly greater than HM.

6.3.3 Total marginal likelihood estimation

The mean log marginal likelihoods from the 100-stepping-stone SS analysis will be taken as truth for purposes of evaluating the performance of all variable-tree-topology analyses. That is, we will assume

$$\log c(T_1|M) = \log \hat{c}(T_1|M) = -529.441, \tag{6.35}$$
$$\log c(T_2|M) = \log \hat{c}(T_2|M) = -532.055, \tag{6.36}$$
$$\log c(T_3|M) = \log \hat{c}(T_3|M) = -532.359. \tag{6.37}$$

The total marginal likelihood $c(M)$, as well as posterior probabilities for each tree topology, can be computed using $c(T_1|M)$, $c(T_2|M)$, and $c(T_3|M)$ as follows:

$$\log c(M) = \log\left(\sum_{i=1}^{3} c(T_i|M)p(T_i|M)\right) = -530.420, \tag{6.38}$$

$$\Pr(T_1|y,M) = c(T_1|M)p(T_1|M)/c(M) = 0.887, \tag{6.39}$$

$$\Pr(T_2|y,M) = c(T_2|M)p(T_2|M)/c(M) = 0.065, \tag{6.40}$$

$$\Pr(T_3|y,M) = c(T_3|M)p(T_3|M)/c(M) = 0.048, \tag{6.41}$$

where $p(T_1|M) = p(T_2|M) = p(T_3|M) = 1/3$.

MCMC analyses in which tree topology was allowed to vary were carried out to generate samples for VHM, VIDR, and VSS. The primary difference in these analyses compared to the single-tree analyses was the insertion of a node slide proposal into each update cycle. The node slide proposal centers a window of width δ over the current position of the single internal vertex. Let the current tree topology be T_i (i=1,2,3). The node can be viewed as being currently placed at a proportion φ along the path from the root to any leaf. That is, the current value equals $\varphi = t_0/(t_0 + t_1)$. A new value φ^* for this proportion is chosen using the current value φ and Uniform(0,1) random deviate u as follows: $\varphi^* = \varphi + \delta(u - 0.5)$. If $\varphi^* < 0$, φ^* is assigned the value $-\varphi^*$. If $\varphi^* > t_0 + t_1$, then a new tree topology T_j is chosen with probability 0.5 from the set of tree topologies $\{T_k, k \neq i\}$, and φ^* is assigned the value $(2(t_0+t_1)-\varphi^*)/(t_0+t_1)$. Like the proposals for updating t_0 and t_1 individually, this proposal is symmetric (Hastings ratio 1.0). While the node slide proposal conditions upon (and thus does not affect) the total tree height t_0+t_1, the fact that t_0 and t_1 are individually updated once per cycle allows the tree height to change.

The total marginal likelihood $c(M)$ was estimated using each method (VHM, VSS, and VIDR) and compared to the "true" value of $c(M)$ using the RMSE (root mean squared error):

$$RMSE = \sqrt{E\left[(\log \hat{c}(M) - \log c(M))^2\right]} \tag{6.42}$$

$$= \sqrt{\text{Var}\left(\log \hat{c}(M)\right) + \left(E[\log \hat{c}(M)] - \log c(M)\right)^2} \tag{6.43}$$

$$\approx \sqrt{\frac{1}{1000}\sum_{i=1}^{1000}\left(\log \hat{c}(M) - c(M)\right)^2}. \tag{6.44}$$

The term $\left(E[\log \hat{c}(M)] - \log c(M)\right)^2$ is the bias. The MCMC standard error and RMSE are thus equal for unbiased estimators. The results (Table 6.2) show that the harmonic mean is clearly biased, with large RMSE (6.827) more than 13 times larger than the SE (0.515). VIDR and VSS, on the other hand, are quite unbiased (RMSE/SE $\approx$ 1.0) and have very low RMSE values (0.030–0.033 for VIDR and 0.002–0.165 for VSS). As with single-topology analyses, increasing the number of stepping-stones improves the MCMC standard error of VSS substantially, and the value of the radius of inflation has little effect on the precision of VIDR estimates.

Table 6.2: Mean, standard error (SE), and root mean squared error (RMSE) of the log total marginal likelihood $\log \hat{c}(M)$ of 1000 variable-topology MCMC simulations from analysis of primate mitochondrial DNA data. Subscripts for VIDR indicate the radius of inflation used. Subscripts for VSS indicate the number of stepping-stones (ratios) used.

Method	Mean	SE	RMSE	RMSE/SE
$\text{VIDR}_{0.01}$	−530.421	0.033	0.033	1.000
$\text{VIDR}_{0.1}$	−530.420	0.030	0.030	1.000
$\text{VIDR}_{1.0}$	−530.423	0.033	0.033	1.006
VSS_1	−530.420	0.165	0.165	1.000
VSS_{10}	−530.420	0.008	0.008	1.000
VSS_{100}	−530.419	0.002	0.002	1.005
VHM	−523.611	0.515	6.827	13.255

Table 6.3: Mean marginal posterior probability of each tree topology based on 1000 replicate MCMC simulations and on mean marginal likelihood estimates from Tables 6.1 and 6.2.

Method	$\Pr(T_1 \vert \text{data})$	$\Pr(T_2 \vert \text{data})$	$\Pr(T_3 \vert \text{data})$
MCMC	0.887	0.065	0.048
$\text{VIDR}_{0.01}$	0.887	0.065	0.048
$\text{VIDR}_{0.1}$	0.887	0.065	0.048
$\text{VIDR}_{1.0}$	0.887	0.065	0.048
VSS_1	0.887	0.065	0.048
VSS_{10}	0.887	0.065	0.048
VSS_{100}	0.887	0.065	0.048
VHM	0.830	0.089	0.081

6.4 Discussion

This chapter is primarily concerned with proving statistical consistency for three marginal likelihood estimators (VHM, VSS, and VIDR) when the tree topology is allowed to vary during MCMC sampling of the posterior distribution. We have also extended the VIDR method to the varying tree topology case: Arima and Tardella (2012) and Chapter 3 address only the case of a fixed tree topology. Finally, we extend the simplest phylogenetic example of Yang (2000a) to Bayesian phylogenetics, providing accurate estimates of both topology-specific marginal likelihoods ($c(T|M)$) as well as the total marginal likelihood ($c(M)$).

The Yang (2000a) example illustrates the strong bias in the VHM method, which has been previously documented in the phylogenetics literature (Lartillot and Philippe, 2006a; Xie et al., 2011; Fan et al., 2011). We used the 100-VSS analysis as our reference, treating these estimated marginal likelihoods as if they were the true values for purposes of evaluating all other methods. One might argue that had we chosen the VHM estimates as the truth, we would conclude that both VSS and VIDR are strongly biased. It is possible to address the question of which method (VHM vs. VSS/VIDR) produces more accurate estimates using estimated marginal likelihoods. Equations 6.39, 6.40, and 6.41 demonstrate how to estimate the marginal posterior probability of tree topology T_i given estimates of $c(T_i|M)$ and $c(M)$. These estimated marginal posterior probabilities can be compared, for each method, to the estimates obtained using the sampled fraction of each tree topology in a variable-topology MCMC simulation. This comparison is presented in Table 6.3. Based on this evidence, VHM is clearly the method that produces incorrect marginal likelihood estimates.

We have demonstrated that all three methods are statistically consistent in the varying tree topology case and will, given an MCMC simulation of sufficient length, accurately estimate the marginal likelihood. Based on our analysis of the Yang (2000a) example, VSS and VIDR are better than VHM at estimating marginal likelihoods in the case of MCMC simulations of finite length. While VSS can produce very accurate estimates, VIDR produces accurate estimates with potentially much less computational effort. VSS is more efficient if a reference distribution is used that is based on the estimated posterior distribution. This requires a preliminary MCMC simulation for purposes of fitting the reference distribution before the VSS analysis can even begin. VIDR has a decided advantage over VSS in that nearly all computation is directed toward sampling from the distribution of primary interest (posterior distribution), whereas VSS spends *all* of its effort (after the reference distribution is established) sampling from power posterior distributions that have no use apart from contributing to an accurate estimate of marginal likelihood. The VIDR method thus appears to manage the tradeoff between computational efficiency and estimating accuracy better than any other approach currently in use. Results from the simple example provided here may be misleading, however, due to the small number of parameters (2) per tree topology and small number of possible tree topologies (3). More work is needed to assess the performance of VIDR and VSS in real-world situations involving much higher dimensional parameter spaces.

6.5 Funding

MHC acknowledges support from National Institutes of Health grants #GM 70335 and #CA 74015. POL acknowledges support from National Science Foundation grant DEB-1036448 (GrAToL).

Acknowledgments

This work made extensive use of the computer cluster maintained by the Bioinformatics Facility (Biotechnology*Bioservices Center), University of Connecticut (http://bioinformatics.uconn.edu/).

7

Bayesian phylogeny analysis

Sooyoung Cheon

Department of Informational Statistics, Korea University, Sejong-city, South Korea

Faming Liang

Department of Statistics, Texas A&M University, College Station, Texas, USA

CONTENTS

7.1 Introduction

The construction of phylogenetic trees is of interest in evolutionary studies. Phylogenetic inference is based on the analysis of hereditary molecular differences, mainly in DNA sequences, to gain information on the evolutionary relationships of organisms. In general, the result of a molecular phylogenetic analysis can be expressed in a phylogenetic tree. Phylogenetic trees have been used for a long time to graphically represent evolutionary relationships among species and genes.

The traditional methods for phylogeny inference select a single "best" tree, either by the neighbor-joining method (Saitou and Nei, 1987) or according to some optimality criterion, e.g., minimum evolution (Kidd and Sgaramella-Zonta, 1971; Rzhetsky and Nei, 1992), maximum parsimony (Fitch, 1971; Maddison, 1991), and maximum likelihood (Felsenstein, 1981, 1993; Kishino et al., 1990; Salter and Pearl, 2001). The neighbor-joining method is a distance-based clustering method, which constructs the phylogenetic tree by successive pairing of taxa that yield the smallest tree length (sum of branch lengths). The minimum evolution, maximum parsimony, and maximum likelihood methods are in practice always combined with a search algorithm looking for the "best" tree. The minimum evolution method seeks the tree with the smallest sum of branch lengths, the maximum parsimony method seeks the tree that requires the minimum number of mutations for reproducing the data, and the maximum likelihood method seeks the tree that is the most likely to have occurred given the observed data and the assumed model of evolution. Although the traditional methods work well for many problems, they do not produce valid inferences beyond point estimates. Alternatively, uncertainty of the phylogeny can be measured by the method of bootstrap resampling (Felsenstein, 1985; Newton, 1996).

Bayesian inference in phylogeny generates a joint posterior distribution for parameters, which include the phylogenetic tree topology and parameters composing the evolutionary model. The posterior distribution can be simulated using Markov chain Monte Carlo algorithms. In Bayesian phylogenetics, the parameters are of the same kind as in maximum likelihood phylogenetics. Thus, typical parameters include tree topology, branch lengths, and substitution model parameters, such as the gamma shape or the transition–transversion ratio parameter. While a maximum likelihood method seeks the best point estimates of parameter values, Bayesian phylogeny inference instead seeks a full probability distribution over all possible parameter values, automatically accounting for uncertainty in phylogenetic trees and model parameters.

7.2 Bayesian phylogeny inference

7.2.1 Tree representation

The relationships among any set of species of all organisms with a common ancestor is called a *phylogeny*, which can be represented by a phylogenetic tree. In this chapter, a phylogenetic tree will be assumed to be a rooted binary tree. Each node represents the most recent common ancestor of its immediate

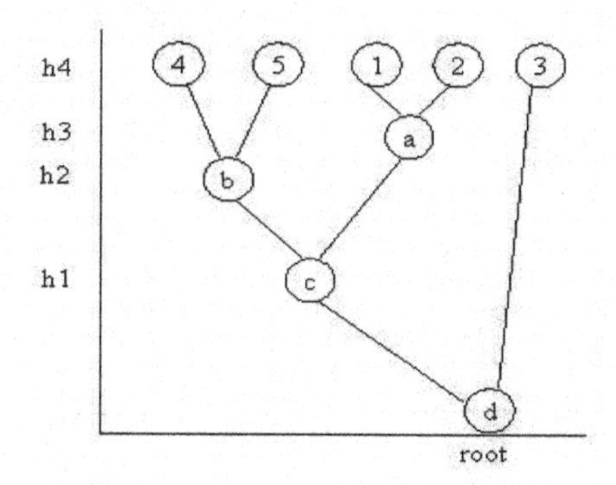

Figure 7.1: Graphical depiction of a sample tree with five taxa.

descendants, and the root represents the common ancestor of all the entities at the leaves of the tree.

In general, a phylogenetic tree of n leaves has $n-2$ internal nodes (excluding the root node) and $2n-2$ branches. A tree is a connected graph with nodes (terminal node and internal node) and branches. Each branch of the tree has a certain amount of evolutionary divergence associated with it. The length of a branch represents the distance between two end node sequences and is often estimated from a model of substitution of residues over the course of evolution. The time separating a child from its parent is its edge weight, also called its branch length, which can be computed by subtraction from the node heights (h_1, h_2, h_3, h_4) in Figure 7.1. For example, node a is at height h_3, node c is at height h_1, and the weight for the edge $c \rightarrow a$ is thus $h_3 - h_1$. We call nodes terminal nodes (or leaves) if they are connected through a single edge and internal nodes otherwise. A true biological phylogeny has a root, or ultimate ancestor of all the sequences. Some algorithms provide information, or at least a conjecture, about the location of the root. Others are completely uninformative about its position and other criteria have to be used for rooting the tree. The placement of the root relative to the leaves determines the direction of time and hence ancestry. The labeled shape of the tree is called the *tree topology*. This is determined by which pairs of nodes coalesce. The topology can be summarized by using parentheses to indicate coalescence. For example, the tree topology in Figure 7.1 is (((4,5),(1,2)),3).

Applications utilize different algorithms for manipulating trees. Many of the different tree algorithms have in common the characteristic that they systematically visit all the nodes in the tree (i.e., the algorithm walks through the tree data structure and performs some computation at each node in the tree). This process of walking through the tree is called a tree traversal (Durbin et al., 1998, pp. 206–210). There are essentially two different methods for systematically visiting all the nodes of a tree: depth-first traversal as a recursive traversal and breadth-first traversal as a non-recursive traversal.

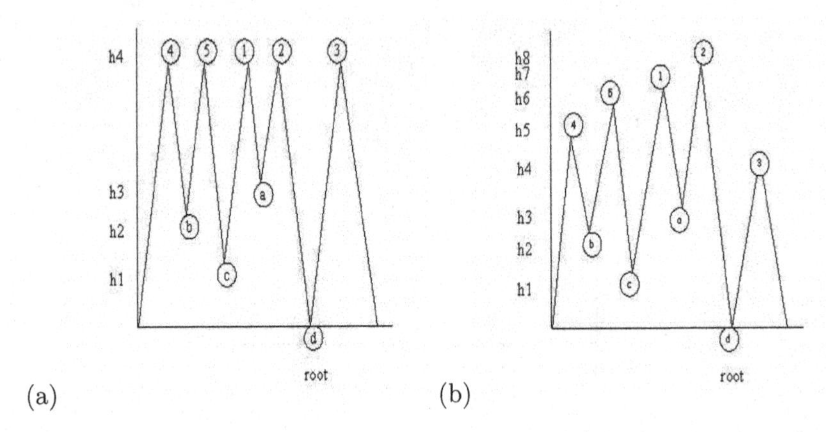

Figure 7.2: (a) Tree with molecular clock assumption and (b) tree without molecular clock assumption.

Certain depth-first traversal methods occur frequently enough that they are given names of their own: pre-order traversal, in-order traversal, and post-order traversal (Drozdek and Simon, 1995). Each taxon (leaf) and internal node appear at a peak and a valley in a graph, respectively. The permutation of taxa is read across the tops of the peaks. Branch lengths and tree topology are determined by $n-1$ (Figure 7.2(a)) or $2(n-1)$ (Figure 7.2(b)) valley depths. There is a left/right choice to make a subtree at each internal node. Based on these choices, there is a unique post-order traversal of the tree. The permutation of leaf labels and the ordered valley depths determine the tree completely.

7.2.2 Evolutionary models

Suppose that n nucleotide sequences (taxa) are arranged as a $n \times N$ matrix, where N is the common number of sites or the common length of the sequences. The data can be viewed as a realization of a stochastic process that has evolved along the branches of an unknown phylogenetic tree.

All evolutionary models deal with the random substitution of one nucleotide for another at individual sites, and most share the following set of underlying assumptions: Markov property, homogeneity, and stationarity (Salemi and Vandamme, 2003). By assuming that evolution among sites is independent conditioned on the given genealogy, modeling is reduced to a single site. Although this assumption greatly simplifies the likelihood calculation, it is quite probably violated by most coding sequence datasets. As explained by Galtier et al. (2005), the sites in a protein (or a RNA sequence) interact to determine the selected three-dimensional structure of the molecule, so the evolutionary processes of interacting sites are not independent. Attempts have been made in the literature to relax the independence assumption by either introducing an autocorrelation parameter for the evolutionary rates of neighboring sites or directly modeling the joint evolutionary process of any two neighboring sites. Significant work in this respect includes Yang (1995a), Felsenstein and Churchill (1996), Thorne et al. (1996), Pollock et al. (1999), Duret and Galtier (2000), and Robinson et al. (2003).

Under the independence assumption, several evolutionary models have been proposed for nucleotides. The one-parameter model (Jukes and Cantor, 1969a) uses nucleotide substitutions with equal rates, and the two-parameter model (Kimura, 1980) allows the rates of transitional and transversional events in nucleotide substitutions to differ. The model of Felsenstein (1981) adds three parameters to the Jukes-Cantor model by allowing the stationary probabilities to be different, and the HKY85 model (Hasegawa et al., 1985) possesses a general stationary distribution of the nucleotides and allows for different rates of transition and transversion events. The HKY85 model is most flexible among the four models.

The transition probability matrix of the HKY85 model is given by

$$Q_{j|i}(T) = \begin{cases} \pi_j + \pi_j\left(\frac{1}{\lambda_j} - 1\right)e^{-\alpha T} + \left(\frac{\lambda_j - \pi_j}{\lambda_j}\right)e^{-\alpha\gamma_j T} & \text{if } i = j, \\ \pi_j + \pi_j\left(\frac{1}{\lambda_j} - 1\right)e^{-\alpha T} - \left(\frac{\pi_j}{\lambda_j}\right)e^{-\alpha\gamma_j T} & \text{if } i \neq j \text{ (transitional event)}, \\ \pi_j(1 - e^{-\alpha T}) & \text{if } i \neq j \text{ (transversional event)}, \end{cases} \tag{7.1}$$

where T denotes time, α denotes the evolutionary rate, $\lambda_j = \pi_A + \pi_G$ if base j is a purine (A or G) and $\lambda_j = \pi_C + \pi_T$ if base j is a pyrimidine (C or T), $\gamma_j = 1 + (\kappa - 1)\lambda_j$, α is the transversion rate, and $\kappa\alpha$ is the transition rate. The stationary probabilities of the four nucleotides are π_A, π_C, π_G, and π_T, respectively. The HKY85 model includes five free parameters, namely, $\alpha, \kappa, \pi_A, \pi_C$, and π_G, which satisfy the constraints $\alpha > 0, \kappa > 0, 0 < \pi_A, \pi_C, \pi_G < 1$, and $0 < \pi_A + \pi_C + \pi_G < 1$.

The most general model used in this chapter , TN93 (Tamura and Nei, 1993), has both HKY85 and F84 (Felsenstein's PHYLIP since 1984) as special cases. The instantaneous rate matrix R is parameterized as follows:

$$R = \alpha \begin{pmatrix} R_{AA} & \kappa\pi_G & \pi_C & \pi_T \\ \kappa\pi_A & R_{GG} & \pi_C & \pi_T \\ \pi_A & \pi_G & R_{CC} & \kappa\gamma\pi_T \\ \pi_A & \pi_G & \kappa\gamma\pi_C & R_{TT} \end{pmatrix}, \tag{7.2}$$

where $R_{AA} = -(\kappa\pi_G + \pi_C + \pi_T)$, $R_{GG} = -(\kappa\pi_A + \pi_C + \pi_T)$, $R_{CC} = -(\pi_A + \pi_G + \kappa\gamma\pi_T)$, $R_{TT} = -(\pi_A + \pi_G + \kappa\gamma\pi_C)$, $\kappa = \frac{4\phi}{\gamma+1}$, and the order (A, G, C, T) for bases is assumed. There are seven parameters, six of which are free. The model is reversible with the stationary distribution given by $(\pi_A, \pi_G, \pi_C, \pi_T)$, where $\sum_{i \in \{A,G,C,T\}} \pi_i = 1$. The parameter α controls the overall mutation rate. The transition–transversion ratio is κ, and γ is the final parameter which affects the ratio of transition–transversion rates among purines and pyrimidines

(Simon and Larget, 2001). When $\phi = \kappa/2$ and $\gamma = 1$, the TN93 model becomes the HKY85 model.

7.2.3 Bayesian approach

A Bayesian analysis requires a likelihood model for sequence evolution through a phylogenetic tree, prior distribution on trees and model parameters, and data. Suppose that x is an alignment of data (e.g., DNA sequences). Let $\omega = (\tau, h, \phi)$ denote a phylogenetic tree, where τ denotes the tree topology, h denotes the vector of branch lengths, and ϕ denotes the vector of parameters of the evolutionary model, describing rates of change among states in the Markov process for a given branch. Modeling is reduced to a single site by assuming that evolution among sites is independent. Thus, the likelihood function $L(\omega|x) = L(\tau, h, \phi|x)$ can be calculated based on observed data x.

The likelihood of the tree can be calculated using the pruning method proposed by Felsenstein (1983). The pruning method produces a collection of partial likelihoods of subtrees, starting from the leaves and working recursively to the root for each site. Let $\mathcal{S} = (\mathcal{A}, \mathcal{G}, \mathcal{C}, \mathcal{T})$ denote the set of nucleotides. For site k of a leaf e, define $L_e^k(i) = 1$ if state i matches the base found in the sequence and 0 otherwise, where i indexes the elements of $\mathcal{S}$. At site k of an internal node ν, the conditional probability of descendant data given state i is

$$L_\nu^k(i) = \left(\sum_{j \in \mathcal{S}} L_u^k(j) Q_{j|i}(T_{\nu u}) \right) \times \left(\sum_{j \in \mathcal{S}} L_w^k(j) Q_{j|i}(T_{\nu w}) \right), \quad i \in \mathcal{S}, \tag{7.3}$$

where u and w denote the two child nodes of ν, T_{ab} denotes the time represented by the branch terminating in nodes a and b, and $Q_{j|i}(T_{ab})$ is a transition probability matrix like (7.1). The likelihood function of the tree can then be written as

$$L(\omega|x) = \prod_{k=1}^{N} \sum_{i \in \mathcal{S}} \pi_0(i) L_\rho^k(i), \tag{7.4}$$

where x denotes the observed sequences of n taxa, ρ denotes the root node, and π_0 is the initial probability distribution assigned to the ancestral root sequence. π_0 is normally set to the stationary distribution.

Let $f(\omega)$ denote the prior distribution of $\omega = (\tau, h, \phi)$. The prior should reflect the scientist's beliefs on how likely particular parameter values are before the data have been observed. The posterior distribution

$$f(\omega|x) \propto L(\omega|x) f(\omega) \tag{7.5}$$

represents the uncertainty about the phylogeny in light of new evidence in the sequence data and is proportional to the likelihood times the prior distribution.

Various samplers can be employed to sample from this posterior. For example, BAMBE (Larget and Simon, 1999) employs the MH algorithm,

MrBayes (Huelsenbeck and Ronquist, 2001) employs the parallel tempering algorithm (Geyer, 1992a), and Cheon and Liang (2008, 2009) employ the stochastic approximation Monte Carlo algorithm (Liang et al., 2007).

7.3 Monte Carlo methods for Bayesian phylogeny inference

7.3.1 Markov chain Monte Carlo algorithms

Since the summation and integrals required in a Bayesian analysis cannot be evaluated analytically, the Markov chain Monte Carlo (MCMC) algorithms have been instead used to approximate the posterior probabilities of trees. The basic MCMC algorithm works as follows.

Metropolis-Hastings algorithm (Metropolis et al., 1953a; Hastings, 1970):

(a) Initialize a phylogeny model ω_0 and set $t = 0$.

(b) Let ω_t denote the current state of Markov chain ω at iteration t.

  (b.1) Randomly propose a new value ω^*. The probability of proposing the new state is $q(\omega_t, \omega^*)$ whereas the probability of proposing the old state ω_t conditional on the new state ω^* (a move that is not actually made) is $q(\omega^*, \omega_t)$.

  (b.2) Calculate the acceptance probability r for the new state ω^* as follows.

$$r = r(\omega_t, \omega^*) = \min\left[1, \frac{f(\omega^*|x)}{f(\omega_t|x)} \frac{q(\omega^*, \omega_t)}{f(\omega_t, \omega^*)}\right]. \tag{7.6}$$

  (b.3) Generate a random variable U that is uniformly distributed on the interval $(0, 1)$. If U is less than r, then accept the proposed state $\omega_{t+1} = \omega^*$. Otherwise, continue with the current state ω_t; i.e., $\omega_{t+1} = \omega_t$.

(c) Set $t \leftarrow t + 1$ and go back to (b).

A minimal requirement for the Markov chain simulation method to work is that the chain should be irreducible, which is needed to balance the desire to move globally through the tree space with the need to make computationally feasible moves in areas of high probability. This process is repeated for a sufficiently large number of iterations. As long as the chain is properly constructed with an irreducible and aperiodic Markov chain, the long-run frequencies of states visited by the Markov chain will approximate the proposed posterior distribution. That is, the Markov chain $\omega_1, \omega_2, \ldots$ formed from an initial state converges in distribution to $f(\omega|x)$ when ω is irreducible (e.g., Tierney, 1994). The important theoretical point is that for almost every realization of the chain,

$$\lim_{K\to\infty} \frac{1}{K} \sum_{i=k_0+1}^{K+k_0} f(\omega_i|x) = E_\pi[f(\omega|x)]. \tag{7.7}$$

The empirical relative frequency of a particular topology in the Markov chain converges to its corresponding (marginal) posterior probability. Credible sets in the Bayesian inference are collections of topologies having high relative frequency in the chain. The topology with the highest relative frequency is called the *best estimate.*

The bias in the Monte Carlo estimates can be reduced by discarding some initial sample points as burn-in ($k_0 > 0$) (Besag and Green, 1993). Burn-in is determined by inspecting time-series plots of the log posterior. It is hard to know whether the iteration K used to approximate is large enough. If K is too small, then the Monte Carlo variance can overwhelm the signal (Geyer, 1992b). Roughly speaking, a chain is said to mix well if it acts like an independent sample. Convergence diagnostics used by Cowles and Carlin (1996) can be employed to check for evidence of poor mixing.

7.3.1.1 Rannala and Yang (1996)

Rannala and Yang (1996) suggested using the maximum posterior probability tree for inferring evolutionary trees. The birth–death process was used in the posterior distribution as a model of speciation and extinction to specify the prior distribution of phylogenies and branching times, and a continuous-time Markov process was taken to model nucleotide substitutions. Parameters of the branching model and the substitution model were estimated by maximum likelihood. By calculating the posterior probabilities of different phylogenies, the phylogeny with the highest posterior probability was chosen as the best estimate of the evolutionary relationship among species, which is called the *maximum posterior probability (MAP) tree.* The posterior probability provides a natural measure of the reliability of the estimated phylogeny.

The n sequences are descended through $n-1$ speciation events, which occurred at times $t_1, t_2, \ldots, t_{n-1}$ in the past, with $t_1 > t_2 > \cdots > t_{n-1}$. The time of the first bifurcation is set to 1 (i.e., $t_1 = 1$) and parameters are then relative to this time scale. Let $\mathbf{t} = \{t_2, \ldots, t_{s-1}\}$.

The birth–death process is a continuous-time process where parameters λ and μ are the branching and extinction rates per lineage, respectively. The substitution model used in Felsenstein's DNAML program (since 1984, version 2.6) was also used for this study. The likelihood function is given by

$$\begin{aligned} L(\lambda, \mu, m, \kappa|x) &= f(x; \lambda, \mu, m, \kappa) \\ &= \sum_{\tau} \int_{t_2=0}^{1} \cdots \int_{t_{s-1}=0}^{t_{s-2}} f(x|\tau, \mathbf{t}; m, \kappa) f(\tau, \mathbf{t}|\lambda, \mu) dt_{s-1} \cdots d_{t_2}, \end{aligned} \tag{7.8}$$

where m is the average substitution rate, κ is the transition–transversion ratio, the summation is over the labeled histories, and the integrations are over the divergence times (Equation (13) in Rannala and Yang (1996)).

Since analytical methods appear quite difficult for evaluating the preceding equations, numerical integration was used to calculate approximate values of the likelihood function. The dimension of the integration for a tree of n sequences is $n - 2$, and the computation is feasible only for a small number of species. Maximum likelihood estimates (MLEs) of parameters λ, μ, m, and κ are obtained by maximizing the log-likelihood function, $l = \log(L)$. A numerical optimization algorithm was used for this purpose. After parameters λ, μ, m, and κ have been estimated by MLEs, the posterior probability of the labeled history (τ), conditional on the observed sequence data, was calculated, which is the "contribution" of tree topology τ to the likelihood function. The tree topology having the maximum posterior probability is the MAP estimate of phylogeny. The maximum posterior probability can be interpreted as the probability that the estimated tree is the true tree under the models, providing a measure of the reliability of the estimated phylogeny.

Calculation of the likelihood function involves evaluation of an $(n - 2)$ dimensional integral for each labeled history. For this calculation, the numerical approach of repeated one-dimensional integration was used, evaluating several algorithms for this procedure (Press et al., 1992, pp. 129–164). The one reason why the calculation does not seem practical with data for more than five species (i.e., a three-dimensional integral) is that the functional form of the integrand in the likelihood function changes with the tree topology, although computer algorithms were devised for performing the integration for any number of species and any tree topology. Thus, the log likelihoods vary greatly for different values of $\mathbf{t}$.

This method was applied to inference of the phylogenetic relationship of human, chimpanzee, gorilla, and orangutan. The estimated best trees were the same as those from the maximum likelihood analysis of separate topologies, but the posterior probabilities were quite different from the bootstrap proportions. The results of the method were found to be insensitive to changes in the rate parameter of the branching process.

Their method was implemented in the PAML (Phylogenetic Analysis by Maximum Likelihood) program package, which is publicly available on the website http://abacus.gene.ucl.ac.uk/software/paml.html.

7.3.1.2 Mau and Newton (1997)

Mau and Newton (1997) described an MCMC method for models satisfying the molecular clock condition, and presented calculations for binary, restriction sites data. This method has different MCMC strategies for the same general problem with Rannala and Yang (1996). A Markov chain was simulated on the space of dendrograms, visiting each labeled history with a long-run relative frequency proportional to its marginal posterior probability, and assessed uncertainty in the estimates by constructing a probability distribution on the space of trees without bootstrapping.

A tree representation in this study was considered to be based on a metric to discover a proposal distribution that nominates candidate trees near the true ones. Following Lapointe and Legendre (1991), binary trees with leaves can be transformed into $n \times n$ cophenetic matrices, whose array elements d_{ij} represent the time to coalescence for each pair of leaf nodes i and j. These matrices are symmetric with zeroes along the diagonal, and the space of matrices is metrizable under the standard Euclidean norm. The cophenetic matrix form is not unique, but depends on the ordering of the labels. Labels can always be rearranged so that the superdiagonal elements, $\{d_{i,i+1}, i = 1, \ldots, n-1\}$, of the corresponding cophenetic matrix, are distinct. A canonical ordering and the corresponding superdiagonal completely determine a dendrogram.

A two-stage proposal distribution was constructed with canonical cophenetic matrices. Starting at a given tree, the first proposal distribution Q_1 produces a random draw from among the 2^{n-1} canonical orderings of that tree by independently flipping a fair coin to determine the orientation of the descendent branches at each internal node, generating a superdiagonal denoted by $\{a_i = d_{i,i+1}, i = 1, \ldots, n-1\}$. In the second stage, Q_2 produces a_i^* on the a_i, which is drawn independently from

$$Q_{2,i}(a_i) \sim \begin{cases} U(a_i - \delta, a_i + \delta) & \text{if } a_i \geq \delta, \\ \dfrac{\delta - a_i}{\delta} U(0, \delta - a_i) + \dfrac{a_i}{\delta} U(\delta - a_i, \delta + a_i) & \text{if } a_i < \delta, \end{cases} \tag{7.9}$$

for a fixed $\delta > 0$. The first case is a uniform density with positive support. The second case is a mixture of two uniform densities with positive support, derived from a single uniform density centered near zero with the probability mass on negative values folded over onto its positive reflection. The complete second stage is

$$Q_2(a_1, \ldots, a_{n-1}) = \prod_{i=1}^{n-1} Q_{2,i}(a_i). \tag{7.10}$$

Note that δ is a tunable parameter which determines how "far" one can jump from the current tree, and hence can be used to modulate the overall acceptance rate of the Markov chain.

Mau and Newton (1997) used a simple stochastic model for the evolution of restriction sites; a two-state Markov process with infinitesimal rates λ and μ, representing the intensity of the instantaneous transition from 0 to 1 and 1 to 0, respectively. The model was reparameterized by its mutation rate $\theta = \lambda + \mu$, and the ratio of its infinitesimal rates, $r = \frac{\mu}{\lambda}$, which yields

$$P_{i,j}(t, \beta = (\theta, r)) = \frac{1}{1+r}\left(r^{1-j} + (-1)^{I\{i \neq j\}} r^t e^{-\theta t}\right), \tag{7.11}$$

where $I\{\cdot\}$ denotes the indicator function.

Ignorance priors are placed on the space of labeled histories, coalescent times, and r, suitably restricted to compact subsets. Thus r is updated (via a uniform proposal distribution centered at its current value) as part of the MCMC cycle. Fortuitously, θ is confounded with time, so the branch lengths are proportional to the amount of evolution, with proportionality constant θ. Hence, θ is fixed throughout the simulation.

Mau and Newton (1997) employed the traditional Metropolis algorithm on the space of dendrograms of trees. The Metropolis algorithm was applied directly to the dendrogram, while a modified copy of the second stage, Q_2, independently updates the coordinates of the component. For monitoring the convergence of the chain, the log posterior and cumulative coalescent times of the current tree printed to the screen after every 1000th iteration, as do the intervening average acceptance rates for candidate trees and generator ratios.

7.3.1.3 Mau et al. (1999) and Newton et al. (1999)

Mau et al. (1999) extended the calculations of Mau and Newton (1997) to problems with more taxa and nucleotide sequence data by considering distances in tree space. Newton et al. (1999) reviewed this approach to quantify the uncertainty in a phylogenetic tree inferred from molecular sequence information. They used a Markov chain to sample from the posterior distribution for a phylogenetic tree given sequence information from the corresponding set of organisms, a stochastic model for these data, and a prior distribution on the space of trees.

A transformation of the tree into a canonical cophenetic matrix form suggests a simple and effective proposal distribution for selecting candidate trees close to the current tree in the chain. This method automatically provides posterior probabilities from which confidence coefficients for a particular tree can be calculated if one so chooses. Since this method averages over local maxima for large numbers of taxa, it has an advantage over other methods that do not guarantee to produce an absolute maximum.

In this study, a uniform prior was considered on the finite set of labeled histories, and a flat prior density was assumed on a compact set of possible coalescent times as well as any propagation parameters from the stochastic model.

In most implementations of the Metropolis-Hastings algorithm, a collection of proposal distributions determine the complete algorithm (e.g., Besag et al., 1995). A single proposal distribution works for the phylogeny problems considered so far. Since this proposal distribution is global, in that a new state ω^* can differ from the current state ω in all respects, this makes the algorithm inefficient. Inefficient algorithms are ones which traverse the parameter space slowly and thus exhibit significant positive correlations on one-dimensional summary. Local, single-parameter updating proposals change parts of the parameter at a time, and are at risk for low efficiency. One risk of a global proposal distribution, on the other hand, is that candidates may be rejected too frequently, and thus results in an inefficient algorithm. This method avoids this by making global changes small in magnitude, and by basing changes on distance within the tree, so that proposed trees are close in posterior density to the current tree.

Mau et al. (1999) considered a two-stage proposal distribution which is somewhat different from Mau and Newton (1997) in terms of Q_2. The first stage randomly selects a canonical representation $\{\sigma, a\}$ for the current tree ω where σ is a canonical ordering and a denotes the times to coalescence between adjacent label pairs in σ, whereas the second stage perturbs the components of a. In particular, the first stage Q_1 samples one of the 2^{n-1} canonical orderings of the current tree by independently flipping a fair coin at each internal node, thus selecting a particular superdiagonal $\{d_{i,i+1} : i = 1, \ldots, n-1\}$ of a canonical cophenetic matrix having times to coalescence $\{a_i = d_{i,i+1}/2\}$. The second stage Q_2 simultaneously and independently modifies the elements of a. Specifically, $a_i^* = |U_i|$ where U_i is uniformly distributed on the interval $(a_i - \delta, a_i + \delta)$ for a tuning constant $\delta > 0$. The tuning constant determines how far one can jump from the current tree and hence can be used to modulate the overall acceptance rate of the chain. That is, when the tunable parameter δ is small, the candidate tree is close to the current tree in terms of pairwise distance between species, and so the likelihood of the candidate tree may be close to that of the current tree. Thus the proposal is to perturb the speciation times of a version of the current tree.

This Markov chain has a property of irreducibility and symmetry, which means that the Metropolis-Hastings ratio reduces to a ratio of posterior densities, and thus to a ratio of likelihoods under a flat prior.

7.3.1.4 Larget and Simon (1999)

Larget and Simon (1999) extended the methods based on MCMC proposed by Yang and Rannala (1997a) and Mau et al. (1999). They developed the Bayesian framework with different update mechanisms for analyzing aligned nucleotide sequence data to reconstruct phylogenies, assess uncertainty in the reconstructions, and perform other statistical inferences. They employed a Markov chain Monte Carlo sampler to sample trees and model parameter values from their joint posterior distribution. All statistical inferences were naturally based on this sample. The sample provides a most-probable tree with posterior probabilities for each clade that is similar to the maximum-likelihood tree with bootstrap proportions, and permits further inferences on the tree topology, branch lengths, and model parameter values.

A composition of two different basic update mechanisms was used to traverse ω. Specifically, an initial tree and model parameter values were randomly chosen, $\omega_0 = (\tau_0, \phi_0)$, from some very dispersed distribution. Given the current state of $\omega_t = (\tau_t, \phi_t)$, a single cycle consists of two stages. In the first stage, while keeping the current tree τ_t fixed, new model parameters ϕ^* were proposed using proposal distribution $q_1(\phi_t, \phi^*)$ in the space of model parameter values

ϕ_t, which are either accepted ($\phi_{t+1} = \phi^*$) or rejected ($\phi_{t+1} = \phi_t$). The second stage modifies the current tree τ_t in a sequence of steps while holding ϕ_{t+1} fixed. One step of the second stage proposes a new tree τ^* according to a proposal distribution $q_2(\tau_t, \tau^*)$ on τ_t, which is accepted or rejected and repeats this process a fixed number of times. The tree τ_{t+1} is the result of a fixed number of Metropolis-Hastings steps according to q_2 from τ_t.

There are two main tree proposal algorithms, each with clock and non-clock versions as follows: global with a molecular clock, which is equivalent to the algorithm presented in Mau et al. (1999); global without the molecular clock; local with a molecular clock, and local without the molecular clock. In the global proposal with a molecular clock; a distance from the root in a depth first traversal of the tree gives a graph with peaks and valleys for any given collection of left/right orientations for subtrees at each internal node. The tree is parameterized by a permutation of the taxa as read from left to right in the representation and the valley depths from left to right. The global proposal without the molecular clock is the same as the global proposal with a molecular clock except that all peaks may be of different distances from the root in the tree representation and thus each valley has two depths, the depths to its left and right peaks (Figure 7.1). The local proposal distribution with a molecular clock modifies the tree only in a small neighborhood of a randomly chosen internal branch, leaving the remainder of the tree unchanged. The local proposal distribution without the molecular clock acts on the unrooted tree. Larget and Simon (1999) also extended these algorithms to being unrestricted by the molecular clock assumption. For both the local and global algorithms, the proposal distribution for the topology and branch times is independent of the model. The acceptance rates do, of course, depend on the model, and these are calculated by integrating over sequences of internal nodes using the pruning algorithm.

This method updates parameters as well. When a parameter κ in the HKY85 model (Hasegawa et al., 1985) is restricted in a range $[0, M]$ where M is any positive real value, a new value $\kappa^* = \kappa + U$ is proposed, where U is chosen uniformly at random between $-\delta$ and δ, reflecting the excess back into the range should κ^* be negative or exceed M. The Hastings ratio is 1 for this proposal distribution.

When a set of parameters is constrained to sum to a constant, a new set of values is proposed according to a Dirichlet distribution centered at the current parameter values. Specifically, if the current values are $\mathbf{z} = (z_1, \ldots, z_k)$, where $\sum_i z_i = c$, then $z^* = cY$, where Y is randomly chosen from a Dirichlet distribution with parameters $(\alpha z_1, \ldots, \alpha z_k)$, with α a tunable parameter. The higher α is, the more likely the proposed parameter values are to remain close to their current values. The Metropolis-Hastings ratio for this proposal distribution is a ratio of two Dirichlet densities. The entire parameter proposal chain is obtained by independently proposing new parameter values for all parameters with these two types of proposals and accepting or rejecting the combined proposal in a single Metropolis-Hastings step. It is critical for proper mixing that the tunable parameters be chosen well for good acceptance rates.

The numeric results of this study showed that their method is much faster than the bootstrap resampling method. Because the tree proposal algorithms are independent of the choice of likelihood function, their method could be used in conjunction with more complex likelihood models.

Simon and Larget (1998, 2001) have developed the software package Bayesian Analysis in Molecular Biology and Evolution (BAMBE), written in ANSI C, which is publicly available on the website http://www.mathcs.duq.edu /larget/bambe.html.

7.3.1.5 Li et al. (2000)

Li et al. (2000) described a Bayesian method based on Markov chain simulation to study the phylogenetic relationship in a group of DNA sequences. This method generates a sequence of phylogenetic trees using the MCMC technique under simple models of mutation events, and provides estimates and corresponding measures of variability for any aspect of the phylogeny under study.

This method is a modification of the local rearrangement strategy used by Kuhner et al. (1995) in a genealogy application. This rearrangement occurs only in the neighborhood of a specific internal node, which cannot be the root of the tree. The neighborhood of an internal node consists of its parent (P), sibling (S), and two children (C_1 and C_2). To define the transition probability function $q(\omega, \omega^*)$, the modified local rearrangement strategy of transition is decomposed into the following three steps: In the first step, a node (target τ) is randomly chosen from among the $n-2$ internal nodes that have both a parent and children. One of C_1, C_2, and S is randomly selected to be the new sibling, which is called S', leaving the remaining two to be the two new children, C_1' and C_2'. The proposed topology is then a modification of the old topology, which has rearranged the positions of the sibling and the two children together with their attached subtrees. In the second step, the time of the new target τ' is picked between the time of the parent node (t_P) and the time of the target's nearest post-arrangement descendant, $\max\{t_{C_1'}, t_{C_2'}\}$, given the topology selected in the first step. In the final step, a nucleotide sequence for the new target is decided after picking $t_{\tau'}$ (see Li et al., 2000, for more details).

The Markov chain generated by applying this local rearrangement strategy will converge to the target distribution because one can move from any topology to any other topology in a finite number of steps, which applies to the local algorithm of Larget and Simon (1999).

This method produces a random initial tree through a distance-based algorithm. First, for each pair of external nodes k and l, the number of sites d_{kl} is computed for which the nucleotides differ. Then a pair of nodes clustered first is drawn with probability proportional to $\hat{p}^{d_{kl}}$, where $\hat{p} = \frac{3}{4}(1 - \exp(-8\widehat{\alpha\tau}))$

is a coarse estimate of the probability that two external nodes are different at a particular site (see Li, 1997, Chapter 3), and $\widetilde{\alpha\tau}$ is our initial estimate of $\alpha\tau$. After a pair of sequences is chosen for clustering, a nucleotide sequence is generated by randomly drawing between the two sequences at each site. This newly generated sequence and the remaining sequences that have not yet been clustered are used to create a new distance matrix. At this point, the number of sequences has been reduced by one. The clustering process is repeated until only two sequences are left. This whole recursive process generates a topology.

Given the topology of a tree, the splitting times of the internal nodes are generated according to the conditional distribution of times given the topology τ proceeding recursively from the time of the root, generating the times of its two children, and so forth, until the times of the parents of the external nodes are picked.

With known topology and branch lengths, the sequences of the internal nodes are generated using the initial estimates. However, in this case it proceeds recursively in postorder sequence until the sequences of the children of the root node are randomly generated.

The proposed proposal for investigating evolutionary histories is a package of model assumptions and movement strategies, together with diagnostics to examine the adequacy of the algorithm's behavior. While Mau et al. (1999), Larget and Simon (1999), and Li et al. (2000) viewed the species as fixed in the model, the Bayesian method proposed by Rannala and Yang (1996) and Yang and Rannala (1997a) considered the species as being sampled from a linear birth-and-death process that has proceeded over time.

This method can change the distribution on topologies to other members of the general class proposed by Aldous (1996), or change the nucleotide substitution model to include rate-variation parameters (Yang, 1993) or non-symmetry parameters (Rodriguez et al., 1990), or specifically model nucleotide substitutions for protein-coding sequences (Goldman and Yang, 1994).

This method allows for the same changes in topologies as in the local algorithm in Larget and Simon (1999), although they have different probabilities. This method keeps track of the sequences of internal nodes. Whereas carrying auxiliary variables may slow down the rate of mixing, this approach has several potential advantages. The sequences themselves may be of interest; for instance, one may wish to determine the probability that certain nucleotide changes are related to clade (subtree)-specific characteristics. Also, the sequences of the internal nodes are helpful in estimating the nucleotide substitution parameters such as α and κ. Second, the proposed changes in node times depend on the evolutionary model (although changes in the topology do not). Targeting the proposal distribution to the posterior is likely to improve statistical efficiency.

This algorithm is very fast per iteration, because it keeps the calculations local to a target node. At the same time, because large subtrees are potentially swapped, substantial changes in the trees are possible within very few moves. Thus there is a good balance between the speed of calculation and the rate of mixing. The algorithm of Mau et al. (1999) strikes a different balance, putting more effort into the calculations at each step, with possible gains in the rate of mixing. A comparison of the overall efficiency of the two algorithms would require using both on the same sets of data. The algorithm of Rannala and Yang (1996) requires the use of Monte Carlo integration at each step, making it impractical for large datasets.

7.3.2 Advanced MCMC algorithms

Although the Bayesian method is very attractive, it suffers from a severe difficulty in applications: the Metropolis-Hastings (MH) algorithm (Metropolis et al., 1953a; Hastings, 1970) tends to get trapped in a local energy minimum, and thus tends to fail to generate correct samples from the posterior distribution. Here, the energy function refers to the negative log-posterior distribution of the phylogenetic trees in terms of physics. The local-trap problem has also occurred in applications of the optimization-based phylogenetic tree construction methods, such as the maximum likelihood and minimum evolution methods. A number of authors have noticed this difficulty and have tried to employ some advanced MCMC algorithms to resolve it. For example, Huelsenbeck and Ronquist (2001), Feng et al. (2003), and Altekar et al. (2004) applied parallel tempering (Geyer, 1992a) to simulate from the posterior distribution. Cheon and Liang (2008, 2009) applied the stochastic approximation Monte Carlo algorithm (Liang et al., 2007) to the problem; Cheon and Liang (2008) utilized a sequential Monte Carlo algorithm focusing on MAP trees, but Cheon and Liang (2009) inferred the phylogeny and the model parameters by dynamically weighted averaging over the samples generated in the simulation. These advanced algorithms can converge much faster than the MH algorithm for the distributions for which the energy landscape is rugged.

7.3.2.1 Huelsenbeck and Ronquist (2001)

Huelsenbeck and Ronquist (2001) developed the MrBayes software, which uses MCMC to approximate the posterior probabilities of trees. MrBayes not only implements the standard MCMC algorithm, but also a variant of MCMC called Metropolis-coupled Markov chain Monte Carlo [or $(MC)^3$ for short; Geyer (1992a)].

The idea of Metropolis-coupled MCMC is to run several chains in parallel (parallel tempering), each chain having a different distribution determined by the value of β, $f(\omega_i|x)^\beta$, and with index swap operations conducted in place of the temperature transition of simulated annealing methods. The chain with distribution $f(\omega_i|x)$ (i.e., $\beta = 1$) is used in sampling and is called the *cool chain.* Other chains are used to improve the mixing of the chains and are called *heated chains.* $(MC)^3$ runs n chains, $n-1$ of which are heated. A heated chain has the steady-state distribution $(f(\omega_i|x))^\beta$. MrBayes uses incremental heating, where the heat applied to the ith chain is $\beta = \frac{1}{1+(i-1)T}$ and T is a heating parameter that must be set by the user. After all n chains have gone

one step, a swap is attempted between two randomly chosen chains. If the swap is accepted, then the two chains switch states. Inferences are based only on the states sampled by the cold chain $\beta = 1$. The heated chains can more easily explore the space of phylogenetic trees; the effect of heating is to lower peaks and to fill in valleys. Importantly, the cold chain can effectively leap across deep valleys in the landscape of trees when a successful swap is made between the cold chain that is perhaps stuck on a local optimum of trees and a heated chain that is exploring another peak. Experience has shown that mixing of the chain is dramatically improved using $(MC)^3$. The generalized Metropolis-coupled MCMC algorithm is as follows.

Metropolis-coupled MCMC algorithm:

(a) Initialize a phylogeny likelihood model ω_0 and set $t = 0$.

(b) Let ω_t denote the current state of Markov chain ω at iteration t.

(b.1) Randomly propose a new value ω^*, for ω_t, according to a proposal distribution $q(\omega_t, \omega^*)$.

(b.2) Accept or reject ω^* with probability r_t.

$$r_t = \min\left[1, \left(\frac{f(\omega^*|x)}{f(\omega_t|x)}\right)^{\beta} \frac{q(\omega^*, \omega_t)}{q(\omega_t, \omega^*)}\right]. \tag{7.12}$$

(b.3) Generate a random variable U that is uniformly distributed on the interval $(0, 1)$. If U is less than r_t, then accept the proposed state ω^*. Otherwise, continue with the current state ω_t.

(c) After all chains have advanced a given number of iterations (say one cycle), two randomly chosen chains j and k are selected to exchange states. The swap of states is accepted with probability

$$r = \min\left[1, \frac{(f(\omega_k|x))^{\beta_j}}{(f(\omega_j|x))^{\beta_j}} \frac{(f(\omega_j|x))^{\beta_k}}{(f(\omega_k|x))^{\beta_k}}\right]. \tag{7.13}$$

(d) A uniformly distributed random number on the interval $(0, 1)$ is generated. If this number is less than the acceptance probability r, then the proposed swap of states is accepted, and thus chains j and k exchange states.

(e) Set $t \leftarrow t + 1$ and go back to (b).

After repeating this process for a sufficiently large number of iterations, the long-run frequencies of states sampled by the cold chain are a valid approximation of the posterior distribution. Successfully swapping states allows a chain that is otherwise stuck on one peak in the landscape of trees to explore other peaks. For example, if the cold chain is stuck on a peak in the posterior distribution of trees, swapping states with another heated chain allows the cold chain to jump to another peak in a single cycle. As a result, the cold chain can more easily traverse the space of trees.

The Metropolis-coupled MCMC algorithm allows multiple peaks in the landscape of trees to be more readily explored and thus is useful in overcoming the local optima problem, but at the cost of increased execution time. $(MC)^3$ takes time proportional to the number of chains being run. The number of heated chains needed for adequate mixing in phylogenetic inference problems is likely to be data dependent. Huelsenbeck and Ronquist (2001) found that running four coupled chains results in successful convergence for several problems where standard MCMC techniques fail, but it is quite possible that adequate mixing will require many more heated chains for difficult datasets. When dealing with large datasets, running even four chains may be computationally prohibitive using traditional algorithms running on single machines.

MrBayes has a command-line interface. The program reads in an aligned matrix of DNA or amino acid sequences in the standard NEXUS format (Maddison et al., 1997), and also implements several methods for relaxing the assumption of equal rates across sites, including gamma-distributed rate variation (Yang, 1993). Also, the program can infer ancestral states while accommodating uncertainty about the phylogenetic tree and model parameters. MrBayes is publicly available on the website (http://mrbayes.sourceforge.net).

7.3.2.2 Feng et al. (2003)

Feng et al. (2003) implemented an MPI (message passing interface) version of the parallel MCMC algorithms for Bayesian phylogenetic inference, referring to this as the parallel Bayesian phylogenetic inference (PBPI) implementation. In this study, entire chains or parts of a chain were distributed to different processors, since in the models used, the columns of the data are independent. They identified a number of important points, including a super linear speedup due to more effective cache usage and the point at which additional processors slow down the process due to communication overhead.

To avoid a computationally intensive process, Feng et al. (2003) employed a parallel implementation. Parallelizing Bayesian phylogenetic inference brings two advantages, which are speeding up the computation and providing the memory space needed for a competitive biological analysis of current data. Since generating the Markov chain requires a large number of iterations, the parallelization will focus on parallelizing the MCMC algorithm. A single-chain MCMC is essentially a sequential program, since the state at time $t + 1$ is dependent on the state at time t. One way to parallelize Metropolis-Hasting MCMC is to parallelize the likelihood evaluation. Another is to run multiple equivalent chains and sample each one after the burn-in stage. This method may involve many random starting points, which provides the advantage of exploring the space through independent initial trajectories, but also has the danger that the burn-in stage may not always be cleared as in Gilks et al.

(1996b). A single Metropolis-coupled MCMC run can be parallelized at the chain level, but chain-to-chain communications are needed frequently. However, just parallelizing on the chain level will not use all the available resources, especially when memory is the limiting factor.

An important issue in the parallelization of Metropolis-coupled MCMC is balancing the load. This issue of load balancing comes from the fact that when a local update scheme is used, different chains may reevaluate over a different number of nodes at each step. More seriously, the local update scheme is often only available for topology and edge length changes; with parameters such as a global rate matrix changing, the likelihood needs to be evaluated across the tree, so all nodes need to be reevaluated. Other issues that must be considered in a parallel algorithm are how to synchronize the processors, how to reduce the number of communications, and how to reduce the message length of each communication.

There are two natural approaches to exploiting parallelism in Metropolis-coupled MCMC: chain-level parallelization and sub-sequence-level parallelization. Chain-level parallelization divides chains among processors; each processor is responsible for one or more chains. Subsequence-level parallelization divides the whole sequence among processors; each processor is responsible for a segment of the sequence, and communications contribute to computing the global likelihood by collecting local likelihood from all processors. These implementations combine two approaches together and map the computation task of one cycle into a two-dimensional grid topology. The processor pool is arranged as a $c \times r$ two-dimensional Cartesian grid. The dataset is split into c segments, and each column is assigned one segment. The chains are divided into r groups, and each row is assigned one group of chains. When $c = 1$, the arrangement becomes chain-level parallel; when $r = 1$, the arrangement becomes subsequence-level parallel.

Two sets of random number generators (RNG1 and RNG2) are used to synchronize the processors in the grid. RNG1 is used for intra-row synchronization. The processors on the same row have the same seed for RNG1, but different rows have a different seed for RNG1. RNG2 is used for inter-row communication. All processors in the grid topologies have the same seed for RNG2. On each row, RNG1 is used to generate the proposal state and draw random variables from the uniform distribution. RNG2 is used to choose which two chains should conduct a swap step and to draw the probability to decide whether or not to accept the swap operation. When the two chosen chains are located on different rows, a two-step communication is conducted on these two rows. The first step is used to exchange the likelihood and temperature of the chosen rows. The second step is used to broadcast the result to other processors located on the grid or the chosen rows. There are several chain swap algorithms available. In each chain swap step, the indices but not the state information of the chains is swapped. An index swap operation changes the temperature of the chains being swapped, and the cool chain may jump from one chain to another. Index swapping reduces the communication contents needed by chain swapping to a minimum.

The processors on the same row always have a balanced load if the differences of the lengths of the subsequences on each column are small enough. The imbalance among different rows is unavoidable, since the instantaneous behavior of a given chain within a step cannot be predicted. However, some techniques can still be used to decrease the imbalance.

The first technique is to synchronize the proposal type on all chains. RNG2 is used to control how to propose a new candidate state. This can prevent the phenomenon of having one chain doing a local update and another chain doing a global update. The second technique is to choose a swap proposal probability to control the interval between two swap steps.

In general, all the parallel strategies may be based on an assumption that any two chains chosen to conduct a swap step need to be synchronized. The symmetric parallel MCMC algorithm is given here (See Feng et al. (2003) for more details).

Symmetric parallel MCMC algorithm:

(a) Problem initialization.

(b) MCMC initialization.

(c) Repeat.

- (c.1) Draw a random variable u_1 from $U(0, 1)$.
- (c.2) If $u_t \leq$ swap probability, do swap step.
- (c.3) Otherwise do parallel step.
- (c.4) If sampling is needed, output the state information of the cool chain.
- (c.5) Set $t = t + 1$.

To further reduce the negative effect of imbalance between different chains, an asymmetric MCMC algorithm can be used. The idea is to introduce a processor as the coordinator node. The coordinator node is used to coordinate the communication between different rows; it does not participate in the likelihood evaluation. After each cycle, the head of each row sends the state information of its chains to the coordinator and retrieves information from it when a swap step is proposed. The asymmetric MCMC algorithm is similar to the shared memory algorithm, but the coordinator can perform other functions such as convergence detection and sampling output. Compared to the symmetric MCMC algorithm, the asymmetric MCMC algorithm wastes one processor. Thus, when the number of rows in the grid topology is not large, the symmetric MCMC algorithm seems favorable.

7.3.2.3 Altekar et al. (2004)

Altekar et al. (2004) introduced a $p(MC)^3$ algorithm that significantly decreases the execution time of a Bayesian analysis using Metropolis-coupled MCMC, $(MC)^3$, for both small and large datasets. $p(MC)^3$ is a variant of MCMC, works by running Markov chains in parallel, and retains the ability to explore multiple peaks in the posterior distribution of trees while maintaining a fast execution time. The algorithm has been implemented using two popular parallel programming models: message passing and shared memory.

Both implementations of this algorithm are evaluated on a cluster of shared memory (symmetric) multiprocessors (SMPs). The message passing implementation makes use of the message passing interface (MPI) (Message Passing Interface Forum, 1994) for sending messages to and from processes. The shared memory implementation uses Cashmere (Stets et al., 1997, pp. 170–183), a software distributed shared memory (SDSM) system developed at the University of Rochester, which provides the illusion of shared memory in software across machines in the cluster. SDSM is able to take advantage of hardware support for sharing within each SMP, which is thereby providing better performance, while allowing seamless expansion without much loss in performance across SMPs. Message passing has the advantage of allowing applications to minimize overhead when communicating across SMPs by sending only the necessary data.

$p(MC)^3$ takes advantage of the random nature of $(MC)^3$, which allows non-swapping chains to proceed to the next generation. The effect is that chains perform useful computation for a majority of the execution. Moreover, this algorithm swaps heats rather than states, thereby decreasing communication overhead significantly. The parallel algorithm has been implemented in both message passing and shared memory programming models. By experimenting with both Ethernet and the Memory Channel networks, Altekar et al. (2004) found that their algorithm scales well to a large number of processors and high scalability opens up the possibility of running a large number of chains for better mixing. Finally, this algorithm performs equally well on both small and large datasets.

The $p(MC)^3$ algorithm significantly decreases the execution time of a Bayesian analysis using $(MC)^3$ and works by running Markov chains in parallel. Specifically, $p(MC)^3$ spreads Markov chains among processes. A process performs all computation associated with its assigned chain(s). This includes calculating the likelihood function with the most computationally intensive operation in any given iteration. In $(MC)^3$, all chains proceed to the next iteration in step with each other. Consequently, swaps are made between chains in the same generation. $p(MC)^3$ breaks away from the notion of all chains proceeding in step with each other. For strict correctness, however, swaps between chains must still take place in the same generation.

1. *Heat swapping.* $p(MC)^3$ reduces communication costs by exchanging heats rather than states. That is, chain states and their associated data structures are never communicated. Rather, the heat values associated with each chain are swapped. Since the heat value associated with each chain is unique to that chain, one can effectively swap the identities of two chains by exchanging the heat values. Once heats are swapped, the swapping chains accept new states based on their newly acquired heat values. Using the heat exchange mechanism, there is no need to communicate the chain states and associated data structures, resulting in interprocessor messages of a few bytes, rather than messages several megabytes in size.

Heat swapping provides a more efficient way to affect swaps than state swapping. Furthermore, heat swapping is particularly easy to implement. A swapping chain can readily access its partner's heat value once both have synchronized and communicated swap acceptance information with each other. Once the swap is accepted, heat values can be swapped without further communication.

State-swapping code requires two steps of communication once chains have synchronized. In the first step, swap acceptance information is conveyed to the swap partner. In the second step, the chain state is communicated if the swap has been accepted. Heat swapping, on the other hand, requires only one step of communication. In the first step, swap acceptance information is conveyed to the swap partner. If the swap is accepted, the current heat value is replaced by the heat value received in the swap acceptance information. By sending the random number used in making the swap decision between the processors along with the rest of the swap information, the processors make identical swap acceptance decisions without having to communicate additional information.

2. *Synchronization.* In $p(MC)^3$, heat exchanges call for synchronization. The exchange rule requires that chains synchronize with one another before swapping heats. A swapping chain must receive swap acceptance data from its swap partner before accepting or rejecting a swap. Altekar et al. (2004) presented two exchange schemes, which are the global exchange scheme and the point-to-point exchange scheme. They address synchronization in $p(MC)^3$ with varying degrees of efficiency and ease of use. Both schemes adhere to the exchange rule and therefore provide results that are identical to the sequential version in terms of phylogenetic data, provided that the number of chains and random number seeds are identical.

In the global exchange scheme, chains are guaranteed to be in the same iteration at the time of a swap, thereby satisfying the exchange rule. In the message passing version, the global exchange scheme is easy to implement and in a shared memory implementation of the global exchange scheme no explicit data communication is required. The major disadvantage of this scheme, however, is that using a barrier can lead to wasted processing time in non-swapping chains due to uneven computation times in an iteration of each chain. While chains involved in the swap are required to wait to communicate and exchange state acceptance information, uninvolved chains must merely wait for the barrier to complete.

The point-to-point exchange scheme minimizes idle time during state exchanges. The scheme works by predetermining which two chains swap in each iteration. Uninvolved chains are allowed to proceed to the next iteration. As a result, little computation time is wasted waiting for swaps to complete. Synchronization in this scheme is achieved through the use of point-to-point synchronization operations. In the message passing programming model, synchronization and data communication are combined in the send and receive operations. The advantage of the point-to-point exchange scheme is that non-swapping chains no longer idle while a swap is taking place. Moreover, the use of point-to-point synchronization rather than barrier synchronization greatly reduces communication cost and synchronization time.

7.3.2.4 Stochastic approximation Monte Carlo

Cheon and Liang (2008) applied the stochastic approximation Monte Carlo (SAMC) algorithm (Liang et al., 2007) to Bayesian phylogeny inference. The SAMC algorithm has the self-adjusting mechanism and thus avoids essentially the local-trap problem suffered by conventional MCMC algorithms in simulating from the posterior distributions of phylogenetic trees. Furthermore, it falls into the category of dynamic importance sampling algorithms; i.e., the phylogeny and the model parameters can be inferred by dynamically weighted averaging over the samples generated in the simulation.

Suppose inference is focused on a distribution,

$$f(x) = \frac{1}{Z}\psi(x), \qquad x \in \chi, \tag{7.14}$$

where χ is the sample space and Z is the normalizing constant. Applying (7.14) to inference on the posterior distribution in (7.5), $\psi(\cdot)$ corresponds to the unnormalized posterior density $L(\omega|x)f(\omega)$ and χ corresponds to the parameter space of ω. Let $U(\omega) = -\log\psi(\omega|x)$, which is called the energy function in terms of physics. $\psi(\omega|x)$ is a non-negative function defined on the sample space with $0 < \int_\chi \psi(\omega|x)d\omega < \infty$. Suppose that the sample space has been partitioned according to the energy function into m disjoint subregions denoted by $E_1 = \{x : U(\omega) < u_1\}$, $E_2 = \{x : u_1 \leq U(\omega) < u_2\}$,..., $E_{m-1} = \{x : u_{m-2} \leq U(\omega) < u_{m-1}\}$, and $E_m = \{x : U(\omega) \geq u_{m-1}\}$, where $u_1, \ldots, u_{m-1}$ are real numbers specified by the user.

SAMC seeks to draw samples from each of the subregions with a pre-specified frequency. Let x_{t+1} denote a sample drawn from an MH kernel $K_{\theta_t}(x_t, \cdot)$ with the proposal distribution $q(x_t, \cdot)$ and the stationary distribution

$$f_{\theta_t}(x) \propto \sum_{i=1}^{m-1} \frac{\psi(x)}{e^{\theta_t^{(i)}}} I(x \in E_i) + \psi(x)I(x \in E_m), \tag{7.15}$$

where $\theta_t = (\theta_t^{(1)}, \ldots, \theta_t^{(m-1)})$ is an $(m-1)$-vector in the space Θ, $\theta_t^{(i)} =$

$\int_{E_i} \psi(\omega|x)d\omega$, and $\theta_t^{(m)} = 0$ for convenience. E_m is assumed non-empty; that is, $\int_{E_m} \psi(x)dx > 0$.

Let $\boldsymbol{\pi} = (\pi_1, \ldots, \pi_m)$ be an m-vector with $0 < \pi_i < 1$ and $\sum_{i=1}^m \pi_i = 1$, which defines desired sampling frequencies for the subregions. Henceforth, $\boldsymbol{\pi}$ is called the desired sampling distribution. Define $H(\theta_t, x_{t+1}) = \mathbf{e}_{t+1} - \boldsymbol{\pi}$, where $\mathbf{e}_{t+1} = \left(e_{t+1}^{(1)}, \ldots, e_{t+1}^{(m)}\right)$ and $e_{t+1}^{(i)} = 1$ if $x_{t+1} \in E_i$ and 0 otherwise.

Let $\{\gamma_t\}$ be a positive, non-decreasing sequence satisfying the conditions,

$$(i) \ \sum_{t=0}^{\infty} \gamma_t = \infty, \qquad (ii) \ \sum_{t=0}^{\infty} \gamma_t^{\delta} < \infty, \tag{7.16}$$

for some $\delta \in (1, 2)$. In the context of stochastic approximation (Robbins and Monro, 1951), $\{\gamma_t\}_{t\geq 0}$ is called the gain factor sequence.

Let $J(x)$ denote the index of the subregion to which the sample x belongs. Let $\{\mathcal{K}_s, s \geq 0\}$ be a sequence of compact subsets of Θ such that

$$\bigcup_{s\geq 0} \mathcal{K}_s = \Theta, \quad \text{and} \quad \mathcal{K}_s \subset \text{int}(\mathcal{K}_{s+1}), \quad s \geq 0, \tag{7.17}$$

where int(A) denotes the interior of set A. Let $\mathcal{X}_0$ be a subset of $\mathcal{X}$, and let $\mathbb{T} : \mathcal{X} \times \Theta \to \mathcal{X}_0 \times \mathcal{K}_0$ be a measurable function which maps a point in $\mathcal{X} \times \Theta$ to a random point in $\mathcal{X}_0 \times \mathcal{K}_0$. Let σ_k denote the number of truncations performed until iteration k. Let $\mathcal{S}$ denote the collection of the indices of the subregions from which a sample has been proposed; that is, $\mathcal{S}$ contains the indices of all subregions which are known to be non-empty. With the above notations, one iteration of SAMC can be described as follows.

The SAMC algorithm:

(a) (Sampling) Simulate a sample ω_{t+1} by a single MH update with the target distribution as defined in (7.15) where $\psi(\omega|x) \propto L(\omega|x)f(\omega)$.

  (a.1) Generate ω^* according to a proposal distribution $q(\omega_t, \omega^*)$. If $J(\omega^*) \notin \mathcal{S}$, set $\mathcal{S} \leftarrow \mathcal{S} + \{J(\omega^*)\}$.

  (a.2) Calculate the ratio

$$r = e^{\theta_{J(\omega_t)}^{(t)} - \theta_{J(\omega^*)}^{(t)}} \frac{\psi(\omega^*)q(\omega^*, \omega_t)}{\psi(\omega_t)q(\omega_t, \omega^*)}. \tag{7.18}$$

  (a.3) Accept the proposal with probability $\min(1, r)$. If it is accepted, set $\omega_{t+1} = \omega^*$; otherwise, set $\omega_{t+1} = \omega_t$.

(b) (Weight updating) For all $i \in \mathcal{S}$, set

$$\begin{aligned} \theta_{t+\frac{1}{2}}^{(i)} &= \theta_t^{(i)} + \gamma_{t+1} \left[I(\omega_{t+1} \in E_i) - \pi_i \right] \\ &\quad - \gamma_{t+1} \left[I(\omega_{t+1} \in E_m) - \pi_m \right]. \end{aligned} \tag{7.19}$$

(c) (Varying truncation) If $\theta_{t+\frac{1}{2}} \in \mathcal{K}_{\sigma_t}$, then set $(\theta_{t+1}, \omega_{t+1}) = \left(\theta_{t+\frac{1}{2}}, \omega_{t+1}\right)$ and $\sigma_{t+1} = \sigma_t$; otherwise, set $(\theta_{t+1}, \omega_{t+1}) = \mathbb{T}(\theta_t, \omega_t)$ and $\sigma_{t+1} = \sigma_t + 1$.

The self-adjusting mechanism of the SAMC algorithm is obvious: If a proposal is rejected, the weight of the subregion that the current sample belongs to will be adjusted to a larger value, and thus the proposal of jumping out from the current subregion will less likely be rejected in the next iteration. This mechanism warrants that the algorithm will not be trapped by local energy minima. The SAMC algorithm represents a significant advance in simulations of complex systems for which the energy landscape is rugged.

The proposal distribution $q(\omega, \omega^*)$ used in the MH updates is required to satisfy the following condition: For every $\omega \in \mathcal{X}$, there exist $\epsilon_1 > 0$ and $\epsilon_2 > 0$ such that

$$|\omega - \omega^*| \leq \epsilon_1 \Longrightarrow q(\omega, \omega^*) \geq \epsilon_2, \tag{7.20}$$

where $|\omega - \omega^*|$ denotes a certain distance measure between ω and ω^* (Roberts and Tweedie, 1996).

SAMC falls into the category of varying truncation stochastic approximation algorithms (Chen, 2002; Andrieu et al., 2005). Following Liang et al. (2007), for all non-empty subregions,

$$\theta_t^{(i)} \to C + \log\left(\int_{E_i} \psi(\omega) d\omega\right) - \log\left(\pi_i + \pi_0\right), \tag{7.21}$$

as $t \to \infty$, where $\pi_0 = \sum_{j \in \{i: E_i = \emptyset\}} \pi_j / (m - m_0)$, $m_0 = \#\{i : E_i = \emptyset\}$ is the number of empty subregions, and $C = -\log\left(\int_{E_m} \psi(\omega) d\omega\right) + \log\left(\pi_m + \pi_0\right)$. In SAMC, the sample space partition can be made blindly by simply specifying some values of $u_1, \ldots, u_{m-1}$. This may lead to some empty subregions.

Let $\widehat{\pi}_i^{(t)} = P(\omega_t \in E_i)$ be the probability of sampling from the subregion E_i at iteration t. Equation (7.21) implies that as $t \to \infty$, $\widehat{\pi}_i^{(t)}$ will converge to $\pi_i + \pi_0$ if $E_i \neq \emptyset$ and 0 otherwise. With an appropriate specification of $\boldsymbol{\pi}$, sampling can be biased to the low-energy subregions to increase the chance of locating the global energy optimizer.

Cheon and Liang (2009) utilized the local moves used in Larget and Simon (1999), which were adopted for updating the trees. The only difference is that the acceptance rate of those moves has been adjusted by the self-adjusting factor $e^{\theta_t^{(J(\omega_t))} - \theta_t^{(J(\omega^*))}}$. Those moves include three parts: the part for updating model parameters, the part for updating branch lengths, and the part for rearranging tree topologies. Larget and Simon (1999) described those moves for both types of trees with and without molecular clocks. Their tree corresponds to the case without molecular clocks. Refer to Larget and Simon (1999) for the details of those moves. Larget and Simon (1999) also prescribed another type of move, the so-called global moves, for updating the trees. Since SAMC is capable of moving across high-energy barriers (due to its self-adjusting ability) and the global moves are much more time-consuming than the local moves, only the local moves were adopted in the implementation of SAMC.

Let $h(\omega)$ denote a quantity of interest for a phylogenetic tree, such as the presence/absence of a branch or an evolutionary parameter. Then $E_f h(\omega)$, the expectation of $h(\omega)$ with respect to the posterior (7.5), can be estimated by

$$\widehat{E_f h(\omega)} = \frac{\sum_{k=n_0+1}^{n} e^{\theta_k^{(J(\omega_k))}} h(\omega_k)}{\sum_{k=n_0+1}^{n} e^{\theta_k^{(J(\omega_k))}}}, \tag{7.22}$$

where $\left(\omega_{n_0+1}, e^{\theta_{n_0+1}^{(J(\omega_{n_0+1}))}}\right), \ldots, \left(\omega_n, e^{\theta_n^{(J(\omega_n))}}\right)$ denotes a set of samples generated by SAMC, and n_0 denotes the number of burn-in iterations. It follows from Geweke (1989) for the usual importance sampling estimate that $\widehat{E_f h(w)} \to E_f h(w)$ as $n \to \infty$.

SAMC was compared with two popular Bayesian phylogeny softwares, BAMBE (Larget and Simon, 1999) and MrBayes (Huelsenbeck and Ronquist, 2001), on simulated and real datasets. Among the three methods, SAMC produced the consensus trees which have the highest similarity to the true trees, and the model parameter estimates which have the smallest mean square errors, but costs the least CPU time.

7.3.2.5 Sequential stochastic approximation Monte Carlo

Cheon and Liang (2009) also applied the sequential Monte Carlo algorithm to the Bayesian approach, focusing on the MAP tree to avoid evaluation of high-dimensional summations and integrals. The traditional MCMC algorithms often suffer from a severe difficulty in convergence. One reason is multimodality. Many techniques to alleviate this problem have been proposed, such as simulated tempering (Marinari and Parisi, 1992), parallel tempering (Geyer, 1992a; Hukushima and Nemoto, 1996), and evolutionary Monte Carlo (Liang and Wong, 2000). However, the slow convergence is not due to the multimodality, but the curse of dimensionality; i.e., the number of samples increases exponentially with dimension to maintain a given level of accuracy.

Cheon and Liang (2008) developed the sequential stochastic approximation Monte Carlo (SSAMC) algorithm. This method is a new phylogenetic tree construction method, which attempts to alleviate these two difficulties simultaneously by making use of the sequential structure of phylogenetic trees in conjunction with stochastic approximation Monte Carlo simulations. The use of the sequential structure of the problem provides substantial help in reducing the curse of dimensionality in simulations, and SAMC effectively prevents the system from getting trapped in local energy minima.

The SSAMC algorithm consists of two steps, buildup ladder construction and sequential SAMC simulation. A buildup ladder (Wong and Liang, 1997; Liang, 2003) comprises a sequence of systems of different dimensions. Typically,

$$dim(\mathcal{X}_1) < dim(\mathcal{X}_2) < \cdots < dim(\mathcal{X}_m), \tag{7.23}$$

where $\mathcal{X}_i$ denotes the sample space of the ith system. The principle of buildup ladder construction is to approximate the original system by a system with a reduced dimension. The solution of the reduced system is then extrapolated level by level until the target system is reached. For phylogeny problems, the buildup ladder can be constructed with an ordering procedure. Let $D_1, \ldots, D_L$ denote L subsets of taxa, $D_1 \subset D_2 \subset \cdots \subset D_L$ where D_i contains the first $|D_i|$ taxa in the buildup order and $D_L = D$ contains all taxa of the dataset. At the latter level of the buildup ladder, the taxa tend to be inserted into different branches of the partial tree, and thus more taxa can be added at one level. For a large dataset, this can make the number of buildup levels substantially smaller than the number of taxa. SSAMC is then employed to sample simultaneously from the following distributions,

$$f(\omega_i|D_i) = \frac{1}{Z_i} L(\omega_i|D_i), \qquad \omega_i \in \mathcal{X}, \quad i = 1, \ldots, L, \tag{7.24}$$

where ω_i is the tree constructed for the taxa contained in D_i, $f(\omega_i|D_i)$ is the posterior distribution of ω_i, and Z_i is the unknown normalizing constant of $f(\omega_i|D_i)$. The SSAMC simulation consists of two stages, the normalizing constant ratio estimation and the target sample generation. One iteration of SSAMC consists of level proposing, tree updating, and estimate updating. Suppose that SSAMC has converged in a run. To generate MCMC samples from the target distribution, one can continue to run SSAMC by fixing the gain factor to zero in the style of simulated tempering (Marinari and Parisi, 1992; Geyer and Thompson, 1995) with the estimated normalizing constants. The complete tree samples generated in this stage can be used for making a Bayesian inference about the phylogeny.

Let i_t denote the current buildup level, $\omega_t^{(i_t)}$ denote the current tree, and $\theta_t^{(i_t)}$ denote an estimate of $\log(Z_{i_t})$ obtained at iteration t. The new tree ω^* generated by the extrapolation or projection operators and the proposed buildup level j are accepted or rejected with the probability

$$\min\left\{1, \frac{e^{\theta_t^{(i_t)}}}{e^{\theta_t^{(j)}}} \frac{L(\omega^*|D_j)}{L(\omega_t^{(i_t)}|D_{i_t})} \frac{T\left(\omega^*, \omega_t^{(i_t)}\right)}{T\left(\omega_t^{(i_t)}, \omega^*\right)} \frac{\widetilde{T}\left(D_j, D_{i_t}\right)}{\widetilde{T}\left(D_{i_t}, D_j\right)}\right\}, \tag{7.25}$$

where $T(x, y)$ denotes the transition probability from x to y for a tree and $\widetilde{T}(x, y)$ denotes the proposal probability of the extrapolation and projection moves from x to y (see Cheon and Liang (2008) for more details). If it is accepted, make the updating $i_{t+1} = j$ and $\omega_{t+1}^{(i_{t+1})} = \omega^*$; otherwise, set $i_{t+1} = i_t$ and $\omega_{t+1}^{(i_{t+1})} = \omega_t^{(i_t)}$.

The SSAMC algorithm:

Let $\widetilde{T}_{mk} = \widetilde{T}(D_m, D_k)$ denote the proposal probability for a transition from level m to level k, where $\widetilde{T}$ is specified as a tridiagonal matrix with the elements

$\widetilde{T}_{1,1} = \widetilde{T}_{L,L} = \frac{2}{3}$, $\widetilde{T}_{1,2} = \widetilde{T}_{L,L-1} = \frac{1}{3}$, and $\widetilde{T}_{m,m+1} = \widetilde{T}_{m,m-1} = \widetilde{T}_{m,m} = \frac{1}{3}$. Initialize the estimates of $\log(Z_m)$'s by setting $t = 0$ and $\theta_0^{(1)} = \cdots = \theta_0^{(L)} = 0$. Generate an arbitrary tree at the first level of the buildup ladder, and set $i_0 = 1$. One iteration of SSAMC consists of the following steps:

(a) (Level proposing) Generate level j according to the stochastic matrix $\widetilde{T}$ and the current level i_t.

(b) (Tree updating)

(b.1) If $j = i_t$, update the model parameters, branch lengths, and tree topology iteratively as in Larget and Simon (1999).

(b.2) If $j = i_t + 1$, propose an extrapolation operation along the buildup ladder, and accept the proposed tree with an acceptance probability.

(b.3) If $j = i_t - 1$, propose a projection operation along the buildup ladder, and accept the proposed tree with an acceptance probability.

(c) (Estimate updating) Set $\theta_{t+1}^{(j)} = \theta_t^{(i_t)} + \gamma_t$, where γ_t is called the gain factor and satisfies the conditions (7.16).

Under mild conditions, Liang et al. (2007) showed that $e^{\theta_t^{(i)}}/e^{\theta_t^{(j)}}$ converges to Z_i/Z_j almost surely. A remarkable feature of SSAMC is that it will not get stuck at a buildup level. This is due to the intrinsic self-adjusting mechanism of the algorithm. If a level transition proposal is rejected, the weight of the current level will increase by the factor e^{γ_t}, and thus a level transition proposal will less likely be rejected in the next iteration. This process can repeat until a success of level transition occurs.

SSAMC is different from the parallel tempering algorithm used in MrBayes (Huelsenbeck and Ronquist, 2001). Parallel tempering works by simulating in parallel a sequence of distributions scaled by different temperatures, while keeping the sample space unchanged at each level. SSAMC works with a buildup ladder with dimension increasing along the ladder. The use of a buildup ladder can provide substantial help to reduce the curse of dimensionality for simulations of complex systems. This has been argued theoretically by Liang (2003) based on the Rao-Blackwellization theorem (Liu, 2001, p. 27), which suggests that one should carry out analytical computation as much as possible in order to improve efficiency of the simulations. Note that the distribution defined at a lower buildup level can be regarded as a marginal distribution of the distribution defined at a higher buildup level. Hence, sampling simultaneously is preferred from a series of partial trees instead of complete trees.

In the second stage, SSAMC seeks to sample from the following mixture distribution:

$$\frac{1}{L}\sum_{m=1}^{L} f(\omega_m|D_m). \tag{7.26}$$

The samples generated at the highest buildup level correspond to complete trees, and are approximately distributed as $f(\omega_L|D_L)$ when the number of iterations becomes large. Other distributions $f(\omega_1|D_1), \ldots, f(\omega_{L-1}|D_{L-1})$ only work as trial distributions for sampling from complete trees. Hence, the choice of the buildup ladder will not affect the equilibrium of the simulation at the highest buildup level, although it may affect the efficiency of the simulation. They suggested choosing the starting taxon randomly. However, the effect of the taxon addition rule can be significant. An arbitrary buildup ladder may cause difficulties for state transitions between some pairs of neighboring levels. Hence, the taxon addition rule should be carefully selected to ensure that the systems on neighboring levels largely resemble each other and thus the state transitions between them can be made smoothly.

SSAMC was compared with two popular Bayesian phylogeny software programs, BAMBE and MrBayes, on simulated and real datasets. At first, for the given tree of 20 nucleotide sequences shown in Figure 7.3(a), a dataset was simulated according to a HKY85 model with the root sequence as shown in Table 7.1 and the parameters $\kappa = 2$, $\alpha = 1$, and $\pi_A = \pi_G = \pi_C = \pi_T = 0.25$. In all simulations, α was restricted to be 1, as the evolutionary rate of each site of the nucleotide sequences was modeled equally.

In SAMC simulation, the taxa were ordered according to the pairwise alignment scores, and a buildup ladder was constructed as follows: $D_1 =$

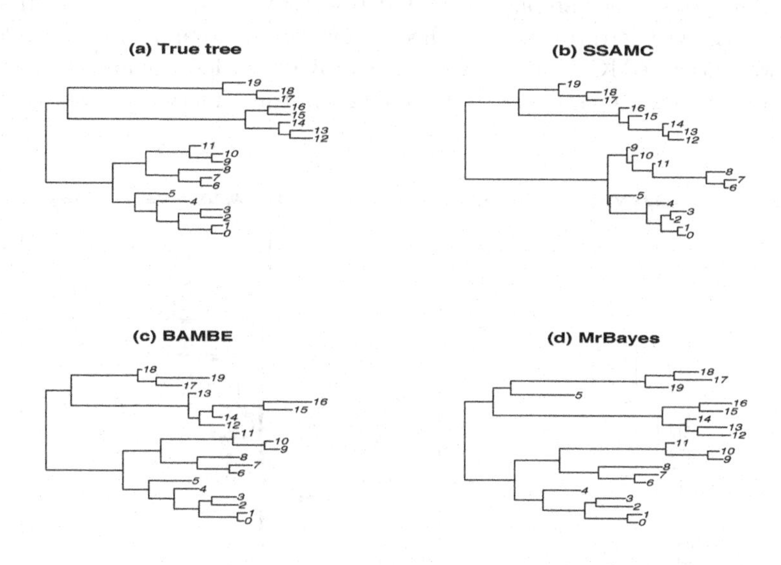

Figure 7.3: Comparison of the phylogenetic trees produced by SSAMC, BAMBE, and MrBayes for the simulated example. The respective log-likelihood values of the trees are (a) −4209.44, (b) −4196.09, (c) −4197.68, and (d) −4198.19.

Table 7.1: The root sequence for the simulated example.

ATGAACCCTT	ACATCCTAAT	AACCCTTCTT	TTCGGACTAG	GTCTAGGAAC
TACAATTACA	TTTGCAAGCT	CCCACTGACT	CCTTGCTTGA	ATAGGCCTTG
AACTAAACAC	CCTCGCTATT	ATCCCACTGA	TAGCCCAACT	CCACCACCCC
CGGGCAGTCG	AAGCTACCAC	AAAATACTTC	CTCACCCAAG	CTGCTGCCGC

$\{1^*,\ldots,4^*\}$, $D_2 = \{1^*,\ldots,5^*\}$,..., $D_7 = \{1^*,\ldots,10^*\}$, $D_8 = \{1^*,\ldots,12^*\}$,..., and $D_{12} = \{1^*,\ldots,20^*\}$, where $1^*,\ldots,20^*$ denote the ordered taxa. MrBayes and BAMBE were run with their default parameter settings. All three methods were run 5 times independently, and each run consists of 2×10^6 iterations. The MAP trees sampled by the three methods indicate that the tree constructed by SSAMC most closely resembles the true tree for this example (Figure 7.3). Figure 7.4 shows the progression curves of the best log-likelihood values produced by the above three methods. It indicates that BAMBE tends to get trapped in a local energy minimum very quickly, and MrBayes and SSAMC perform better in this respect. However, during these runs MrBayes fails to produce better trees than those produced by SSAMC.

The Bayesian analysis has been done for the parameters of the HKY85 model for this example. In SSAMC, the same parameter settings as those used above were used except that the gain factor was set to zero during the last 10^6 iterations. The samples generated by SSAMC, BAMBE, and MrBayes during the last iterations were collected for the Bayesian analysis. Table 7.2 shows that SSAMC produces the highest averaged log-likelihood value and the most accurate estimate for κ among the three methods. Note that the

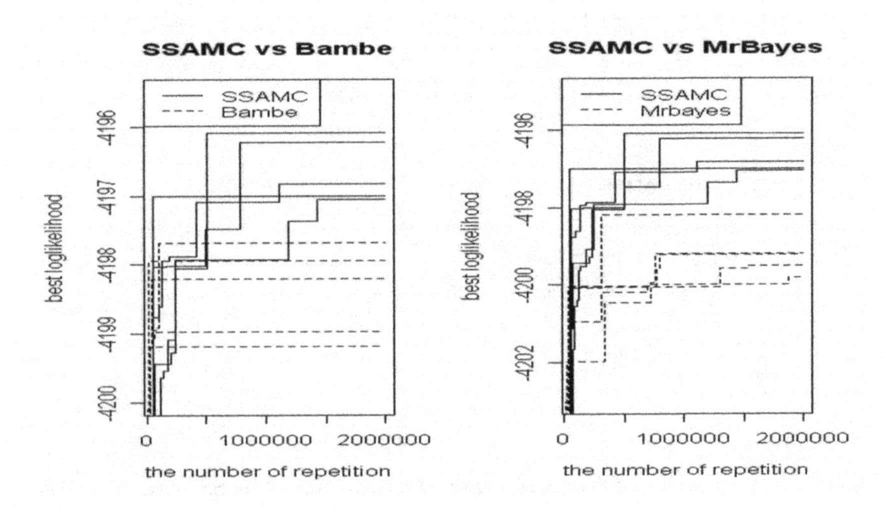

Figure 7.4: Comparison of the progression curves of the best log-likelihood values produced by SSAMC, BAMBE, and MrBayes.

Table 7.2: Bayesian analysis for the parameters of the HKY85 model used for the simulated study. CPU: CPU time (in minutes) cost by a single run of the algorithm on an Intel Pentium III computer. Each of the other entries of the table is calculated by averaging over five independent runs, and the number in the parentheses represent the standard deviation of the corresponding average.

	Methods		
	SSAMC	BAMBE	MrBayes
CPU	19.2	25.6	138.9
Averaged			
Log likelihood	−4202.06 (0.11)	−4208.46 (0.09)	−4202.64 (0.05)
κ	1.85 (0.03)	1.77 (1.9×10^{-3})	1.77 (3.2×10^{-4})
π_A	0.254 (5.5×10^{-5})	0.253 (3.8×10^{-5})	0.255 (4.6×10^{-5})
π_G	0.225 (4.8×10^{-5})	0.225 (2.8×10^{-5})	0.227 (3.8×10^{-5})
π_C	0.262 (3.9×10^{-5})	0.263 (3.7×10^{-5})	0.261 (6.2×10^{-5})
π_T	0.258 (4.1×10^{-5})	0.259 (2.6×10^{-5})	0.257 (8.5×10^{-5})

improvement made by SSAMC is significant in terms of averaged log-likelihood values and estimates of κ, although the absolute differences are small. However, the estimates of the nucleotide frequencies produced by the three methods are similar, as they are mainly determined by the data. Cheon and Liang (2009) also compare SSAMC with BAMBE and MrBayes on the sequences generated for the trees with 30 taxa and 40 taxa. The results are similar.

Later, MrBayes and BAMBE were re-run independently 10 times, and each run consists of 2×10^7 iterations, a 10-fold increase of the total number of iterations performed in previous runs. However, neither MrBayes nor BAMBE can produce a better log-likelihood value than that produced by SSAMC in the previous runs.

SSAMC costs the least CPU time per iteration among three methods. BAMBE has the simplest program structure among the three algorithms. It is a single Markov chain algorithm, and the state (or tree) undergoes a single MH update per iteration. MrBayes is a population MCMC algorithm in which five Markov chains are run in parallel. At each iteration, the state of each chain is updated once besides the state-swapping operations between chains. Hence, the CPU time cost by an iteration of MrBayes is much more than that cost by an iteration of BAMBE. Like BAMBE, SSAMC is a single-chain algorithm, and its state only undergoes a single update at each iteration. Since in most iterations it works on partial trees, on average SSAMC costs less CPU time per iteration than BAMBE.

For the real data analysis, aligned protein coding mitochondrial DNA sequences were obtained from 32 species of cichlid fishes (Kocher et al., 1995; Cheon and Liang, 2008). Taxa 1–5 form a flock from Lake Malawi. The remainder from Lake Tanganyika (6–31 taxa) constitute a Tanganyikan flock.

The Malawi, Ectodini, and Lamprologini tribes are represented by the letters A, C, and D, respectively. Class B consists of taxa 6–9, a combination of most of Tropheini and one species of Limnochromini. Classes $E = \{22, 23, 24, 26, 27\}$ and $F = \{28, 29, 30, 31\}$ are convenient conglomerations (pseudoclades) of the remaining tribes. Taxa $\{25\}$ is not grouped. Taxon 32 is an outgroup from cichlid America. Each DNA sequence consists of 1044 sites. Identical nucleotides are observed on 567 sites, and the nucleotides on the remaining sites are used for phylogenetic tree construction.

SSAMC was first applied to this example. The buildup ladder was constructed as follows: $D_1 = \{1^*, \ldots, 4^*\}$, $D_2 = \{1^*, \ldots, 5^*\}, \ldots$, $D_7 = \{1^*, \ldots, 10^*\}$, $D_8 = \{1^*, \ldots, 12^*\}, \ldots$, and $D_{18} = \{1^*, \ldots, 32^*\}$, where $1^*, \ldots, 32^*$ denote the ordered taxa. SSAMC was run five times and each run consists of 10^6 iterations. The best log-likelihood value found among the five runs is -7726.51. For comparison, MrBayes and BAMBE were also applied to this example. Each software was run five times, and each run consists of 10^6 iterations. The best log-likelihood values produced by them are -7876.98 and -7888.57, respectively. Figure 7.5 compares the MAP trees found by the three methods. As reported below, the best log-likelihood value produced by the maximum likelihood method is -7881.18. These values are all significantly lower than that produced by SSAMC.

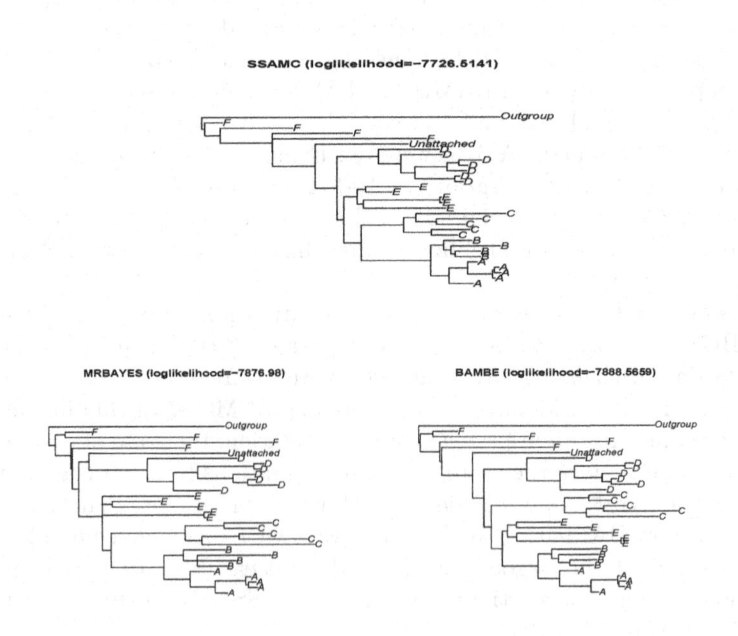

Figure 7.5: Comparison of the MAP trees produced by SSAMC, MrBayes, and BAMBE for the African cichlid fish example.

For a thorough comparison, non-Bayesian methods, neighbor joining, minimum evolution, maximum parsimony, and maximum likelihood, were also tried for this example. These methods have been implemented in many software programs, e.g., PHYLIP (http://evolution.genetics.washington.edu/phylip.html), MEGA (http://www.Megasoftware.net), PAUP (http://paup.csit.fsu.edu/index.html), and APE (http://cran.r-project.org). The tree for the maximum likelihood method was produced using the software APE with the HKY85 model, while others were produced using the software PHYLIP with default parameter settings. Figure 7.6 shows the trees constructed by them. All trees shown in Figures 7.5 and 7.6 have a fair degree of similarity. Each has clades A, B, C, D, and F in common. The greatest disparity between estimates involves the attachment of taxa from clade E. Interestingly, all methods occur in placing clade B closer to clade A.

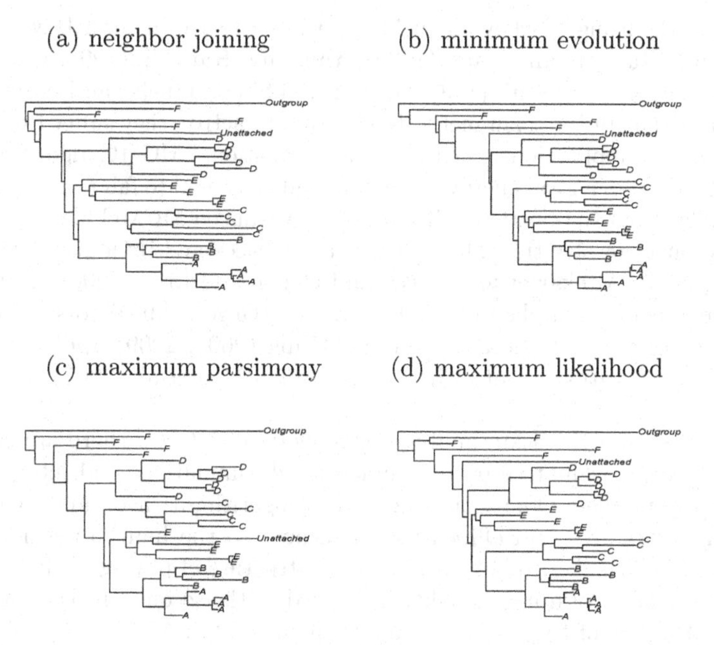

Figure 7.6: Best trees produced by (a) neighbor joining, (b) minimum evolution, (c) maximum parsimony, and (d) maximum likelihood (with the log-likelihood value −7881.18) for the cichlid fishes example.

7.4 Summary

Bayesian inference has been applied to a wide range of problems in evolutionary biology. In general, Bayesian approaches have been used to express the uncertainty in the phylogeny and parameters of the sequence mutation model with a posterior probability distribution. Bayesian analysis should prove useful in addressing some of the outstanding problems in phylogenetics, such as detecting a process that complicates phylogenetic analysis of evolutionary relationships, performing phylogenetic analyses and understanding evolutionary relationships to appropriate interpretation of bioinformatics results, and constructing large trees by combining the results of smaller and overlapping analyses.

Related works using Markov chain Monte Carlo algorithms in a Bayesian context include Rannala and Yang (1996), Yang and Rannala (1997a), Mau and Newton (1997), Mau et al. (1999), Larget and Simon (1999), and Newton et al. (1999). Although Bayesian methods are very attractive, they suffer from a severe difficulty in applications, such that the conventional MCMC algorithms tend to get trapped in a local energy minimum, and thus tend to fail to generate correct samples from the posterior distribution. Advanced MCMC algorithms developed so far to resolve this difficulty are Huelsenbeck and Ronquist (2001), Feng et al. (2003), Altekar et al. (2004), and Cheon and Liang (2008, 2009). The first three methods applied parallel tempering (Geyer, 1992a) to simulate from the posterior distribution. Cheon and Liang (2008, 2009) applied the stochastic approximation Monte Carlo algorithm (Liang et al., 2007) to the problem.

Furthermore, in a Bayesian analysis of phylogenetic trees, there are two difficulties with the evaluation of high-dimensional summations and integrals and a local trap problem. The sequential SAMC algorithm (Cheon and Liang, 2008) is very attractive and efficient with respect to the high-dimensional problems because it makes use of the sequential structure of phylogenetic trees to overcome the curse of dimensionality. In regard to the second problem, it is also efficient because of the self-adjusting mechanism of SAMC.

In future work, one may consider more complex phylogenetic models and investigate possible ways to improve the performance of MCMC and advanced algorithms, such as multiple-try Metropolis and population-based MCMC algorithms.

8

SMC (sequential Monte Carlo) for Bayesian phylogenetics

Alexandre Bouchard-Côté
Department of Statistics, University of British Columbia, Vancouver, British Columbia, Canada

CONTENTS

Until recently, Markov chain Monte Carlo (MCMC) has been the lone, faithful workhorse of Bayesian phylogenetics. MCMC methods have turned an abstract decision theoretic framework into a practical toolbox that has been applied to most phylogenetic inferential questions.

The theory behind MCMC is now a mature field offering a flexible suite of methods to approach approximate posterior computations. However, as we will discuss shortly, there are strong motivations for developing computational methods that can complement MCMC. The goal of this chapter is to give an accessible introduction to one promising complementary approach, *Sequential Monte Carlo* (SMC, also known as *particle filter methods*).

Before delving into the theory and practice of SMC methods, we start this chapter with an overview of two motivations for applying SMC to Bayesian phylogenetic inference.

The first motivation is the growing gap between the amount of phylogenetic data and the computational resources available to analyze these data. It is not atypical to find in the literature examples of a sampler running for weeks (Hackett et al., 2008), and these examples can be viewed as symptoms of a larger problem. Simply put, advances in processor speed no longer seem able to catch up with the advances in sequencing technologies (Mardis, 2011; Wetterstrand, 2012). The phylogenetic data storm is coming from many fronts: more sequenced species (Valentini et al., 2009), more data per species (Wapinski et al., 2007), and measurements of variations at the population level (Li et al., 2008; Shah et al., 2009). The main response to this deluge has been to exploit advances in parallelization, either at the level of processing units (Altekar et al., 2004; Suchard and Rambaut, 2009) or clusters (Feng et al., 2003). However, the intrinsically serial nature of MCMC makes it nontrivial to adapt it to parallel architectures. In contrast, we will argue that SMC is an architecture able to easily tap into these new computational resources.

The second motivation for developing MCMC complements is the growing gap between the resolution at which we understand molecular evolution, and the models we use on a day-to-day basis to perform phylogenetic inference. There are many pieces of information that are currently excluded from phylogenetic analyses, not from a lack of reasonable models, but from a lack of practical computational tools. Morrison (2009) discusses the example of *slipped strand mispairings* (SSMs). The repeat patterns left by SSMs form a cue commonly used when manually preprocessing alignments. However, this information is often discarded from subsequent phylogenetic analyses. Another example comes from RNA or protein structural constraints, which are known to affect evolutionary inference (Nasrallah et al., 2011). Not all of these cues will necessarily improve reconstruction accuracy (Pachter, 2007), but it is unlikely that the cues that are easy to accommodate within MCMC precisely coincide with the phylogenetically informative ones.

In this chapter, we will focus on the first motivation, the computational gains brought by SMC methods. By using familiar models in our examples, we can keep the exposition easier to follow. We will return to the second motivation in the last section of this chapter, where we discuss future directions.

There is a large and healthy literature on SMC methods, and excellent books, reviews, and tutorials have been written on the subject (Doucet et al., 2001; Kotecha et al., 2003; Doucet and Johansen, 2009). However most of this previous work has focused on either the special case of state space models (Friedland, 2005) (models with the structure of a hidden Markov model), or on general setups which are non-trivial to apply to phylogenetics (Moral et al., 2006). Recent work has started filling this gap, and the main goal of this chapter is to summarize these results (Teh et al., 2008; Görür and Teh, 2008; Bouchard-Côté et al., 2012; Wang, 2012).

This chapter is organized into three parts. In the first part, we describe SMC from the user's perspective, focusing on how to use the output of SMC for Bayesian phylogenetic analyses. In the second and third parts, we describe the core algorithm behind SMC and extensions of SMC, respectively.

8.1 Using phylogenetic SMC samplers

8.1.1 Overview

In its most general form, SMC methods cover a wide range of algorithms: at one end of the spectrum, we end up with a close cousin of MCMC, and at the other end, with an algorithm with very different computational properties. In most of this chapter, we will focus on the latter category to give a concrete example of a different approach to phylogenetic posterior inference. We will come back to less "radical" SMC flavors in Section 8.3.5.

The most striking difference between the SMC methods discussed here and MCMC lies in the way they represent hypotheses over phylogenies. At every intermediate iteration of an MCMC chain, a complete state (fully resolved tree over the set of taxa under study X) is kept in memory. By fully resolved, we mean that all the taxa under study are connected through an hypothesized phylogenetic tree. In contrast, the SMC methods discussed here will maintain a partially resolved tree in intermediate steps, or *partial state* for short.

What do we mean by partially resolved? There is again considerable flexibility here, but the most fundamental example of a partially resolved tree is a forest. Recall that a *forest* is an acyclic graph, but in contrast to a tree, it is not required to be connected. Equivalently, a forest can be described as a collection of disjoint trees. See Figure 8.1(a).

For concreteness, assume for now that inference is performed over clock trees (how to do SMC inference over non-clock trees is discussed in Section 8.2.5). Assume also that the evolutionary parameters are known—we will discuss how this can be relaxed in Section 8.3.5. In this setup, the partial states are sets of rooted trees such that the set of species $X_i \subset X$ at the leaves of each tree form a partition of X. By this, we mean that the X_i's satisfy the following two conditions: (1) the union of the X_i's is the set of species under study, X; (2) the X_i's are disjoint, i.e., $X_i \cap X_j = \emptyset$ for all $i \neq j$.

Two special cases are worth mentioning. First, the case of the completely disconnected forest, where each observed taxon forms a trivial tree with one node and zero edges. This extreme case, which we call a *trivial forest*, denoted $\perp$, will be used to initialize our SMC algorithms. The second special case consists of forests containing a single tree connecting all the observed taxa, i.e., a standard phylogeny. We call this state *complete*.

The partial state representation is close in spirit to neighbor joining, or more generally, to agglomerative clustering methods (Teh et al., 2008). As in neighbor

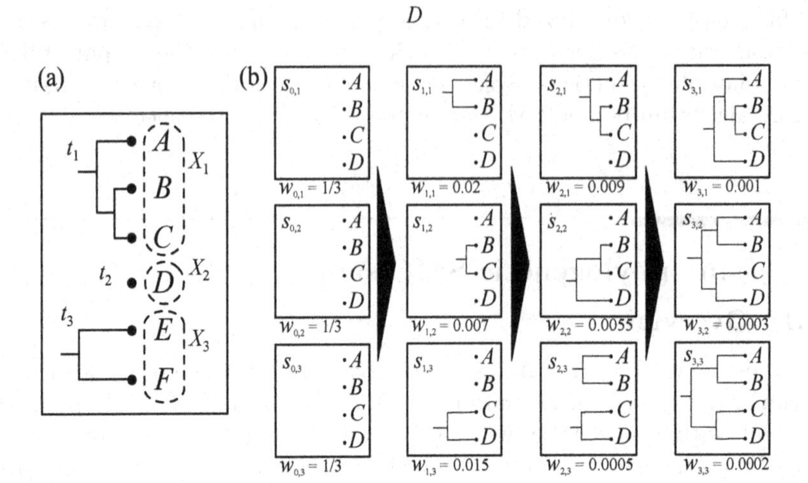

Figure 8.1: (a) An example of a partial state s, a clock forest in this case, i.e., a collection of clock trees. The set X_i is the set of leaves for tree $t_i \in s$. Note that the X_i's are disjoint, and their union is X, the set of all leaves. The tree t_2 is an example of a degenerate tree, with a topology having a single node and no edge. (b) Examples of particle populations at each generation of an SMC algorithm, with $K = 3$. Each particle is a forest $s_{n,k}$ associated with a positive weight $w_{n,k}$. Note that the weights do not sum to one within a population; they need to be normalized before creating the distribution $\hat{\pi}_{n,k}$ associated with each population. The details of how one population is produced from the previous one (denoted by black arrows in this figure), is explained in more detail in Section 8.2.

joining, SMC progresses by "merging" pairs of trees, i.e., by creating a new tree with one new root node connecting two trees from the previous iteration. However, in contrast to neighbor joining, SMC is based on a likelihood model rather than summary distance statistics. SMC is also less greedy than neighbor joining, entertaining several hypotheses simultaneously at every iteration. This reflects the fact that the goal of SMC is different than that of neighbor joining, aiming to sample from the posterior distribution rather than approximating the maximum of an objective function.[1]

Each of the parallel hypotheses is called a *particle* (see Figure 8.1(b)). A particle is composed of a forest as described above, along with a weight, which is simply a positive number reflecting the algorithm's current assessment of the quality of the corresponding hypothesis. This number does not directly reflect a probability, but we will see shortly how it can be transformed into

[1]But by being less greedy, SMC can also have an edge in maximization tasks where local optima are expected (Görür and Teh, 2008).

a posterior probability estimate. We use the term *particle population* for a collection of competing hypotheses represented at one SMC generation. Each round, consisting of selecting one merging operation for each particle, is called a *generation*. Note that since we merge exactly two trees at each SMC generation, all the forests found in a given particle population have the same number of trees.

The output of a phylogenetic SMC algorithm is a particle population containing complete states. We will describe shortly how this population is computed, but for now, let us look at how we can use this population to answer phylogenetic questions.

As a first example, consider the problem of assessing the support of a clade (the posterior probability that a subset of taxa $X' \subset X$ are the leaves of a subtree of the phylogeny). To approximate this quantity from the population of complete states output by SMC, proceed as follows. First, compute the sum of the weights of the particles where the clade of interest is present; second, divide this number by the sum of all the weights. This gives a number between zero and one that approximates the posterior clade support. An important asymptotic result from SMC theory (Crisan and Doucet, 2002) is that as the number of particles K goes to infinity, this estimate converges to the posterior clade support. Since asymptotic theory does not guarantee the quality of approximations based on finite K, Bouchard-Côté et al. (2012) used simulations to assess the performance of SMC. Instead of evaluating a single clade posterior at a time, this previous work summarized all of the clades posteriors by constructing a consensus tree (Felsenstein, 2004). Refer to Figure 8.2, where two of their simulation results are shown. These results demonstrate that as the number of particles increases, the consensus tree constructed from the approximate clade posteriors becomes closer in average to the "true tree," i.e., the tree used to generate the data. The results also show that for a range of computational budgets, SMC achieves a tree reconstruction error lower than MCMC.

As a second example, let us suppose that we wish to estimate a divergence time between a pair of species $x, x' \in X$ using a simple clock model. Given a phylogenetic tree t, let us define the divergence time $d_{x,x'}(t)$ as half of the sum of the branch length separating x to x' in t. To estimate this quantity from a weighted population t_k, w_k, we start by looking at the divergence time $d_{x,x'}(t_k)$ for each particle t_k in turn, and we multiply each of these times by the corresponding weight w_k. We then divide the sum of the quantities computed in the first step by the sum of the weights, obtaining:

$$\frac{\sum_{k=1}^{K} d_{x,x'}(t_k) w_k}{\sum_{k'=1}^{K} w_{k'}}.$$

Again, this estimate will converge to the correct posterior mean as the number of particles K goes to infinity.

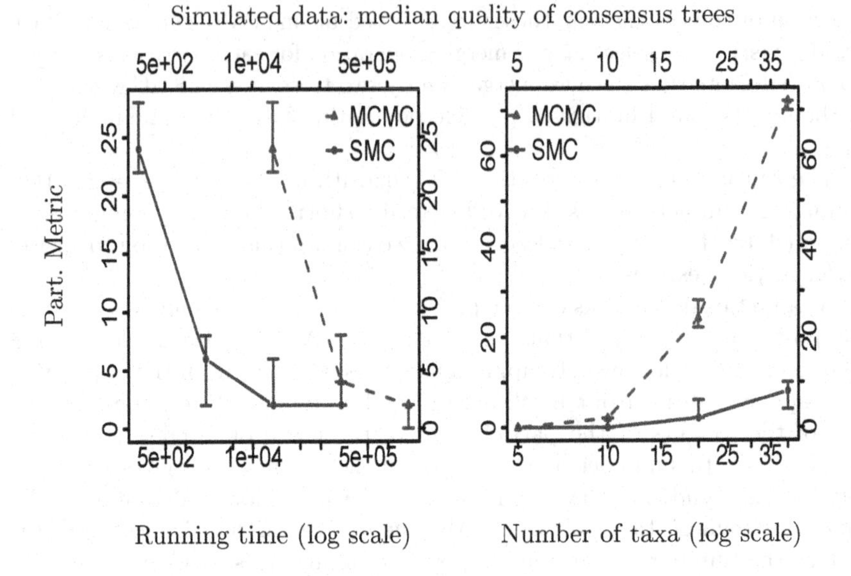

Figure 8.2: Assessment of the quality of the posterior clade estimates from Bouchard-Côté et al. (2012). Left: the rate at which the partition metric between the held-out tree used to simulate the data and the consensus tree decreases as a function of the running time. Running time is measured by the number of times the Felsenstein recursion is computed (see Section 8.3.1). For MCMC, the running time is controlled by the number of MCMC iterations, for SMC, by the number of particles K. Right: a computational budget is fixed, and the number of taxa is varied, creating increasingly difficult posterior inference problems. See Bouchard-Côté et al. (2012) for detailed experimental conditions.

8.1.2 A general framework for understanding the output of Monte Carlo algorithms

Before going over the details of how the particle populations are computed, it is worth discussing in more generality how the output of SMC algorithms can be used to compute almost any expectation of interest. In particular, we will be able to put the two examples given at the end of the last section under the same umbrella.

To achieve this, we start by making the observation that the normalized weights, denoted by

$$\bar{w}_k = \frac{w_k}{\sum_{k'=1}^{K} w_{k'}}, \tag{8.1}$$

can be viewed as discrete probabilities (as they form a list of positive numbers summing to one). Each of these probabilities is assigned to a location, namely the sampled tree associated with it. This view gives us a discrete probability distribution over the space of trees, called the distribution induced by a particle population, $\hat{\pi}$. As the number of particles increases, this discrete distribution can get arbitrarily close to the posterior distribution over trees.

This suggests the following procedure for approaching generic inference problems. First, identify a function f such that the quantity of interest is expressed as an expectation under the posterior distribution, $\mathbb{E}_\pi f(t) = \mathbb{E}[f(t)|\mathcal{Y}]$, where π denotes the posterior distribution of the phylogenetic tree random variable t given the data $\mathcal{Y}$. Second, we simply replace the posterior distribution π in the above expectation by the discrete approximation $\hat{\pi}$ induced from the particle population output by SMC. This yields the estimator:

$$\mathbb{E}_\pi f(t) \approx \mathbb{E}_{\hat{\pi}} f(t) \tag{8.2}$$

$$= \sum_{k=1}^{K} \bar{w}_k f(t_k). \tag{8.3}$$

For instance, in the first example, f can be taken as the function that is equal to one if the input tree has the clade of interest X', and zero otherwise. With such f, (8.3) reduces to the sum of the normalized weights corresponding to the trees consistent with the clade of interest X'.

In the second example, f can be taken as the function returning the divergence time of two species in an input phylogeny, i.e., $f = d_{x,x'}$. This also yields the weighted average described before.

Note that if one is interested in $d_{x,x'}$ for more than one pair of taxa x, x', only one posterior approximation $\hat{\pi}$ needs to be computed, and can be reused to compute all the required posterior expectations.

The technique described in this section is closely related to the way MCMC approximations are computed. With MCMC, the same basic equation (8.2) is used, but the discrete approximation $\hat{\pi}$ is formed differently. The support of $\hat{\pi}$ is taken to be the set of distinct phylogenies $\{t_k\}$ visited by MCMC, and the weight w_k, to be the number of times t_k was visited by the sampler (because of rejections, it is possible to find identical phylogenies in the MCMC output).

8.2 How phylogenetic SMC works

In this section, we elaborate on how the particles are computed. We start with some background on two related, simpler algorithms: *importance sampling* (IS), and *sequential importance sampling* (SIS). We then present SMC itself, which adds resampling on top of SIS. These three ideas, importance sampling, sequential proposals, and resampling, form the core of the SMC methodology.

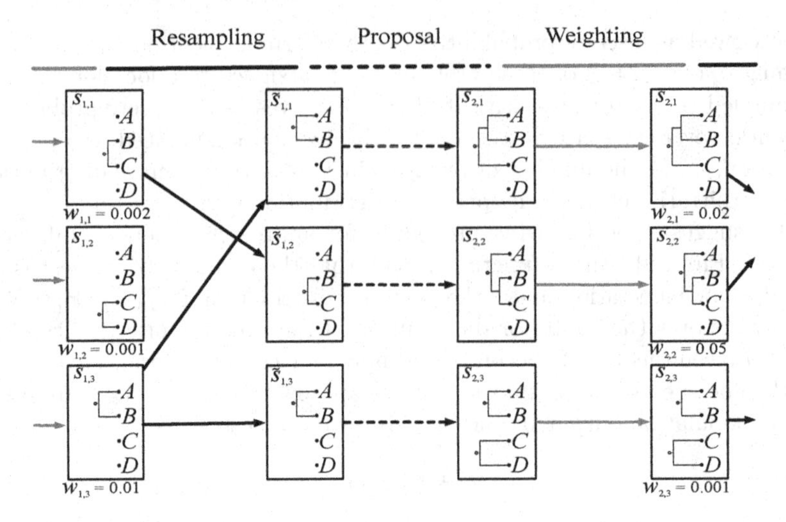

Figure 8.3: A schematic example from Bouchard-Côté et al. (2012) of the tree steps used in SMC to produce one particle population to the next population.

We then give a description of the techniques involved in implementing scalable samplers.

As in the previous section, π is the posterior distribution over trees given the data. From Bayes' theorem, this posterior can be written as a density of the form $\gamma(t)/Z$, where γ is the product of the prior density times the likelihood, and Z is the marginal probability of the observed data, $Z = \mathbb{P}(\mathcal{Y})$. Note that the function γ can be efficiently evaluated pointwise. This involves summing over the internal nucleotides while fixing the tree, an operation that can be carried efficiently using Felsenstein's peeling recursion (Felsenstein, 1981), a special case of the sum-product or junction tree algorithm (Bishop, 2006).

Computing Z, on the other hand, is intractable, as it would involve a sum over all possible tree topologies, as well as an iterated integral over all possible branch lengths.

8.2.1 The foundations: importance sampling

We start this section by describing importance sampling, an algorithm that is simpler than SMC but that also shares important similarities with its more advanced cousin. In particular, the output of an importance sampling has the same form as the output of an SMC algorithm, i.e., it is a particle population as described earlier. The main difference with SMC (and SIS) is that this population is computed in a single step. This generally creates a less reliable approximation than the one produced by SMC, which, as we will see, proceeds in many incremental steps interspersed by resampling steps.

Before using importance sampling, the user is required to design a proposal distribution q_{imp}. This proposal distribution, defined over the space of phylogenies, should be chosen according to two criteria. The first criterion concerns the support of q: if a tree t has positive posterior density, then q_{imp} should assign positive density to t as well. Note that this condition says nothing about the relative magnitudes of the posterior and proposal. Having the regions of high probability in the posterior and proposal roughly align will decrease the number of particles needed to get to a certain level of accuracy, but is not required to prove asymptotic results. The second criterion is that it should be computationally easy to sample from q_{imp}, and to evaluate the density $q_{\text{imp}}(t)$ of this sample.

In a Bayesian context, a simple (but often inefficient) way to construct a proposal for an importance sampling algorithm is to use the prior as the proposal. This automatically satisfies the first criterion as the support of the posterior is always a subset of the support of the prior. The second criterion will also be satisfied for most prior distributions—how this is done is explained in more detail in the next section. Computing the density under the prior is also possible using the known closed-form formula for the number of topologies on $|X|$ leaves.[2]

Importance sampling creates each particle in the population independently and in two steps. In the first step, we sample from the proposal distribution with density q_{imp} to obtain a tree t_k. In the second step, we correct the discrepancy between the proposal and the posterior distribution by assigning a weight w_k to the proposed tree t_k. This is done using the following ratio:

$$w_k = \frac{\gamma(t_k)}{q_{\text{imp}}(t_k)}. \tag{8.4}$$

We now have the ingredients needed to form a particle population. As described in the first part of the chapter, this population can be used to answer a wide range of phylogenetic questions.

Note that the form of (8.4) is motivated by the law of large numbers. It ensures that the approximation $\hat{\pi}$, defined in Section 8.1.2, converges to the target posterior distribution π as the number of particles K goes to infinity (see Bishop (2006)).

8.2.2 Toward SMC: sequential importance sampling (SIS)

Sequential importance sampling (SIS), is an equivalent way of viewing importance sampling, but reparameterized in a way that naturally leads to SMC. Instead of computing each particle in one step, as in IS, SIS builds them in stages or generations. In our phylogenetic setup, the number of stages is equal

[2]Technically, for importance sampling the normalization of the proposal is not required. This contrasts with SMC, where being able to compute the normalization of the proposal is important and often overlooked.

to the number of observed taxa minus one. To avoid confusion with MCMC iterations, we use the terminology of generation for these stages.

It is often natural to break the proposal into smaller steps. For example, a uniform topology is usually sampled by merging rooted trees in stages. Formally, assume that at generation $n \in 1, 2, \ldots, N$, $N = |X| - 1$, each particle is a forest of $|X| - n$ rooted trees. We denote the forest or partial state at generation n and particle index k by $s_{n,k}$. Proposing a successor of this particle $s_{n,k}$ is done by sampling one of the $\binom{n}{2}$ pairs uniformly at random, that is, by sampling a subset of size two without replacement from the set of trees in the forest.

Proposing branch lengths incrementally is also generally straightforward. Consider, for example, the task of sampling from phylogenies equipped with a Yule prior, and let us define the height of a forest as the height of the tallest tree in the forest. Sampling branch lengths in this case is equivalent to sampling increments δ between two forests. To mirror the definition of the Yule prior, we let this increment be distributed according to an exponential distribution with rate $|X| - n + 1$.

In order to formalize the proposal used in SIS and SMC, we first need to introduce some notation. We denote the set of partial states (introduced in Section 8.1.1) by $\mathcal{S}$, a superset of the set of trees of interest $\mathcal{T}$. For example, if $\mathcal{T}$ is the set of clock trees, one natural choice for $\mathcal{S}$ is the set of forest containing clock subtrees (we call these clock forests).

The incremental proposals q used for SIS (and SMC) are specified by a transition probability or density on $\mathcal{S}$. We denote the conditional density of proposing $s_{n,k}$ from $s_{n-1,k}$ by $q(s_{n,k}|s_{n-1,k})$. Note that proposals for SIS and SMC are different than those used in IS—the former being conditional densities, while the latter was just a density over trees.

Since we now propose states incrementally, we also need to discuss how to weight the intermediate states. In order to do this, we start by rewriting the monolithic proposal q_{imp} of importance sampling into a sequence of small conditional probabilities. This is done using the chain rule of probability:

$$\begin{aligned} q(t_k) &= q(s_{1,k}, s_{2,k}, \ldots, s_{N,k}) \\ &= q(s_{1,k}|\perp)\, q(s_{2,k}|s_{1,k}) \ldots q(s_{N,k}|s_{N-1,k}), \end{aligned} \tag{8.5}$$

where $\perp$ denotes a completely disconnected forest, i.e., a forest where each tree is degenerate and consists of a single observed taxon and no edge. Note that (8.5) holds because in the ultrametric case, a tree t_k uniquely corresponds to a sequence of partial states $s_{1,k}, s_{2,k}, \ldots, s_{N,k}$—the correspondence is given by the sequence of speciation events in their chronological order. In the case of non-clock tree, the discussion is slightly more involved (see Section 8.2.5).

To obtain intermediate weights $w_{n,k}$, we rewrite the importance sampling weights, (8.4), into a product of N factors. We combine (8.5) with a telescoping product to obtain:

$$
\begin{aligned}
&\frac{\gamma(t_k)}{q_{\text{imp}}(t_k)} \\
&= \frac{\gamma(t_k)}{q(s_{1,k}|\perp)q(s_{2,k}|s_{1,k})\dots q(s_{N,k}|s_{N-1,k})} \\
&= \underbrace{\underbrace{\underbrace{\left(\frac{\gamma(s_{1,k})}{\gamma(\perp)q(s_{1,k}|\perp)}\right)}_{w_{2,k}}\left(\frac{\gamma(s_{2,k})}{\gamma(s_{1,k})q(s_{2,k}|s_{1,k})}\right)}_{w_{2,k}}\dots\left(\frac{\gamma(s_{N,k})}{\gamma(s_{N-1,k})q(s_{N,k}|s_{N-1,k})}\right)}_{\substack{\vdots \\ w_{N,k}}}.
\end{aligned}
\tag{8.6}
$$

Note that all factors of the form $\gamma(\cdot)$ cancel out, except for the last one, which is the target $\gamma(s_{N,k}) = \gamma(t_k)$, and the first one, $\gamma(\perp)$, which is constant for all particles, and therefore disappears after normalizing the weights in (8.3). Note also that these weights can be updated incrementally as follows:

$$
w_{n+1,k} = w_{n,k}\frac{\gamma(s_{n+1,k})}{\gamma(s_{n,k})\ q(s_{n+1,k}|s_{n,k})}. \tag{8.7}
$$

Before going further, let us pause and look at (8.6) more closely. The careful reader may have noticed that we have not yet formally defined some of the symbols in this equation: recall that γ is a density over phylogenetic trees, while in several of the factors in (8.6), we feed γ with forests. We therefore need to generalize our definition of γ, to allow not only trees but forests as arguments.

How should this be done? We show a concrete example in Section 8.2.4, but surprisingly, the only restriction for the asymptotic correctness of the algorithm is that γ should assign a positive density to all of the proposed forests. The intuition behind why the numerical value of γ for intermediate forests does not affect asymptotic correctness is that these values all cancel each other in the telescoping product of (8.6). However, when we will add the resampling step in the next section, the choice of values γ assigned to intermediate forests will have an effect on the quality of the approximation since a finite number of particles is used in practice. But fortunately, even with resampling, the asymptotic results still hold under weak conditions.

To summarize, SIS proceeds in N generations, maintaining a population of particles at each generation. To obtain a new population from the previous one, the incremental proposal is used to propose a new particle $s_{n+1,k}$ from each previous particle $s_{n,k}$. Equation (8.7) is then used to compute a new weight for each particle.

8.2.3 Resampling

It is apparent from the derivation in (8.6) that the distribution of the final particle population in SIS is the same as in IS. In particular, both SIS and IS have serious limitations when the target distribution π is defined over a large space, for example, over the space of phylogenies. The main symptom of this problem is a large weight imbalance: one gets a population where most particles in the population have a negligible contribution to expectation estimates in (8.3). This problem is called *particle degeneracy*.

In this section, we show how this problem can be alleviated by pruning unpromising particles. Remarkably, this can be done while preserving the asymptotic properties of the approximation, meaning that as the number of particles increases, the approximation can still become arbitrarily close to the true posterior.

It is important to realize that not all pruning strategies would preserve these asymptotic properties. For example, a deterministic scheme that would naively pick the k particles of highest weights and create K/k duplicates of each would not preserve asymptotic correctness.

The solution is to use a randomized scheme instead. In this section, we describe a simple-to-implement randomized scheme called multinomial resampling (alternatives do exist however, see Douc et al. (2005)). This scheme is an asymptotically correct method for addressing degeneracy.

Multinomial resampling consists of sampling K times independently from the probability distribution induced by the particle population from the previous generation, namely the discrete distribution giving probability $\bar{w}_{n,k}$ to the particle $s_{n,k}$. To get more intuition on this procedure, assume that the weights have been normalized already, and that we place the particles in an arbitrary but fixed order. Having done that, we can think of the weights as line segments covering the unit interval (see Figure 8.4 (left)). Multinomial resampling consists of throwing K darts on this one-dimensional target, and looking at how many darts fall on each segment (particle).

From this picture, we can see that particles with a high weight will generally get resampled several times, while particles with low weights will generally get pruned (i.e., get a weight of zero). By construction, this scheme also has the property that the expected number of times each particle will be resampled is proportional to the particle weight. This property is key when establishing that the resampling step preserves the asymptotic results.

After throwing the darts, the weight of each particle is modified. It is set to the number of darts falling on the corresponding segments, divided by the total number of darts K.

Note that multinomial resampling can be viewed as a generalization of the classical bootstrap procedure: if all the weights before resampling were equal to $1/K$, multinomial resampling is equivalent to resampling uniformly with replacement.

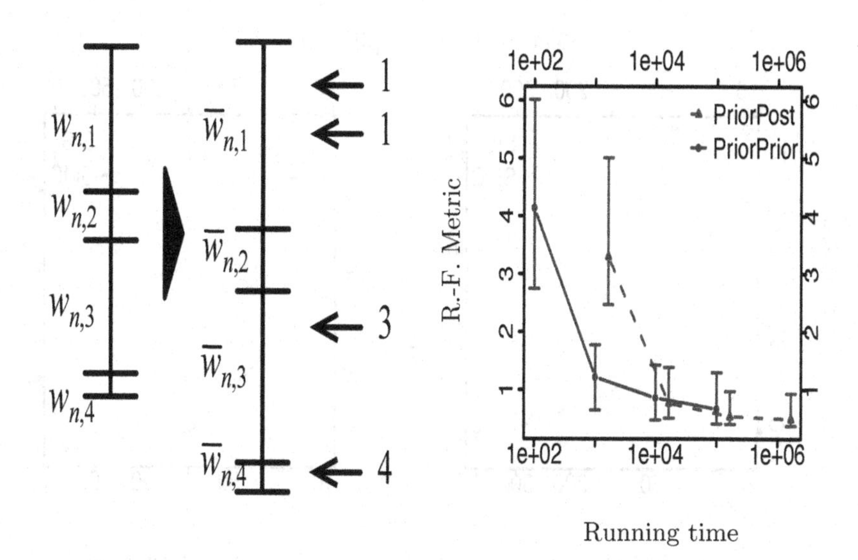

Figure 8.4: Left: schematic representation of a multinomial resampling step. After arbitrarily ordering the weights $w_{n,k}$, we can view them as segments on a stick. We then normalize these weights using (8.1) to obtain $\bar{w}_{n,k}$. Finally, we simulate K iid uniform random variables ($K = 4$ in this example) to get a new list of particles (1,1,3,4 here). Right: comparison of two SMC proposals, Prior-Prior and Prior-Post. Experimental conditions are as in Figure 8.2. Adapted from Figure 6 in Bouchard-Côté et al. (2012).

After adding a resampling step between the proposal steps of SIS, we get a first example of an SMC algorithm. We show in Figure 8.5 that this addition has a big impact in practice. For the SIS algorithm, increasing the number of particles only improves the performance of the reconstructed tree at a very slow rate. In contrast, the improvement is substantial with SMC, thanks to the resampling step.

We summarize in Algorithm 1 all the steps involved in a simple phylogenetic SMC algorithm for ultrametric trees. In this pseudo-code, we denote the distribution assigning probability $\bar{w}_{n-1,k}$ to $s_{n-1,k}$ by $\hat{\pi}_n$. As a special case, we get that $\hat{\pi}$ as defined in Section 8.1.2 coincides with $\hat{\pi}_N$.

Note that there is an important difference between the weight equation for SIS, (8.7), and the counterpart for SMC, (8.8), in the pseudocode of Algorithm 1. In the latter equation, the weight from the previous generation is not multiplied. This is occasioned by the resampling step: after this step, each state in $\tilde{s}_{n-1,1}, \tilde{s}_{n-1,2}, \ldots, \tilde{s}_{n-1,K}$ is equally weighted.

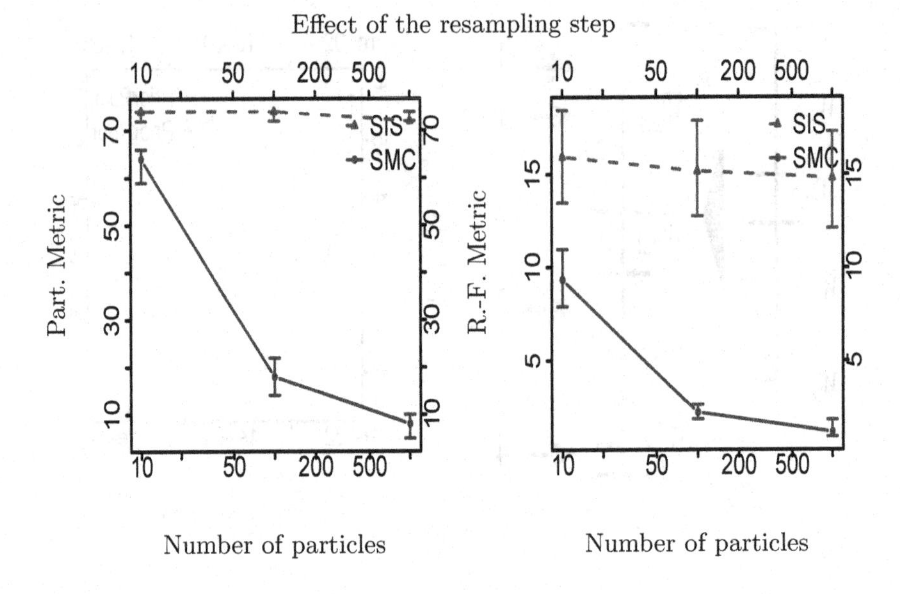

Figure 8.5: Error rates of SMC and SIS as a function of the number of particles, averaged over ten random forty-taxon trees and ten executions per tree. Left: error is measured using the partition metric. Right: error is measured using the Robinson-Foulds metric. It is clear from these results that the resampling step plays a crucial role in the performance of phylogenetic SMC samplers. The experimental conditions are otherwise identical to those of Figure 8.2.

Algorithm 1 SMC for ultrametric trees.

for all $k \in \{1, \ldots, K\}$ **do**
  initialize $s_{0,k}$ to $\perp$
  initialize $w_{0,k}$ to $1/K$
end for
for generation $n = 1, 2, \ldots, N$ **do**
  for all $k \in \{1, \ldots, K\}$ **do**
    resample $\tilde{s}_{n-1,k}$ from $\hat{\pi}_{n-1}$
    propose $s_{n,k}$ from the proposal with density $q(\cdot|\tilde{s}_{n-1,k})$
    weigh:

$$w_{n,k} = \frac{\gamma(s_{n,k})}{\gamma(\tilde{s}_{n-1,k})\, q(s_{n,k}|\tilde{s}_{n-1,k})} \tag{8.8}$$

  end for
end for
return $\hat{\pi} = \hat{\pi}_N$

8.2.4 Example: Yule trees

As a concrete example, we show in this section the detailed form of the proposal, extension, and weight updates in the case of inference over Yule trees. To do this, we use a notation similar to Bouchard-Côté et al. (2012), viewing each particle as a set of trees $s = \{t_j\}$. Each tree has a set of leaves corresponding to a subset of the observations, denoted by $X(t_j) \subset X$. If s' was proposed from s, then s' is obtained from s by removing two trees, $t_l(s')$ and $t_r(s')$ and adding a new tree $t_m(s')$ formed by merging t_l and t_r. In particular, $X(t_m) = X(t_l) \cup X(t_r)$. We also use $L(t)$ to denote the likelihood of the observations corresponding to $X(t)$ given t.

As mentioned in Section 8.2.2, the form of the weight update used in Algorithm 1 assumes that each tree t over X corresponds to a unique sequence of particles $s_0, s_1, \ldots, s_N$ starting from the degenerate forest $s_0 = \perp$ and ending in tree over all the leaves $s_N = t$. The method we have discussed so far to achieve this is to ensure that the height of the forest increases at each generation. Formally, we denote by $h(s) = (h_0, h_1, \ldots, h_{n(s)})$ the list of the heights of the forests leading to s. From this, we can derive a list of increments, $\delta_i(s) = h_i - h_{i-1}$. With this notation, the height increase condition is simply that $\delta_i(s) > 0$. The set of successor forests that can be reached from s by merging a pair and increasing the height will be denoted by $S(s)$.

We can now express the extended density over forests as follows:

$$\gamma(s) \propto \left(\prod_{i=1}^{n(s)} f(\delta_i(s); |X| - i + 1) \right) \left(\prod_{t \in s} L(t) \right),$$

where $\propto$ denotes proportionality up to a constant that can depend on s only through $|s|$, and $f(x; \lambda)$ denotes the exponential density with rate λ.

As for the proposal, it can be written as

$$q(s'|s) \propto \mathbf{1}[s' \in S(s)] \; f(\delta_{n(s')}(s'); |X| - i + 1).$$

The weight is therefore equal to

$$\begin{aligned} w_{n,k} &= \frac{\gamma(s')}{\gamma(s) q(s'|s)} = \frac{f(\delta_i(s); |X| - i + 1) \; L(t_m(s'))}{L(t_l(s')) \; L(t_r(s')) \; q(s'|s)} \\ &\propto \frac{L(t_m(s'))}{L(t_l(s')) \; L(t_r(s'))}. \end{aligned}$$

8.2.5 Inferring non-clock trees

We have assumed so far that the trees being sampled are ultrametric. We review in this section results from Wang (2012) showing how this assumption can be relaxed. For concreteness, we assume in this section a simple non-clock model where the prior distribution over topologies is uniform, and the prior distribution over branch lengths is exponential with rate λ.

To infer non-clock trees, there are two changes to make to Algorithm 1: a different proposal, and a different weighting equation.

For non-clock proposals, the simplest construction, denoted q_{NC}, is obtained by adding one degree of freedom to a clock proposal. This extra degree of freedom allows the incremental construction of rooted non-clock trees[3] More precisely, q_{NC} first picks a pair of rooted non-clock trees to merge, and then samples from two independent exponential distributions to determine the lengths of the two edges joining the pair of trees to a new internal node.[4]

As for the non-clock weight update, it consists of a generalization of the ultrametric weight formula. The new equation is given by:

$$w_{n,k} = \frac{\gamma(s_{n,k})}{\gamma(\tilde{s}_{n-1,k})} \frac{c(\tilde{s}_{n-1,k}|s_{n,k})}{q_{\text{NC}}(s_{n,k}|\tilde{s}_{n-1,k})}, \tag{8.9}$$

where c is an *over-counting function*. The theoretical justifications and constraints on c are described in more detail in Wang (2012). A simple choice that works well in practice is to set $c(s|s')$ to the number of non-degenerate trees in the forest s', i.e., the number of trees with at least two leaves.

Note that (8.9) superficially looks like a Metropolis-Hastings ratio, but has the important difference that in general $c \neq q_{\text{NC}}$.

8.3 Extensions and implementation issues

After having described the core ideas behind phylogenetic SMC, we now look at various ways in which the basic algorithm can be made more efficient, and can be applied to more general setups.

8.3.1 Efficiently computing and storing the particles

The most expensive operation in phylogenetic SMC algorithms is generally the weight update, shown in (8.8) in Algorithm 1. As shown in Section 8.2.4, this amounts to evaluating the likelihood $L(t)$, for different subtrees t in a forest, which in turn involves summing over the internal nucleotides. This is done using Felsenstein's tree-peeling algorithm. Note that this computation is also the bottleneck of MCMC tree sampling algorithms, where it has to be computed for each proposed tree in order to determine the Metropolis-Hastings acceptance ratio.

Although tree-peeling is needed in both SMC and MCMC algorithms, an advantage of SMC is that we can reuse subtree recursion computations across particle generations. In contrast, even the simplest MCMC proposal, perturbing

[3]Even in models where the root is not identifiable, it is advantageous to construct rooted trees for computational reasons described in Section 8.3.1. The artificial sampled rootings can always be discarded.

[4]Except for the merging operation at the last generation N, where only one edge is added, and consequently only one branch length is sampled.

a single random branch length, requires recomputing the tree-peeling recursion for a large portion of the tree (more precisely, recomputation will be needed for all the edges along the path from the first perturbed edge to the next).

In SMC, the order in which the trees are incrementally constructed mirrors the order in which the tree-peeling recursion is computed. This means that by storing the recursion in each tree in the forest, we can avoid redundant tree-peeling computations.

The storage cost for each recursion is in the order of the number of sites times the number of characters in the alphabet. When the number of particles is large, the storage requirements can become problematic. In fact, while MCMC is time-bound, the simple phylogenetic SMC algorithm described so far is generally memory-bound. Several implementation strategies can be used to alleviate this memory limitation.[5]

Additionally, it is possible to exploit the resampling step to trade memory for space. The key insight is that because of resampling, a large fraction of the particles will be pruned. It is therefore useless to keep the peeling recursions in memory for these particles.

In order to do this, we can use a two-pass method: in a first pass, we compute the weight of each particle in the population, but only store the random seed used by the proposal. In other words, the forest (and peeling recursion arrays) are stored implicitly in the first pass. This is very memory efficient since only two variables are needed per particle at this point. After this step, since we have the weights for all particles, the resampling step can then be performed. Finally, the actual particles are reconstructed from the random seeds, but only for the particles that survived the resampling step, and once per group of identical particles (since resampling is done with replacement, there is a positive probability of having duplicate resampled particles).

This scheme takes at most twice the time needed for the standard SMC implementation, but can significantly reduce the memory requirements. Moreover, since memory writes are slower than floating point operations in modern architectures, we have observed that in practice the two-pass scheme can actually be both faster and more memory efficient.

If further memory reductions are needed, one can compose two stages of multinomial sampling to control the number of distinct particles present in the final population. This is done by doing a first round of sampling, sampling K' particles with replacement with $K' < K$, and then doing a second round on the output of the first round, this time, sampling K times (also with replacement) from these K' particles. Given a memory limit of L distinct particles, the number of particles K' in the first generation can be chosen using the formula for the expected number of distinct particles sampled from a population with

[5]First, identical subtrees will be shared by several particles (because of resampling), so particles should have pointers to recursion values rather than the values themselves. Second, when a node is not a root in any particles in the current generation, it can be safely discarded.

weights w_k, i.e., by finding via line search the largest K' such that

$$K' - \sum_{k=1}^{K} (1 - w_{n,k})^{K'} \leq L.$$

Another option available to reduce the memory footprint is to use Particle MCMC (PMCMC) methods (Andrieu et al., 2010).

8.3.2 More SMC proposals

So far we have used the simplest possible proposal: picking a pair of trees in the forest uniformly at random, and a height increment of the forest from an exponential distribution. In this section, we review some alternative proposals that have been described in the literature.

The seminal paper by Teh et al. (2008) introduces three proposals, named "Prior-Prior," "Prior-Post," and "Post-Post." The proposal discussed so far is the equivalent, adapted to Yule trees, of the Prior-Prior proposal used in this previous work on coalescent trees. The two other proposals, Prior-Post and Post-Post, are computationally more expensive, but the hope is that these particles will be of better quality. In other words, the goal is to construct a proposal closer to the target distribution by using the information provided in the observations.

With Prior-Post, $q_{\text{pr-po}}(s'|s)$, the topology proposal is improved, keeping the branch length proposal simple and inexpensive. This is done as follows. First, an increment on the height of the forest, δ, is sampled from the same distribution as in the Prior-Prior proposal. Second, for all unordered pairs of trees in the forest, $\{t, t'\} \subset s$, the likelihood of merging t and t' to form a new tree $t_{t,t',\delta}$ is computed. Third, a multinomial distribution over the possible pairs is sampled, with each outcome $\{t, t'\}$ having probability proportional to

$$p_{t,t'} = \frac{L(t_{t,t',\delta})}{L(t)\ L(t')}.$$

The weight update (assuming resampling at every step) is equal to

$$w_{n,k} = \sum_{\{t,t'\} \in s} p_{t,t'}. \tag{8.10}$$

See Bouchard-Côté et al. (2012) for a derivation of (8.10), as well as an empirical evaluation of this proposal, reproduced in Figure 8.4 (right) for convenience. As in previous similar plots, the running time is measured by the number of times the peeling recursion was computed, with the four points in each series obtained using different numbers of particles in $\{10, 100, 1000, 10000\}$. It can be seen that in this dataset, for a given number of particles, Prior-Post slightly outperforms Prior-Prior, but since the particles of Prior-Post are more expensive to compute, Prior-Prior achieves better performances on a fixed computational budget. One possible explanation is that Prior-Post spends too much computational resources for the shallow merging operations relative to the deeper ones. It might be possible to alleviate this issue by allocating different numbers of particles at different generations.

The third proposal introduced in Teh et al. (2008), Post-Post, uses the information in the observation to inform the sampling of both the height of the new forest, and the pair to merge. At a high level, Post-Post works as follows. First, a pair to merge is sampled using a process similar to the one described above for Prior-Post, except that δ is marginalized in the weight calculation. Once a pair is picked, the height of the forest is then sampled proportionally to π. The difficulty in this scheme is the first step. The integral involved is intractable when the likelihood is a general continuous time Markov chain. However, if the likelihood model is a Brownian motion (Felsenstein, 1973), the integral can be identified as the tail of a generalized inverse Gaussian integral Jorgensen (1982). Even in this case, the calculation is non-trivial to implement in practice.[6]

One benefit of Post-Post is that its proposal over height increments can have fatter tails than the exponential proposal of Prior-Prior and Prior-Post. This can be important when the branch lengths of a dataset significantly deviate from the expected branch lengths of the prior. Note that there might be simpler ways to achieve these fatter tails, as the height increment proposal in Prior-Prior is not required to mirror the prior over branch lengths, as the weights will correct the discrepancy.

In general, there is more flexibility in designing SMC proposals than in designing MCMC proposals, in part because the reversed move, $q(s|s')$, does not need to be considered. The literature has only scratched the surface of the benefits of this flexibility. One exception is the work of Görür et al. (2012), which uses an efficient heuristic to propose pairs to merge in a more informed fashion.

8.3.3 More on resampling

In this section, we discuss the resampling step in more detail. We start by describing how to efficiently implement multinomial resampling.

When the number of particles is small to moderate, the naive scheme described in Section 8.2.3 (throwing a dart on the unit length), works reasonably well. In this regime, most of the computational budget is spent in the proposal step. However, for a large number of particles, the naive resampling algorithm can become a bottleneck.

This is because in the naive implementation, each of the K darts requires as much as K operations to look up into which particle the dart falls. This means that this implementation can take time on the order of K^2, while the cost of proposals grows on the order of K (but with a much larger constant).

[6]In the first generations of the algorithm, this integral can be shown to be equivalent to a modified Bessel function of the second kind, but in general, numerical integration computed in log-space to avoid underflows is necessary.

Fortunately, multinomial resampling can also be implemented in time linear in K, by using classical methods for simulating order statistics (Ripley, 1987; Doucet, 1997; Pitt and Shephard, 1999; Carpenter et al., 1999). The basic idea is to sample $K+1$ independent exponential random variables with an equal but arbitrary rate. Since these points can be viewed as a realization of a Poisson process, their locations after normalization to $[0,1]$ are uniformly distributed. We can then traverse the sorted list once to determine the number of particles of each type to keep in the next generation.

Beyond multinomial resampling, other schemes that preserve the asymptotic correctness of SMC exist. Examples include residual resampling, stratified resampling, systematic resampling, and dynamic resampling (see Douc et al. (2005) for a review). Some of these alternatives have theoretically been shown to be an improvement over multinomial sampling (Douc et al., 2005).

It may also be advantageous in certain contexts to only do resampling in a subset of the generations. For example, one may choose to do no resampling between generation n and $n+1$, but to resample between generation $n+1$ and $n+2$. From an algorithmic point of view, this is easy to implement: skipping resampling can be reduced to the case where resampling is done at every step but with a transformed proposal. Suppose, for example, that we only want to skip resampling between generations n and $n+1$. We can construct a transformed proposal q'_n equal to the composition of the proposals q_n and q_{n+1} at generations n and $n+1$ in the original list of proposals. By using this transformed proposal q'_n, we can construct a new SMC algorithm that requires one fewer generation but that is equivalent to the original SMC algorithm. The advantage is that this transformed SMC scheme can be handled with Algorithm 1.

The most widely used method for determining when to resample is based on a quantity called the *effective sample size* (ESS). ESS is typically approximated using the formula

$$\frac{1}{\sum_k (\bar{w}_{n,k})^2},$$

where $\bar{w}_{n,k}$ denotes the normalized weights. ESS is maximized when all the weights have the same value, in which case the effective sample size is equal to the number of particles K, and minimized when one particle has all the weight, an extreme case of particle degeneracy. A popular heuristic selecting the subset of the generations on which to do resampling is to compute the ESS at each particle generation, and to resample when it falls under a threshold.

8.3.4 Estimating the marginal likelihood

In addition to computing expectation, another quantity of interest in Bayesian phylogenetic inference is the marginal likelihood, $\mathbb{P}(\mathcal{Y})$, also known as the evidence. For example, a popular way to compare two probability models $\mathbb{P}$ and $\mathbb{P}'$ in the Bayesian framework is to look at their Bayes factor, defined as the ratio of the marginal likelihood of the data under each model, $\mathbb{P}(\mathcal{Y})/\mathbb{P}'(\mathcal{Y})$.

SMC provides an easy-to-implement estimator for the marginal likelihood. When multinomial resampling is done at each step, this estimator is simply the product over generations of the weight averages:

$$\hat{\mathbb{P}}(\mathcal{Y}) = \prod_{n=1}^{N} \frac{1}{K} \sum_{k=1}^{K} w_{n,k}. \tag{8.11}$$

This estimator is consistent, i.e., as the number of particles goes to infinity, $\hat{\mathbb{P}}(\mathcal{Y})$ converges to $\mathbb{P}(\mathcal{Y})$. Compared to the naive marginal likelihood harmonic estimator from MCMC samples (Newton and Raftery, 1994b), or to other importance sampling methods, the variance of the SMC estimator is generally better behaved, owing to the resampling step (Doucet and Johansen, 2009). However, we do not know this at the moment of empirical or theoretical studies that compare the behavior of the estimator $\hat{\mathbb{P}}(\mathcal{Y})$ from SMC against more sophisticated MCMC estimators (Chib, 1995; Gelman and Meng, 1998; Lartillot and Philippe, 2006a; Xie et al., 2011).

8.3.5 Combining SMC and MCMC

In this section, we review a method for using SMC in combination with MCMC algorithms. Using such combinations can be motivated by tight memory constraints (as discussed in Section 8.3.1), or to jointly sample evolutionary parameters. Both cases are discussed in Wang (2012), and have a firm theoretical foundation based on *particle MCMC* (PMCMC) methods (Andrieu et al., 2010).

We limit the discussion in this section to the second motivation, jointly inferring evolutionary parameters. In this chapter, we have assumed so far that the parameters are fixed and known. This assumption is clearly unrealistic, and it defies one of the main motivations behind using Bayesian methods for phylogenetics: modeling our uncertainty over evolutionary model parameters. Fortunately, PMCMC provides a solution to this issue by alternating between evolutionary parameter resampling and SMC tree reconstruction. Formally, a PMCMC algorithm is an MCMC chain with a powerful proposal distribution constructed using an SMC algorithm. In this section, we describe one of the simplest versions of PMCMC, *particle marginal Metropolis Hastings* (PMMH). We refer the reader to Andrieu et al. (2010) and Wang (2012) for many other possibilities of potential interest in Bayesian phylogenetics.

In PMMH, each step in an MCMC chain is computed as follows. First, we propose a new value θ^* for the evolutionary parameter, using a proposal that can depend on the current value θ of the parameter. We then use Algorithm 1 to create a new set of weights $w^*_{n,k}$, partial states $s^*_{n,k}$, and associated approximation $\hat{\pi}^*$ targeting $\pi(t) = p(t|\theta^*, \mathcal{Y})$. We select a proposed tree t^* by sampling from $\hat{\pi}^*$. Because the SMC algorithm is run with a finite number of particles, t^* is only approximately distributed according to the posterior given θ^*, however in the following, we show how to compute an acceptance ratio to correct this discrepancy.

Computing the acceptance ratio is accomplished by making use of the approximation of the marginal likelihood introduced in (8.11). More precisely, we will construct a ratio reminiscent of a Bayes factor comparing the current population of particles to the previous population—i.e., to the population produced by the SMC algorithm during the last accepted MCMC step.

We denote the unnormalized weights of this previous population by $w_{n,k}^{(i-1)}$: the index i indexes MCMC iterations (which can be thought as the outer loop of the algorithm), the index n and k denote, as before, the particle generation and index (which become indices for inner loops in PMCMC). Similarly, $w_{n,k}^*$ denotes the weights of the proposed population. The population is accepted if a uniform random number u satisfies:

$$u \leq \frac{\prod_n \frac{1}{K} \sum_k w_{k,n}^*}{\prod_n \frac{1}{K} \sum_k w_{k,n}^{(i-1)}} = \prod_n \frac{\sum_k w_{k,n}^*}{\sum_k w^{(i-1)}}.$$

If the proposal is accepted, we set the next population, evolutionary parameter, and tree to the proposed values, $w_{k,n}^{(i)} = w_{k,n}^*$, $\theta^{(i)} = \theta^*$, $t^{(i)} = t^*$. Otherwise, we keep the old values, $w_{k,n}^{(i)} = w_{k,n}^{(i-1)}$, $\theta^{(i)} = \theta^{(i-1)}$, $t^{(i)} = t^{(i-1)}$.

For all K, this scheme was shown in Andrieu et al. (2010) to be a valid MCMC algorithm. In other words, for a function of interest ϕ on the trees and evolutionary parameters, as the number of MCMC iterations I goes to infinity,

$$\lim_{I \to \infty} \frac{1}{I} \sum_{i=1}^{I} \phi(t^{(i)}, \theta^{(i)}) = \int \phi(t, \theta) \pi(\,dt,\, d\theta),$$

if weak regularity conditions are satisfied; see Andrieu et al. (2010) for a precise statement. Note that we do not have to assume an increase of K as the MCMC iteration index increases. This is important since higher values of K require more memory, while the cost of higher I is only an increase in time.

While they are not required by the basic consistency result, higher values of K do help obtain a faster mixing. In fact, as K goes to infinity, the acceptance probability converges to one (Andrieu et al., 2010). At the other end of the spectrum, the case where $K = 1$ reduces to a standard MCMC algorithm.

8.4 Discussion

We have reviewed in this chapter various SMC techniques based on incremental construction of a forest. These techniques differ from standard MCMC methods in interesting ways. In particular, new types of proposals can be considered, and likelihood calculations at a given generation reuse calculations from previous generations. Even more importantly, multi-core parallelization in SMC algorithms is easy to implement. The computational bottleneck in most cases is to sample from the proposal K independent times, so for a large K, SMC can be qualified as an "embarrassingly parallel algorithm" (Foster, 1995). This type of parallelization has been tested for phylogenetic SMC samplers in Wang (2012), yielding promising results on small numbers of cores. Scaling this technique to hundreds of cores is theoretically possible using the same techniques.

Some of the benefits of SMC, easy parallelization in particular, can also be brought to samplers using more traditional state representations and proposals. In *population Monte Carlo* samplers, each particle is a complete state (Cappé et al., 2004), just as in a standard MCMC sampler. Note that this type of sampler requires a weight update similar to (8.9), based on a backward kernel. See Cappé et al. (2004) and Moral (2004) for details.

We have focused so far on purely *computational* considerations for motivating phylogenetic SMC: computing a posterior on bigger datasets but using existing models. However, there is also a very promising *statistical* potential. We conclude this chapter by giving an overview of this future direction.

Let us turn our attention to the continuous time processes used to model the evolution of a biological sequence along one edge of a phylogenetic tree. Since the pioneer work by Jukes and Cantor (1969a), there has been tremendous progress in making evolutionary models more realistic. These advances include using generalized rate models (Tavaré, 1986), local correlations (Goldman and Yang, 1994), rate variation (Yang, 1996; Thorne et al., 1998), and indel operations (Thorne et al., 1991).

However, current models are still lacking many known components of evolution. For example, the work of Nasrallah et al. (2011) shows that ignoring structural constraints, as most current models do, can lead to biased tree estimates. This issue was shown to be especially severe in the case of RNA sequences. This motivates the development of more sophisticated likelihood models that take structural constraints into account.

Serious computational challenges arise in the development of these new likelihood models, especially within the Bayesian framework. The key difficulty in going beyond independent site models lies in the computation of the marginal probability of the observed sequences given a tree. This step requires complicated high-dimensional data augmentations or auxiliary variables (Tanner and Wong, 1987) that can be difficult to resample within the standard MCMC methodology. Examples of these auxiliary variables include partially or fully resolved internal sequences, as well as non-local evolutionary events localized on a phylogeny.

By sequentially constructing trees, SMC can be used to jointly propose a population of possible values for the high-dimensional auxiliary variables. More precisely, in the SMC framework, the problem of proposing auxiliary values can be reduced to proposing values along a single branch at a time. To approach these smaller, single branch problems, techniques from endpoint conditioned sampling setups can be used (Hobolth and Stone, 2009a).

9

Population model comparison using multi-locus datasets

Michal Palczewski
Department of Scientific Computing, Florida State University, Tallahassee, Florida, USA[1]

Peter Beerli
Department of Scientific Computing, Florida State University, Tallahassee, Florida, USA

CONTENTS

9.1 Introduction

Bayesian inference has changed the study of phylogenetics and population genetics. Just a few years ago, researchers using probabilistic methods had to justify using such methods rather than parsimony-based tree inferences in phylogenetics and allele-frequency-based methods in population genetics. Molecular phylogenetics seems to be more progressive than population genetics in accepting Bayesian or maximum likelihood methods because today it is common to find phylogenetic reports that only employ probabilistic methods; in contrast, population genetics reports that do not report summary statistics alongside probabilistic methods are rare. We assume this is mostly based on the fact that in phylogenetics usually only one marker, a long stretch of DNA, was collected from many different species; this made it rather simple

[1] Current affiliation is Google Inc.

to develop statistical methods and focus on the mutation model that changes the sequence data over evolutionary time, leading in turn to development of a large number of different mutation models and variants. These models considered, for example, site rate variation and coding versus non-coding sequences. Population genetics, on the other hand, focused on allele frequencies among many sampling locations of a single species. Once sequencing was feasible for many individuals, however, it became obvious that sequencing the same stretch of DNA from many individuals in a single population contributes little additional information because most individuals are identical by descent. The allozyme era of the '80s revealed, however, that populations show many differences if we are willing to look at many loci. This led to a search for cheap markers, such as microsatellites and single nucleotide polymorphisms (SNPs). Recently, studies on non-model organisms that use many stretches of DNA sequence have emerged.

Approaches in biogeography have blurred the boundaries between phylogenetics and population genetics. More within-species sampling for strict phylogenetics purposes has led to examination of the effects of variability within and between species samples. In population genetics, methods for explicitly modeling population divergence and combinations of other population genetic forces such as changes of population size through time and migration patterns among populations, are beginning to emerge.

For many of these analyses, the use of more data improves the accuracy of the inference. Typically, one can increase the number of individuals, the number of populations or species, the lengths of the sampled sequences, and the number of unlinked loci. Felsenstein (2005), Pluzhnikov and Donnelly (1996), and Carling and Brumfield (2007) have shown that in a population genetic framework, the information gained from increasing the number of individuals is limited: a sample of many individuals from the same population will reveal many close relatives — the samples are not independent from each other. Increasing the number of populations and species often does not help because we also increase the size of the model that our inference needs to solve. Increasing the sequence lengths may help but eventually the assumption that the sites in the sequence all have the same evolutionary history is violated because of recombination. We would need to treat the left and the right ends of a large stretch of a sequence as different, unlinked loci. This leaves the last option, increasing the number of unlinked loci, as a natural way to improve the analysis. In phylogenetics, many sequences are now partitioned to allow different mutation models, but few methods allow independent analysis of each partition, for example, by running independent Markov chains for each partition to sample independent groups of trees.

In this chapter, we will focus on population genetics analyses, but believe that the same problems and solutions will hold for phylogenetic inferences with multilocus datasets.

In current applications in population genetics, the number of independent loci that can be used has increased considerably, many of the new genome-scale approaches use single nucleotide polymorphisms and summary statistics either based on allele frequencies or other summarizing tools such as principal components or similar approaches. Coalescence-based Bayesian approaches have not yet been tested with thousands of loci, but in principle, programs like MIGRATE (Beerli, 2006), or LAMARC (Kuhner, 2006) can be run with an unlimited number of loci. Likelihood inference for a multilocus dataset can be run in parallel because, assuming the loci are independent, the calculation for each locus can be easily parallelized, and the final result is a simple combination of the individual results. We have run MIGRATE successfully using 10,000 loci using a two-population model on a computer cluster with 256 nodes within 2 hours.

In Bayesian inference, the independent calculations of the posterior distributions for each locus are simple, but, in contrast to the combination of maximum likelihood estimates over loci, the product of these posterior distributions leads to an overuse of the prior. Correction for this overuse allows us to calculate the posterior distributions independently on different computers or CPU cores, therefore improving the speed of analysis.

9.2 Bayesian inference of independent loci

Bayesian inference is the process of making statements of belief about parameters of interest based on a set of data. At the core of Bayesian inference is Bayes' formula

$$\mathrm{P}(\theta|D) = \frac{\mathrm{P}(\theta)\mathrm{P}(D|\theta)}{\mathrm{P}(D)}. \tag{9.1}$$

The goal of the calculations is to make a statement about the parameter θ in light of the data D: the posterior distribution of the parameter. We use θ as a placeholder for a single parameter or a vector of parameters. The $\mathrm{P}(\theta)$ is the prior belief about the parameter θ. For example, if we are interested in estimating the probability of tossing a head with a particular coin, we could assume that all probabilities in the range of 0 and 1 are feasible and equally likely. Alternatively, we could believe that usually a coin is not manipulated and so the estimate should be close to 0.5, in which case we could use a distribution that peaks at 0.5 and has lower probability at 0 and 1, for example, a beta distribution with parameters $\alpha = \beta = 2$. The denominator is the probability of the data, which is equivalent to the integral of the numerator over the range of θ. For most inference purposes, this is simply a scaler of the posterior probability. In a typical Bayesian analysis, the posterior distribution is presented as the result. Often we cannot calculate the posterior analytically and we resort to stochastic methods, for example, Monte Carlo. For tree-based inferences in phylogenetics and population genetics we sample parameter values from this posterior using Markov chain Monte Carlo (MCMC). These samples are collated and a histogram representing the final posterior is created. The popularity of MCMC stems from the fact that to find the relative posterior distribution we can ignore the denominator in Formula (9.1) because MCMC uses the ratio of previous and current state in the Markov chain (Metropolis et al., 1953c).

Unfortunately for model comparisons we need to calculate the value of the denominator, which is commonly referred to as the "the probability of the data." However, placing a probability value on the data can be counterintuitive; therefore, it is sometimes called "the probability of simulating the data." To gain further insight we rephrase (9.1) as

$$\mathrm{P}(\theta|D) = \frac{\mathrm{P}(\theta)\mathrm{P}(D|\theta)}{\int_\theta \mathrm{P}(\theta)\mathrm{P}(D|\theta)d\theta}, \tag{9.2}$$

or we can be even more explicit by adding the inference model M

$$\mathrm{P}(\theta|D,M) = \frac{\mathrm{P}(\theta|M)\mathrm{P}(D|\theta,M)}{\mathrm{P}(D|M)}. \tag{9.3}$$

The probability of the data given the model which is equivalent to the expectation of the likelihood under the prior:

$$\mathrm{P}(D|M) = \int_\theta \mathrm{P}(\theta|M)\mathrm{P}(D|\theta,M)d\theta = \mathrm{E}[\mathrm{P}(D|\theta,M)]_{\mathrm{P}(\theta|M)}. \tag{9.4}$$

Incorporating the model M explicitly into the marginal likelihood emphasizes the implicit assumption of a particular model under which all terms are calculated. We estimate the probability of the data given a model. If one can calculate this marginal likelihood under model A and then calculate it again under model B, we can directly compare model A and model B. The model with the higher marginal likelihood is preferred because it explains the data better. Unlike other methods of model inference, such as the likelihood ratio test, this method can give a higher likelihood and show preference for a model with fewer parameters than a more complicated model.

Jeffreys (1961a) quantified the degree of support for one model over another by calculating the Bayes factor

$$\mathrm{BF} = \frac{\mathrm{P}(D|M_1)}{\mathrm{P}(D|M_2)}. \tag{9.5}$$

He also indicated an interpretation of the values (Table 9.1). These values assume that the prior on selecting models is uniform, and that they all have equal prior probability.

The previous exposition assumes that the data D are a single piece of information, but often the dataset is structured in one way or another. For example, it is common in phylogenetics to partition the dataset into different

Table 9.1: Bayes factors using the cutoff values devised by Jeffreys (1961a) for two models M_1 and M_2.

BF	log BF	Strength of evidence in favor of M_1
$< 1{:}1$	< 0	Negative (Supports M_2)
1:1 to 3:1	0 to 1.1	Barely worth mentioning
3:1 to 10:1	1.1 to 2.3	Substantial
10:1 to 30:1	2.3 to 3.4	Strong
30:1 to 100:1	3.4 to 4.6	Very Strong
$> 100{:}1$	> 4.6	Decisive

subsets and use different mutation models for each subset of the partition; in population genetics it is common to assume that different loci are independent of each other: every locus is an independent replicate of the evolutionary process (model) that lead to the data observed, thus every locus can be thought of as an independent dataset. Since estimates from each locus can be independently obtained, it makes sense to parallelize the analysis for high-performance computing and run each locus on a different processor in parallel. Beerli (2004) used such an approach for calculating maximum likelihood estimates by running large numbers of loci independently and then finding the maximum likelihood using

$$\mathrm{P}(D_1, D_2, \ldots, D_n|\theta) = \prod_{i=1}^{n} \mathrm{P}(D_i|\theta). \tag{9.6}$$

Simply translating this procedure to Bayesian inference would lead to

$$\mathrm{P}(\theta|D_1, D_2, \ldots, D_n) = \frac{\prod_{i=1}^{n} \mathrm{P}(\theta|D_i)}{\mathrm{P}(D_1, D_2, ..., D_n)}. \tag{9.7}$$

Unfortunately this is incorrect.

Although each of our estimates of the parameter is independent, they are thought to be estimating the same parameter value. Overuse of the prior is a concern. Instead, correctly evaluate the following.

Theorem 1. *The posterior*

$$\mathrm{P}(\theta|D_1, D_2, ..., D_n) = \frac{\mathrm{P}(\theta) \prod_{i=1}^{n} \mathrm{P}(D_i|\theta)}{\int_\theta \mathrm{P}(\theta) \prod_{i=1}^{n} \mathrm{P}(D_i|\theta) d\theta} \tag{9.8}$$

with independent locus data $D_1, D_2, ..,, D_n$, and a set of parameters θ can be calculated by

$$\mathrm{P}(\theta|D_1, D_2, ..., D_n) = \frac{\mathrm{P}(\theta)^{1-n} \prod_{i=1}^{n} \mathrm{P}(\theta|D_i)}{\int_\theta \mathrm{P}(\theta)^{1-n} \prod_{i=1}^{n} \mathrm{P}(\theta|D_i) d\theta}. \tag{9.9}$$

Proof. Expanding $\mathrm{P}(\theta|D_i)$ in (9.9) leads to

$$\mathrm{P}(\theta|D_1, D_2, ..., D_n) = \frac{\mathrm{P}(\theta)^{1-n} \prod_{i=1}^{n} \frac{\mathrm{P}(\theta)\mathrm{P}(D_i|\theta)}{\int_\phi \mathrm{P}(\phi)\mathrm{P}(D_i|\phi)d\phi}}{\int_\theta \mathrm{P}(\theta)^{1-n} \prod_{i=1}^{n} \frac{\mathrm{P}(\theta)\mathrm{P}(D_i|\theta)}{\int_\phi \mathrm{P}(\phi)\mathrm{P}(D_i|\phi)d\phi} d\theta}.$$

The integrals over ϕ cancel, so that

$$\mathrm{P}(\theta|D_1, D_2, ..., D_n) = \frac{\mathrm{P}(\theta)^{1-n} \prod_{i=1}^{n} \mathrm{P}(\theta)\mathrm{P}(D_i|\theta)}{\int_\theta \mathrm{P}(\theta)^{1-n} \prod_{i=1}^{n} \mathrm{P}(\theta)\mathrm{P}(D_i|\theta)d\theta}. \tag{9.10}$$

Moving the $\mathrm{P}(\theta)$ in (9.10) out of the products results in equivalence of (9.8) and (9.9). ***Qed***

Bayes factors offer a convenient tool for comparing different population models without requiring that models be nested. In usual MCMC-based Bayesian inference, the marginal likelihoods are not computed because these normalizing weights cancel in comparisons during the run. They need to be computed and recorded, however, when the combined marginal likelihoods need to be reported. We must therefore evaluate the denominator of (9.8)

$$\mathrm{P}(D_1, D_2, ..., D_n|M_i) = \int_\theta \mathrm{P}(\theta|M_i) \prod_{i=1}^{n} \mathrm{P}(D_i|\theta, M_i)d\theta. \tag{9.11}$$

It would be tempting to calculate the marginal likelihood for each D_i independently, but this is incorrect. Even though each dataset is an independent sample, they are thought to come from the same set of parameters

$$\mathrm{P}(D_1, D_2, ..., D_n|M_i) \neq \mathrm{P}(D_1|M_i)\mathrm{P}(D_2|M_i)...\mathrm{P}(D_n|M_i). \tag{9.12}$$

The interdependence of the loci based on the same set of parameters must be taken into account.

Theorem 2. *The combined marginal likelihood over all independent data blocks can be calculated as a product of independently calculated marginal likelihoods for each data block and a term that depends on the model and the data.*

Proof. The combined estimator of the posterior distribution is

$$\mathrm{P}(\theta|D_1, ..., D_n, M) = \frac{\mathrm{P}(\theta|M) \prod_{i=1}^{n} \mathrm{P}(D_i|\theta, M)}{\mathrm{P}(D_1, ..., D_n|M)}. \tag{9.13}$$

Converting the likelihoods using posteriors on the right:

$$\begin{aligned}\mathrm{P}(\theta|D_1, ..., D_n, M) &= \frac{\mathrm{P}(\theta|M) \prod_{i=1}^{n} \mathrm{P}(\theta|D_i, M)\mathrm{P}(D_i|M)}{\mathrm{P}(\theta|M)^n \mathrm{P}(D_1, ..., D_n|M)} \\ &= \frac{\prod_{i=1}^{n} \mathrm{P}(\theta|D_i, M)\mathrm{P}(D_i|M)}{\mathrm{P}(\theta|M)^{n-1} \mathrm{P}(D_1, ..., D_n|M)},\end{aligned}$$

moving $P(D_1, ..., D_n|M)$ to the left and $P(\theta|D_1, ..., D_n, M)$ to the right results in

$$P(D_1, ..., D_n|M) = \left(\prod_{i=1}^{n} P(D_i|M)\right)\left(\frac{\prod_{i=1}^{n} P(\theta|D_i, M)}{P(\theta|M)^{n-1}P(\theta|D_1, ..., D_n, M)}\right).$$

The fraction has to be a constant with respect to θ because both the product of the individual marginal likelihoods and the combined marginal likelihood on the left are constants with respect to θ:

$$K = \frac{\prod_{i=1}^{n} P(\theta|D_i, M)}{P(\theta|M)^{n-1}P(\theta|D_1, ..., D_n, M)}.$$

Moving the combined posterior and integrating both sides with θ leads to a re-expression of K:

$$P(\theta|D_1, ..., D_n, M)K = P(\theta|M)^{1-n}\prod_{i=1}^{n} P(\theta|D_i, M)$$

$$\int_\theta P(\theta|D_1, ..., D_n, M)K d\theta = \int_\theta P(\theta|M)^{1-n}\prod_{i=1}^{n} P(\theta|D_i, M)d\theta,$$

and because

$$\int_\theta P(\theta|D_1, ..., D_n, M)d\theta = 1,$$

we can evaluate the scaling factor

$$K = \int_\theta P(\theta|M)^{1-n}\prod_{i=1}^{n} P(\theta|D_i, M)d\theta. \tag{9.14}$$

This allows the calculation of the combined marginal likelihood using independent inferences

$$P(D_1, ..., D_n|M) = K\prod_{i=1}^{n} P(D_i|M).$$

Qed

9.2.1 What K represents qualitatively

K represents the "agreement" that multiple loci have about the parameters. For example, one locus favors strong migration while another favors weak migration in the same model. We would expect that there is less evidence for that particular model than when the two loci would be in better agreement. Figure 9.1 showcases this "agreement" for four different situations with two loci. The subfigures show two posterior distributions, one for each locus. In

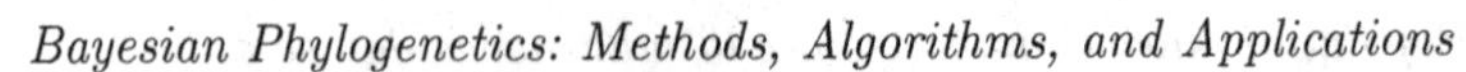

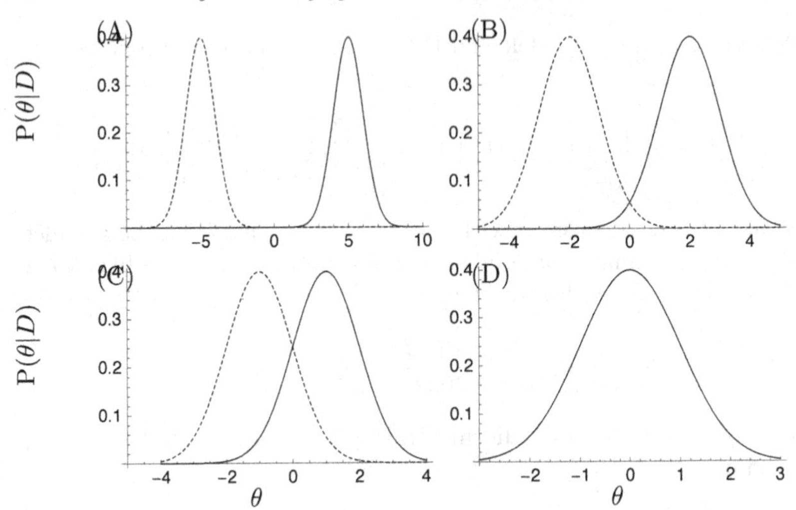

Figure 9.1: Four graphs of possible posterior distributions and their associated K value (9.14) needed to combine the single-locus marginal likelihoods. (A) $\mu_1 = -5, \mu_2 = 5$, $\log K = -23.2698$; (B) $\mu_1 = -2, \mu_2 = 2$, $\log K = -2.26978$, (C) $\mu_1 = -1, \mu_2 = 1$, $\log K = 0.73022$; (D) $\mu_1 = 0, \mu_2 = 0$, $\log K = 1.73022$.

each scenario, the posterior for each locus is a normal distribution with one parameter μ with variance 1.0.

$$\mathrm{P}(\theta|D, \mu) = \frac{e^{-\frac{1}{2}(\theta-\mu)^2}}{\sqrt{2\pi}}. \tag{9.15}$$

We will truncate this distribution by assuming that the prior used is uniform from -10 to 10. The parameter μ can vary among loci. Therefore, the posteriors can range from almost no overlap to being right on top of each other. When the means are very different, K is very small; when the means are similar, K becomes larger. This suggests that with many loci, where we may have more potential for disagreement of the posterior distribution estimate among the loci, K will be high. By extension, with many parameters, we also expect more disagreement and higher K values.

9.2.2 Calculating K

Although (9.14) gives us a formula for K, it is not an easy quantity to estimate. If θ were a one-dimensional parameter, the estimation of K would be a trivial one-dimensional integration. However in population genetics, K can range from a 2-dimensional θ in the case of two similarly sized populations exchanging similar numbers of migrants to high-dimensional θ, such as 400 in the case of a 20-population model where every population can exchange varying numbers of migrants (Beerli and Palczewski, 2010). In a multi-dimensional case, Equation

9.14 can be rewritten like this

$$K = \int_{\theta_1}\int_{\theta_2}\cdots\int_{\theta_m} \psi^{1-n}\prod_{i=1}^{n} \mathrm{P}(\theta_1, \theta_2, \ldots, \theta_m|D_i, M_1)d\theta_1 d\theta_2 \ldots d\theta_m, \quad (9.16)$$

where

$$\psi = \mathrm{P}(\theta_1, \theta_2, \ldots, \theta_m|M_1)$$

is the prior distribution. The naive way to estimate this would be to take the estimated samples from an MCMC run, bin them to create a large multi-dimensional histogram, and then sum over all the terms.

$$K \approx \sum_{\theta_1}\sum_{\theta_2}\cdots\sum_{\theta_m} \psi^{1-n}\prod_{i=1}^{n} \mathrm{P}(\theta_1, \theta_2, \ldots, \theta_m|D_i, M_1)(\Delta\theta_1\Delta\theta_2\ldots\Delta\theta_m)$$

Unfortunately, estimating this term by using a full histogram would be prohibitive. The default number of histogram bins that MIGRATE uses is 1500. In a 400-parameter case, this would mean that $1500^{400} = 2.73 \times 10^{1270}$ terms would be used for this sum. This is far greater than the number of atoms in the observable universe ($\sim 10^{80}$). This equation uses $O(nh^m)$ terms, where n is the number of loci, m is the number of parameters, and h is the number of histogram bins. Although MIGRATE can estimate accurate individual parameter estimates with far fewer samples, it will be unlikely to get multiple observations for each histogram bin during an MCMC run. Therefore we are forced to use a simplification. Although we know that parameters have the potential to be correlated, we make the following adjustment:

$$\mathrm{P}(\theta_1, \theta_2, \ldots, \theta_m|D_i, M_1) \approx \mathrm{P}(\theta_1|D_i, M_1)\mathrm{P}(\theta_2|D_i, M_1)\ldots\mathrm{P}(\theta_n|D_i, M_1).$$

Thus we assume that all parameter posteriors are independent. We also make an additional adjustment of the prior distribution

$$\mathrm{P}(\theta_1, \theta_2, \ldots, \theta_m) \approx \mathrm{P}(\theta_1)\mathrm{P}(\theta_2)\ldots\mathrm{P}(\theta_m).$$

Although, it is possible to formulate non-independent priors, we only consider this case.

Our Equation (9.16) simplifies to the parameter-unlinked

$$\begin{aligned} K_u &= \prod_{j=1}^{m}\int_{\theta_j} P(\theta_j)^{1-n}\prod_{i=1}^{n} P(\theta_j|D_i, M_1)d\theta_j \\ &\approx \prod_{j=1}^{m}\Delta\theta_j\sum_{\theta_j}\left[P(\theta_j)^{1-n}\prod_{i=1}^{n} P(\theta_j|D_i, M_1)\right]. \end{aligned} \quad (9.17)$$

This equation is $O(nhm)$. Although there is an accuracy tradeoff, this equation is possible to calculate. The density for each parameter is calculated individually.

Each parameter has a term that contribute to the scaling factor. If parameters are highly correlated, this could pose a problem. We are using this K_u in our program MIGRATE.

The worst-case scenario for our approximation is the situation where all parameters are completely linked. This is very unlikely with real data, in particular with our main interest: the inference of population sizes and migration rates in structured populations. Assuming this worst-case scenario for two parameters θ_1 and θ_2, we can express one parameter as a linear combination of the other, for example:

$$\theta_2 = c_1\theta_1 + c_2$$

so that

$$\mathrm{P}(\theta_1, \theta_2|D) = \mathrm{P}(\theta_1|D)I(\theta_2 = c_1\theta_1 + c_2),$$

then (9.16) becomes the parameter-linked

$$K_l = \int_{\theta_1} \mathrm{P}(\theta_1)^{1-n} \prod_{i=1}^{n} \mathrm{P}(\theta_1|D_i, M_1)d\theta_1.$$

MIGRATE uses a combination of the above formulae. In the case of asymmetric parameters and population sizes that differ among populations, MIGRATE uses the independent version of the formula. In the case of symmetric migration rates or populations with the same size, the parameters are fully linked and the second formula is appropriate.

This is equivalent to computing K for just one parameter. K_l and K_u are at different ends of the spectrum with regards to the true value of K. The square of the correlation (r^2) between two parameters explains how much of the variance of one variable can be explained by the variance of the other variable. Thus we could construct a weighted average

$$K \approx K_l(r^2) + K_u(1 - r^2).$$

For more than two parameters, one could imagine an analysis of variance–based approach, but we have not investigated this option.

9.3 Model comparison using our independent marginal likelihood sampler

Bayes factors make it easy to test nested or non-nested models. Accuracy of the marginal likelihood approximation from inferences using MCMC is a great concern. Estimators of the marginal likelihood based on the harmonic mean (cf

Kass and Raftery, 1995) have been proven to be unreliable (Fan et al., 2011; Beerli and Palczewski, 2010; Neal, 2008). Our own thermodynamic integration is less sophisticated than those introduced by Xie et al. (2011) and Fan et al. (2011) and may need more computation, but does accurately judge models (Beerli and Palczewski, 2010). Analyzing population structure can involve rather complicated models that may have a wide range of parameters. We can count all "unidirectional" migration models ignoring population sizes using

$$\sum_{i=0}^{n(n-1)} \binom{n(n-1)}{i}. \tag{9.18}$$

For 4 populations, this results in 4096 models. The number of models increases considerably if we allow bidirectional migration and also take into account that some of the sampling locations could be part of the same panmictic population, for example, locations 1–3 are one population and location 4 is the second.

With population genetics or phylogenetic data, we rarely know the detailed history, and therefore we may never know the truth. We use models to help us to understand the history of our samples and the hidden truth. Current research is commonly testing a priori defined hypotheses. Bayesian model comparison will make it easy to compare these a priori models and thus reduces the number of interesting models. Whether the best model is close to the truth is often easy to answer with "no." But this model selection process may allow researchers to formulate new models and test those with even more or better data.

We explore the problem of model selection and comparison with a small example that is sufficiently complex. Beerli (2006) used simulated datasets that were generated using a model of 4 populations that exchange migrants in a round-robin scheme, where population P_1 receives migrants from population P_2, P_2 receives migrants from P_3, P_3 receives migrants from P_4, and finally, P_4 receives migrants from P_1; we abbreviate this scenario as $P_1 \to P_4 \to P_3 \to P_2 \to P_1$. We picked the first 10 single-locus datasets and generated 3 new datasets, one with two, five, and ten loci, respectively. For each dataset we evaluate six models (Table 9.2). Model H_0 is the model used to simulate the datasets, and we call it the true model. Models H_1 and H_2 add an additional

Table 9.2: Different models.

Model	Parameter	Explanation
H_0	8	True model: $P_1 \to P_4 \to P_3 \to P_2 \to P_1$
H_1	9	Same as H_0 and addition of $P_4 \to P_2$
H_2	9	Same as H_0 and addition of $P_3 \to P_1$
H_3	1	One panmictic population
H_4	4	Two populations $(P_1, P_3) \leftrightarrow (P_2, P_4)$
H_5	4	Two populations $(P_1, P_4) \leftrightarrow (P_2, P_3)$
H_6	16	Full model, all migration routes

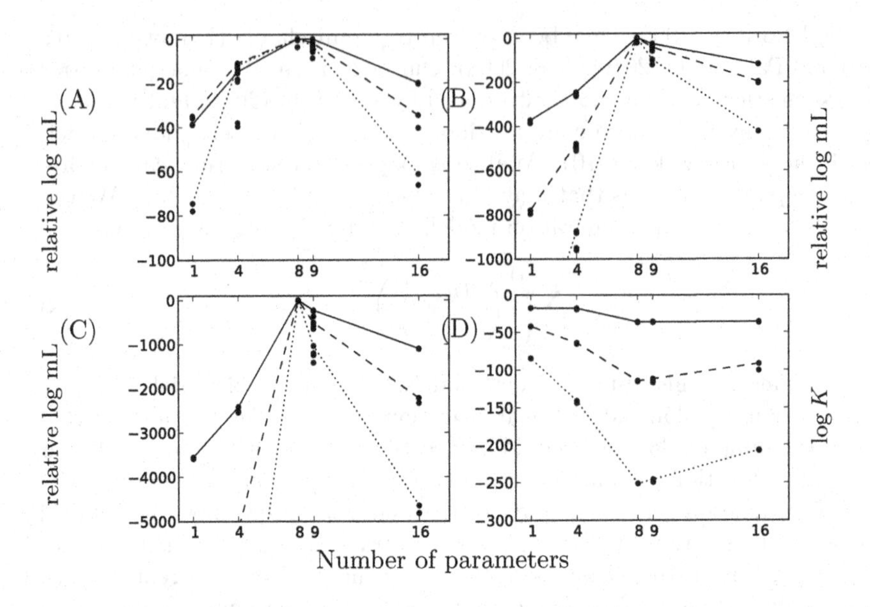

Figure 9.2: Relative log marginal likelihoods (Log Bayes factors) of the models shown in Table 9.2. Each model was run twice. The lines connect the highest scores for each number of parameters and 2 (solid line), 5 (long dashes), and 10 loci (short dashes). (A) dataset had 100 bp per individual; (B) 1,000 bp; (C) 10,000 bp; and (D) scaling factors for (C).

migration route. They are similar in structure to each other and also should be similar in results because the round-robin scheme does not have a particular source population. Models H_3, H_4, and H_5 are rather different from the truth. Model H_6 uses 16 parameters and represents the most complex model. With sufficient data, this model is capable of representing the parameters correctly (Beerli, 2006). We ran each model twice, using relatively short runs (about two hours) on the high-performance cluster at Florida State University using 32 nodes.

Figure 9.2 summarizes these runs. The difference between all the models is smallest for the two-locus 100-bp dataset and largest for the ten-locus 10,000-bp dataset. With large amounts of data (10,000 bp) the log Bayes factor difference among the models is very large and the true model H_0 wins even with a small number of loci. Models H_1 and H_2 are most similar to the true model H_0 because they only differ by one additional parameter. This is reflected in the model order from best to worst: $H_0 > H_1 \sim H_2 > H_6 > H_4 \sim H_5 > H_3$. The simulated scenario is clearly structured. This leads to low marginal likelihoods for models that lump populations (models H_3, H_4, and H_5). The full model H_6 has a lower marginal likelihood than model H_0 because 16 varying parameters predict the data poorly. Even with poor data

(Figure 9.2A), model H_0 is superior to complex or very simple models, but the marginal likelihood difference from models H_1 or H_2 are small, suggesting that we cannot distinguish H_0 from these models with certainty and we need more data than two loci with only 100 bp.

With ten loci and very short sequences, we may be able to distinguish complex models from each other, but more loci or longer sequences would improve the distinction considerably. For example, using 1000 bp and two loci delivers clearer results than ten loci of 100 bp each.

Our evaluations suggest that even if we omit the true model from the set of tested models, our results still favor models that explain the structured nature of the dataset (models H_1 and H_2).

More data with models that are closer to the true model lead to lower magnitudes of the scaling factors K or $\log K$ (Figure 9.2D), suggesting that low $\log K$ indicate a better model fit than high $\log K$, but we have not explored this relationship in more detail.

9.4 Conclusion

The calculation of Bayes factors for inferences that need MCMC are complex and time-consuming. Current approaches for the approximation of the marginal likelihood employ multiple chains with different, static temperatures or a single chain that dynamically changes the temperature to collect likelihood samples. We provide a framework for combining independently calculated estimates of marginal likelihoods under the assumption that the parameters are independent of each other. This permits analyses of large-scale biogeographic or population genetic datasets on computer clusters.

Acknowledgments

Our work was supported by grants DEB-0822626 and DEB-1145999 from the National Science Foundation. We acknowledge the use of the high-perfomance computing facility at Florida State University.

10

Bayesian methods in the presence of recombination

Mary K. Kuhner

Department of Genome Sciences, University of Washington, Seattle, Washington, USA

CONTENTS

10.1 Introduction to non-treelike phylogenies

Phylogenetics began with the concept of a single branching tree representing ancestral relationships among its taxa. Reality is not always so straightforward. Non-treelike phylogenetic signal can arise in several ways. While it can be mimicked by error obscuring an underlying treelike phylogeny, it can also represent genuine reticulations in the ancestry relationships: cases in which a single taxon or sequence has multiple ancestors. Reticulations can arise within populations via recombinational mechanisms including crossing over, gene conversion, and chromosome segregation, and between populations or species by ancestral polymorphism (incomplete lineage sorting), hybridization, or horizontal gene transfer. For example, when a patient is infected with two strains of HIV-1, recombination can produce a hybrid molecule with two distinct ancestors, each contributing one or more segments of the viral chromosome. Such a sequence cannot be correctly placed on a normal phylogenetic tree, because it did not arise by a treelike process.

Phylogeneticists are slowly developing approaches to these tangles. This chapter will focus on reticulations that can be modeled as genetic recombination; some of the techniques discussed may be relevant to other forms of reticulation as well. Recombination is easier to model than most other forms of reticulation because it represents a recurrent process with fairly well-understood dynamics. The genetic map gives us an idea of how likely two loci are to be separated by a recombination in each generation. It is much more difficult to know how often a locus might be, for example, horizontally transferred. This impedes phylogenetic analysis, which will want to weight the possibility of horizontal transfer against the alternative of parallel evolution.

A recombination, in phylogenetic terms, is the creation of a new lineage by a single interchange between segments from two parental lineages. Over short distances it seems reasonable to model recombinations as happening one at a time, so that the new lineage will contain the left-hand portion of one of its progenitors and the right-hand portion of the other, with a definite "breakpoint" between them. While more than one recombination might happen in a single generation, over evolutionary time it does little harm to assume that they happened close together in time but not quite simultaneously, and single events are easier to model. The exception to this general principle is that if the genome is circular, viable recombinations necessarily happen in pairs; this resembles gene conversion.

Gene conversion normally produces a new lineage which contains the flanking regions of one progenitor with a segment of the other progenitor sandwiched between them: a conversion is similar to two simultaneous recombinations. Most current approaches lump conversions in with recombinations, although this may bias rate estimates as conversion, a single event, will have different kinetics than the two independent events of a double recombination.

Phylogenetic consideration of recombination ignores the fact that a recombination event generally produces two reciprocal products: it is deemed unlikely that both products found their way into offspring and both resulting lineages survived.

It is useful to contrast recombination with hybridization. As hybridization has generally been defined, if we specify that a lineage is a hybrid, we assert that its genetic material is a mix of two progenitors but make no statement about the spatial relationships between pieces with different ancestry. When we specify that a lineage is a recombinant we are making a stronger claim: the left segment has one ancestry and the right segment has another, and in principle we could determine where the breakpoint lies.

A hypothesis of recombination is easier to handle phylogenetically than the less-restrictive hypothesis of hybridization. In a fully resolved recombinant graph, each individual sequence position is assigned a treelike phylogeny, and the probability of that sequence position given its phylogeny can be calculated as usual. In contrast, if a sequence is hypothesized to be a hybrid but no attempt is made to assign its parts to the two progenitors, assessing the fit of data to phylogeny is much more challenging. For a review of the hybridization problem, see Nakleh (2010).

We can also contrast a recombinational model of reticulation with one in which each individual locus has a branching tree, but these trees may differ between loci. This can arise by incomplete lineage sorting (also known as ancestral polymorphism or deep coalescence), recombination, hybridization, or horizontal transfer. The simplifying assumption is usually made that separate loci are unlinked, and that thus the only correlation among gene trees is that imposed by the population or species tree. Several Bayesian methods approach gene trees in this way (Liu and Pearl, 2007; Heled and Drummond, 2008).

Except in the case of putatively non-recombining genomic regions such as mtDNA, the genes trees themselves may be reticulate, since recombination can happen within as well as between loci. Models which attempt to reconcile the gene tree with the species tree normally ignore this possibility. In the long run, gene tree/species tree methods may need to be augmented with methods that can handle within-locus recombination.

Ignoring recombination. What happens if we apply standard phylogenetic techniques to data containing a recombinant lineage? The result depends on how much data supports each of the component non-recombinant phylogenies — a recombination that occurs near one end of the sampled region will have little effect on the inferred phylogeny compared to one that occurs in the middle — and on the relationship of the two progenitors relative to the rest of the phylogeny. A recombinant lineage carrying mainly material from one of its progenitors may cluster with that progenitor's other offspring. A recombinant lineage with substantial input from two dissimilar sequences often seems to move down the tree toward the common ancestor of its two progenitors, as was observed for inter-specific hybrids by McDade (1992); she also observed that the presence of a hybrid between distantly related lineages often pulls those lineages closer to one another in the inferred tree. The incompatibilities introduced by recombination will tend to lengthen the branch leading to the recombinant taxon, which may distort the tree further; this effect is particularly marked for phylogeny methods that are vulnerable to branch-length effects. In McDade's study, which was based on parsimony, inclusion of hybrids between distantly related species sometimes led to major disruptions of the phylogeny.

It is therefore dangerous to assume a branching tree when none in fact exists. Doing so guarantees a wrong inference about the recombinant or hybrid taxon, often leads to wrong inferences about its progenitors, and can even misplace "innocent" taxa.

When branch lengths are being inferred, the incompatibilities introduced by recombination will need to be explained by recurrent mutation, and branch lengths will therefore be biased upward. This is a problem for phylogenetics, but an even more severe one for within-population inference, where branch lengths provide key information about parameters such as mutation rate and effective population size. In both within-species and between-species cases, biased branch lengths also jeopardize inference of dates: a recombinant taxon is likely to be inferred as older than it actually is, and it may also disrupt dating of other lineages.

Many studies have experimented with culling the data to render them more treelike. One approach is to remove any taxa suspected of being recombinants or hybrids. This may reduce distortion to the tree, but of course reduces the number of taxa that can be studied. Also, because reticulate ancestry is easier to detect in more diverse sequences, removal of visibly recombinant taxa may tend to remove the most diverse parts of the tree, biasing any inference relying on tree shape or branch length distribution. This problem is especially worrisome in coalescent analyses, where inference of population size may be severely distorted by selective removal of the most variable taxa.

Alternatively, sequences can be kept short to reduce the risk of them containing a recombination breakpoint, and gene tree/species tree methods can be used to deal with discordance among the trees from the various sequences. However, this approach tolerates only completely linked and completely unlinked sequences. Such data are not available for all organisms. For example, in HIV-1, which has a small genome and a high recombination rate, unlinked loci do not exist, and sequences short enough to be reliably recombination-free may be too short to be informative.

Inference of reticulate phylogenies. It would clearly be preferable to include recombinations in our phylogenies rather than ignoring them, but to do so is challenging. The phylogenetic search space is already absurdly large; the recombinant-phylogeny search space is much larger. At the limit, each site position might have its own phylogeny. In practice, runs of adjacent sites often share the same phylogeny; the distance between ancestral recombination breakpoints along the chromosome has been estimated to be between 250 and 2500 bp in humans (Robertson and Hill, 1983). Nonetheless, perfect reconstruction of a recombination phylogeny will often be impossible, because some of the breakpoints will be closely spaced and there simply will not be enough variable sites between one breakpoint and the next to allow the phylogeny to be determined.

This situation cries out for Bayesian phylogenetics, as Bayesian methods move the emphasis from finding the single best tree — often hopeless in the presence of recombination — to finding a collection of representative trees. This chapter will review various approaches to phylogenetic inference in the presence of recombination. Surprisingly, while there are recombinant phylogeny inference programs and Bayesian methods incorporating recombination, no program currently available can quite be described as a Bayesian recombinant phylogeny inference method. Much remains to be done in this area.

10.2 Describing the ARG

Biologists contend that trees grow with their leaves in the air and their roots in the ground. Mathematicians and computer scientists have the opposite opinion.

I will follow biologists here, and call the direction toward the tips "upward" and toward the root "downward."

Griffiths and Marjoram (1997) coined the term "Ancestral Recombination Graph" (ARG) for a clocklike recombinant phylogeny or genealogy with specified recombination sites. The ARG is a special case of the Directed Acyclic Graph (DAG) or phylogenetic network used to represent generalized reticulations: it is augmented with information about which sites traverse which lineages.

An ARG can be decomposed into branching trees by breaking the sequence at each recombination point, as seen in Figure 10.1. I will call these recombination-free sequence segments "intervals" and the tree governing an interval the "interval tree" (they are also called "marginal trees" in the literature). Adjacent interval trees may be very similar or quite dissimilar, as a single breakpoint between intervals may be the site of more than one recombination. Adjacent interval trees may also be identical due to cryptic recombinations such as the one shown in Figure 10.2(3).

An ARG is not uniquely specified by its interval trees. For example, two ARGs with identical sets of interval trees can differ in the number and location of their cryptic recombinations. However, if sites evolve independently, the probability of the data given the ARG is completely determined by the interval trees; therefore, most standard methods for evaluating this probability work for ARGs as well. ARG inference can therefore use the rich array of mutational models developed for maximum-likelihood and Bayesian phylogeny inference. The only exceptions are models which allow non-independence between sites in a way that depends on the character at that site, such as models for tRNA stem-loop structures (Tillier, 1994). We may know that an A at one site predicts a T at another, but if those two sites derived from different ancestral lineages, they would not have been interacting as pairing partners. However, other forms of non-independence can still be accommodated: for example, recombination is not a barrier to the use of hidden Markov models for autocorrelation of mutation rate along the sequence (Yang, 1995b). Such models assume that adjacent sites will tend to be fast-evolving or slow-evolving in a correlated fashion, but not that the actual states of these sites are correlated; the correlations are therefore not disrupted by recombination.

The ARG is not Markovian along the sequence. A subtle but important feature of the ARG is that not only are adjacent interval trees correlated, but *non-adjacent* interval trees are not even conditionally independent. That is, if we are moving along the sequence and observe interval tree 1, then interval tree 2, our speculations about the topology of interval tree 3 should be informed by tree 1 as well as tree 2. In particular, tree 3 has a higher probability of being identical to tree 1 than would be expected if non-adjacent subtrees were conditionally independent. An example is shown in Figure 10.3. The ARG is thus non-Markovian, in that you cannot correctly generate an interval tree using only its immediate predecessor, or indeed any finite number of immediate

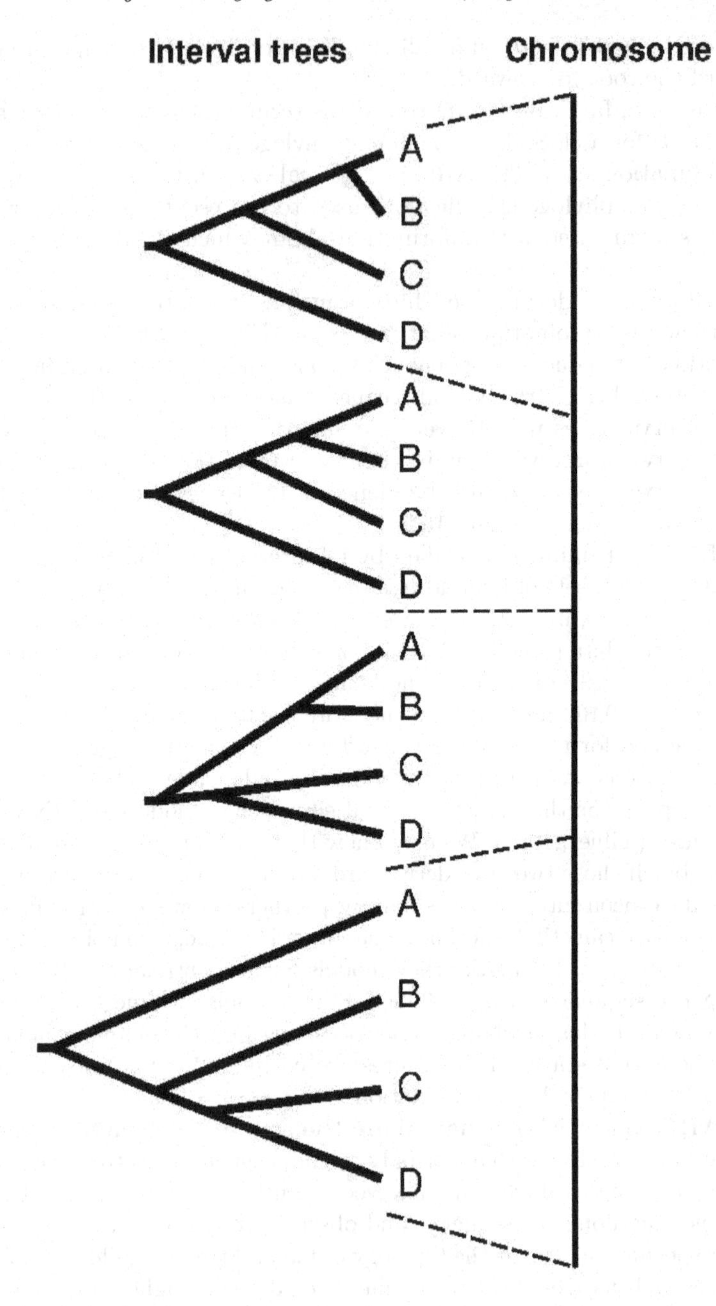

Figure 10.1: Interval trees along a chromosome. In this figure the interval trees of an ARG are shown spread out along the chromosome. Branches are drawn proportional to time, illustrating that adjacent areas of the genome may or may not be the same age (the bottommost interval tree represents a greater age than the others).

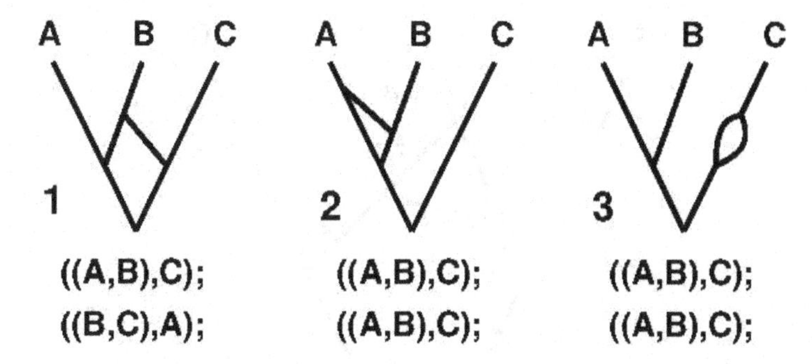

Figure 10.2: ARGs and their interval trees. (1) A recombination which changes the topology of the interval trees. (2) A recombination which changes branch lengths, but not topology. (3) A *cryptic* recombination which does not change interval tree topology or branch lengths.

predecessors. (A Markovian process is one in which you need only consider a specified number of predecessors to know the probability of the current state.)

The following example may make this idea more intuitive. Suppose that we are analyzing the ancestry of one of my chromosomes. The first segment happened to come from my maternal grandfather Peter, but as we move along the chromosome we reach an area that came from my maternal grandmother instead. The Markovian principle would say that as we continue further, the chance that additional intervals will derive from Peter is no higher than that they will derive from any other member of his generation. This is clearly false: possessing a given set of ancestral relationships in one segment of the genome increases the probability that they are also present elsewhere. To correctly assess the chance that the next interval will come from Peter, we would want to know whether *any* previously examined intervals came from Peter. The inheritance probabilities are thus non-Markovian for this very close relative, and although it is less obvious for more distant relatives, they remain non-Markovian.

It is easy and tempting to write simulators for the Markovian ARG rather than the actual coalescent ARG, but their behavior is slightly different. A correct algorithm for simulating the ARG by proceeding along the chromosome, avoiding the erroneous assumption of Markovian behavior, is given in Wiuf and Hein (1999). More commonly the ARG is simulated by working downward from the tips, considering the whole sequence at once, as is done by the *ms* simulator (Hudson, 2002).

McVean and Cardin (2005) explore use of the Markovian ARG as an approximation to the full recombinant coalescent distribution, trading off statistical correctness for ease of inference. They find that the properties of the Markovian approximation are very similar to those of the full coalescent. In particular, the bulk attributes of the interval trees (such as mean time to

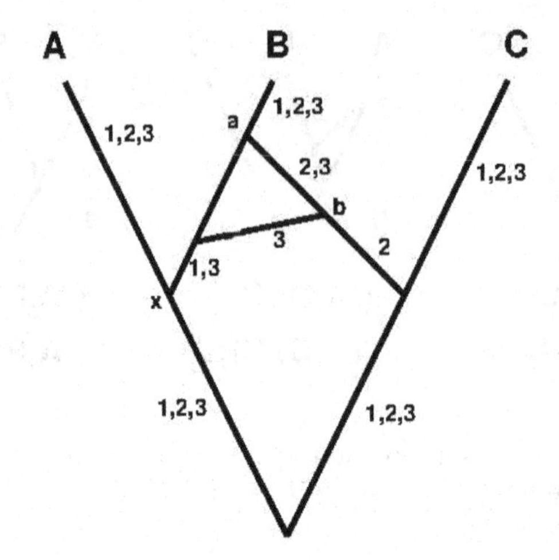

(a) An ARG spanning three regions

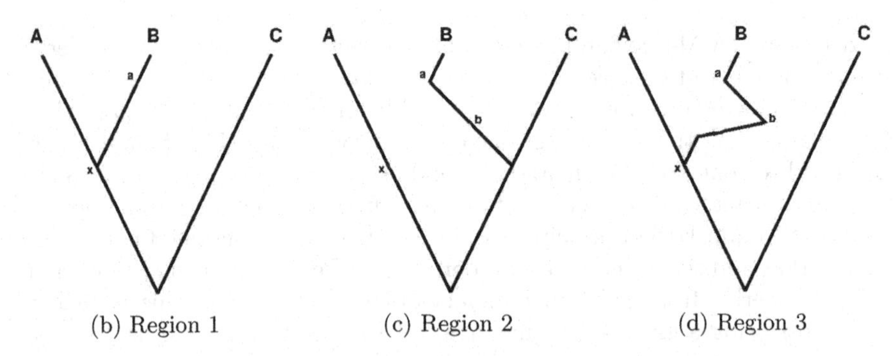

(b) Region 1 (c) Region 2 (d) Region 3

Figure 10.3: The ARG is not Markovian along the sequence. This figure shows an ARG spanning regions 1, 2, and 3 of a sequence. If we were simulating the ARG one interval tree at a time starting from the left, we would first insert recombination (a). In the resulting interval tree there is no coalescence corresponding to point X. Doing an independent simulation to produce recombination (b) would then have a very low probability of independently generating a coalescence at point X and thus returning to the original interval tree. However, on the actual ARG, this sequence of events is fairly probable.

the most recent common ancestor) are the same, but generated ARGs have far fewer recombinations, so the state space is greatly reduced. This idea of the sequential Markovian coalescent has been extended by Marjoram and Wall (2006) but much remains to be done.

Which lineages should be included in an ARG? As we trace a lineage back in time from the observed sequences, it is broken up by recombinations. A lineage in the past will often have parts which are ancestral to observed sequences and parts which are not. Repeated recombination can lead to past lineages which contain no ancestral material at all. The original definition of the ARG (Griffiths and Marjoram, 1997) includes only lineages containing ancestral material. Lineages giving rise solely to non-ancestral material are considered irrelevant, as no hypothesis about them can influence our interpretation of the sampled data. However, O'Fallon (2011) has explored the possibility of including non-ancestral lineages in the ARG, finding that the resulting simplification of ARG rearrangement may outweigh the computational costs.

A related issue appears at the root. Any given interval tree has a root (most recent common ancestor) in the usual fashion. However, different interval trees may have different roots. It is possible to discontinue tracing a line of descent as soon as it reaches the root of the interval tree, or trees, to which it is relevant. In this case, the ARG does not necessarily have a global root but may instead resemble a mangrove thicket with multiple roots.

A more treelike ARG results if one tracks all ancestral lineages until the global root is reached, but the resulting ARG may be much larger. The size of the global-root ARG can be reduced by regarding sites which have reached their interval tree root as no longer capable of "interesting" recombinations. Kuhner et al. (2000) named this the "final coalescence" (FC) optimization since a site becomes recombinationally uninteresting when it has reached its final (rootmost) coalescence, but it was discovered much earlier by Richard Hudson (personal communication). The fit of a dataset to an ARG, under any model with independence of site states, is unaffected by use or non-use of the final coalescence rule, or indeed by the decision to have a global root at all.

A multiple-root simulation or analysis implies the FC optimization, but FC can also be used on its own. If the count of recombinations per ARG is a parameter of interest, it is critical to know whether FC is in effect, as it can have a dramatic impact. The most commonly used coalescent/recombination simulator, *ms* (Hudson, 2002), uses the multiple-root approach and therefore also the FC optimization. In practice, non-FC samplers or simulators tend to bog down when the recombination rate is high as the number of recombinations becomes enormous.

Cryptic and silent recombinations. A troublesome feature of real ARGs is the presence of cryptic recombinations or "bubbles." These represent situations where a recombination conjoins two sequences immediately derived from a common ancestor. This produces a recombinant lineage all of whose sites derive from the same ancestor, and though they derive via two separate paths, nothing in the data can tell us so (see Figure 10.2(3)): the two interval trees are identical. More complex multi-recombination structures may behave in the same way.

Another category of recombinations is not inherently indetectable — they change the branch length or topology of the interval tree — but in a specific realization of the ARG as data they are indetectable because of a lack of informative sites. For example, a recombination between two identical sequences can never be detected. These could be termed silent recombinations. Their frequency depends on the density of informative sites and can be quite high. Hudson (1985) repeatedly resimulated a scenario with 44 segregating sites and 11 sampled sequences, using parameters inspired by a real dataset. They found that the mean number of actual recombinations was 118 (using the FC definition of the ARG) and that the mean number of detectable recombinations (neither cryptic nor silent) was 5.

Cryptic and silent recombinations are a severe nuisance. They really occur, but they are completely invisible in data. If they are included in a reconstructed ARG, they bloat it for no apparent gain. However, cryptic recombinations are very difficult to exclude. ARGs are usually simulated or redrawn working downward from the tips, and at the time when the recombination is proposed, the fact that it is going to be a bubble is not yet determined. Bubbles could be removed from the ARG when detected, but the formula for the coalescent probability of an ARG with its bubbles removed is not known. As a result, existing ARG simulators and samplers generally include bubbles in their view of the ARG. Attempts to find the single best ARG, of course, do not.

Users of recombination-inference programs should always remember that many, often most, real recombinations are cryptic or silent. Inference based on a stochastic search of ARG space may suggest a quantity of recombinations broadly consistent with the truth, but there is no way to recover the actual number or location of many of the events. A program based on best-fit inference of the ARG will lack cryptic and silent recombinations, and it is highly likely that the best-fit inference will therefore have fewer recombinations than the truth, possibly far fewer: in the case explored by Hudson (1985), a "perfect" best-fit reconstruction of the ARG would reveal only 4% of its actual recombinations.

If a site is not a single nucleotide polymorphism (SNP) (does not vary in our sample), we have negligible power to determine whether it falls to the left or the right of a postulated recombination breakpoint. We might therefore simplify the ARG by supposing that only non-silent recombinations — recombinations separating two or more SNPs — are of interest, and that recombinations occur at arbitrary points between adjacent SNPs (say, midway between). Such an approximate view of the ARG preserves the interval tree of every SNP, though not of non-SNPs, and will tend to have substantially fewer recombinations than the full ARG. This reduction of the search space may improve search efficiency, especially for human-like data where SNPs are widely spaced. It was probably first used by Wang and Rannala (2008) in their program *InferRho*, though it has been independently rediscovered by my lab and probably others.

The molecular clock. Unlike inference of topology and branch lengths, inference of recombination demands that the direction of time be known. If we see that sequences A and B are similar at the right end but dissimilar at the left end, we may suppose that one of them descends from a recombinant, but there is no way to know which. If we further note that many sequences are similar to A but few are similar to B, we might suppose that A is the ancestral non-recombinant type, but we could easily be mistaken—perhaps the recombination event is old and its descendants numerous. Only by making time-based arguments can we infer which type is ancestral and therefore which lineage has undergone a recombination.

It should not be necessary to assume a strict molecular clock in order to infer genealogies with recombination, however. The relaxed-clock approach implemented in *BEAST* (Drummond et al., 2006) seems applicable to recombinant cases, though currently no published software exists for this combination.

10.3 Inference of the ARG

Single-estimate methods. Hein (1993) attempted to infer the best ARG using parsimony, applying separate weights to transitions, transversions, and recombinations, and using a heuristic search to find an ARG minimizing the resulting score. A similar approach could be built around maximum likelihood evaluation. In both cases, assumptions must be made about the relative probability of recombination and mutation.

There are also several methods which attempt to detect the presence of recombinations but do not reconstruct the ARG. For example, the bootscanning algorithm (Salminen et al., 1995) looks for sudden changes in the bootstrap support of competing phylogenies along the length of the sequence, detecting a recombination breakpoint at an abrupt change from one supported phylogeny to another. Another class of methods recovers an undirected network or graph, rather than an ARG. Many such methods are collected in the *SplitsTree* program (Huson and Bryant, 2006). (This publication also gives an excellent overview of network-recovery algorithms and concepts.)

Attention has moved away finding a single best estimate of the ARG because the task is nearly impossible. Even if cryptic recombinations are ignored, few datasets contain enough variation to allow all interval trees to be correctly inferred. For example, two sibling lineages may be identical at one end of the sequence. If a copy of this region is found in a recombinant, it will be completely unclear whether one or the other sibling, or their parent, was the donor. Even in highly polymorphic data, short interval trees often cannot be correctly reconstructed.

Inference of a cloud of ARGs. Rather than attempting to determine the single best ARG, one could make a cloud of plausible ARGs. This approach was pioneered by the *recom* sampler of Griffiths and Marjoram (1996), a coalescent genealogy sampler similar in approach to *GENETREE* (Griffiths and Tavaré, 1994).

These samplers are based on developing an expression for the likelihood of the observed samples as a functional of a Markov chain: a massive summation over potential coalescences, mutations, and recombinations at each step in the tree. The functional begins with consideration of the observed data — a certain number of sequences bearing certain mutations — and works backward in time to consider sequences of events that could explain the observed data. Any path through the summation — choosing, for example, a particular coalescence at the first step, a particular mutation at the second step, and so on — represents a single ARG. The functional is too large to evaluate directly, but Markov chain Monte Carlo (MCMC) can be used to make a large sample of ARGs from it. At each step, a coalescence, mutation, or recombination is chosen from among all legal events according to their relative probabilities. These probabilities are defined for trial values of the parameters, the driving values. Once a large collection of ARGs has been generated, the sampler makes a maximum likelihood estimate of the parameters (mutation rate, population size, and recombination rate) under which the collected ARGs are most probable, weighting to compensate for the influence of the driving values.

Each sampled ARG is made independently: these methods are therefore called independent-sample or IS samplers (the acronym can also be taken to refer to importance sampling, but in my view this is misleading as all genealogy samplers use importance sampling).

In their non-recombinant implementations, IS samplers are appealing because they mix well and relatively few samples are needed for a good estimate. Independent sampling means that peaks are unlikely to be missed. However, this independence comes at the cost of requiring a mutational model without multiple hits or back mutation (an infinite-sites model), as otherwise the number of terms to be summed would potentially be infinite. In practice, an IS model with back mutation will try out innumerable reversions, but is not directed toward the small subset of possible reversions which could actually enhance fitting of the genealogy to the data. Most genealogy realizations will contain far too many mutations, with a correspondingly low value of the functional.

A similar problem has been observed in IS algorithms with recombination. While the number of possible mutational histories is finite with an infinite-sites mutation model, the number of recombinational histories is not. As found by Griffiths and Marjoram (1996), this leads to heavy sampling of hyper-recombinant ARGs: they collectively represent common paths through the functional, but they contribute almost nothing to its evaluation, merely eating up computer time and storage. The authors of *recom* recommend breaking off evaluation of any ARG which acquires too many recombinations, setting the functional value to zero. In a run on 50 sequences Griffiths and Marjoram (1996) found that 89% of their two million attempted samples were discarded for this reason.

The fundamental difficulty is that at any step in the ARG, many recombinations are possible, but most of them do nothing to explain the observed data. We would like to focus attention on recombinations which separate discordant regions of the sequence. The IS strategy with recombination was strengthened by Fearnhead and Donnelly (2001) by using an approximate sampling distribution inspired by the optimal sampling distribution (which is unfortunately intractable). In brief, recombinations generating a haplotype similar to other haplotypes present in the genealogy are preferentially sampled. In order to avoid bias, the sampler must downweight such "popular" haplotypes, which might seem to make the whole strategy futile, but the space of haplotypes that could be proposed is so large that pulling likely haplotypes into the search set results in huge improvements in search effectiveness even in the face of this weighting. A haplotype might have an enormously favorable value of the functional, easily able to overcome adverse weighting, but unless this haplotype is proposed, the sampler will never discover how good it is. The Fearnhead and Donnelly sampler is implemented as program *fins* available at `http://www.maths.lancs.ac.uk/~fearnhea/software/Rec.html`. This direction shows promise for further work.

Another family of genealogy samplers uses a correlated-sample rather than independent-sample approach. The first member of this family was the *RECOMBINE* program of Kuhner et al. (2000). In contrast to *recom*, *RECOMBINE* does not construct genealogies independently, but makes small changes to the current genealogy. This leads to an autocorrelated sample which may not explore the full state space, but no longer requires an infinite-sites mutational model for efficient use. Samplers of this kind can be termed CS or correlated-sample samplers. Some authors refer to this family as MCMC samplers, but this term is misleading as IS samplers also use MCMC, although they use it very differently. IS samplers use their Markov chain to generate successive events in a single tree; the next tree will be a new, independent chain. CS samplers use theirs to generate successive, correlated trees.

The basic search strategy of CS samplers is to make a starting realization of the genealogy (or ARG) and calculate the probability of the observed data, using a specified mutational model, on that genealogy. (This quantity can be written $P(D|G)$.) A small change is made to the genealogy and $P(D|G)$ is recalculated. If the new genealogy is superior it is always accepted. If it is inferior, a stochastic decision to accept or reject is made based on the ratio of $P(D|G)$, so that the worse the new genealogy is, the less likely it is to be accepted. This is the MCMC algorithm of Metropolis et al. (1953b). If the changes made to the genealogy involve any bias, this must be compensated for by an additional factor, due to Hastings (1970), in the acceptance/rejection formula, yielding Metropolis-Hastings sampling. Genealogy samplers based on Metropolis-Hastings sampling were first introduced with *COALESCE* (Kuhner et al., 1995).

CS samplers were originally developed in a maximum likelihood context. They can also be used in a Bayesian fashion by intermixing proposals to modify the genealogy with proposals to change the current parameter values. This was first done by Beerli and Felsenstein (1999).

The CS approach to the coalescent was inspired by Felsenstein's suggestion (Felsenstein, 1992a) that uncertainty about the genealogy could be handled by summing over bootstrap replicates (the "bootstrap Monte Carlo" algorithm). Felsenstein's original suggestion proved to have a severe bias because bootstrapping imposes an inappropriate prior on the distribution of genealogies, as was first pointed out by Richard Hudson (personal communication) and noted in Kuhner et al. (1995), but attempts to overcome this problem led directly to the development of the first CS genealogy sampler. Bayesian CS samplers are very similar to Bayesian phylogenetic algorithms, but developed fairly independently.

The advantage of the CS approach is that genealogies close to the current one have a reasonable chance of fitting the data well even in the vastly increased search space induced by a finite-sites mutational model. The corresponding disadvantage is that areas of the search space distant from the current point may be missed unless the sampler is run for an extremely long time. In the worst case, different "peaks" in the search space might be separated by "valleys" where $P(D|G)$ was zero; such valleys would be an impassable barrier to the MCMC search, which would be trapped on its initial peak. As most mutational models give a non-zero $P(D|G)$ to any genealogy with non-zero branch lengths, this worst case should not be encountered in practice. However, near-worst cases with deep valleys do occur with real data.

A detailed look at the ARG-rearrangement algorithm used by *RECOMBINE* is presented in Section 10.4 of this chapter. *RECOMBINE* is no longer distributed as it has been supplanted by *LAMARC* (Kuhner, 2006), which uses the same search strategy but allows a wider range of other evolutionary forces. *LAMARC* is capable of both maximum likelihood and Bayesian inference, and in its Bayesian mode could almost be considered a Bayesian ARG inference tool. (*recom* and *RECOMBINE* are likelihood based.) However, while all of these programs sample ARGs, none of them readily make the ARGs available to the user. (*LAMARC* can be made to do so but the capability is undocumented.) Instead, they focus on inference of population parameters such as the rate of recombination. There is no conceptual reason why the ARGs sampled by these programs could not be treated as a cloud of ARGs for phylogenetic-inference purposes. However, practical tools needed for this undertaking are underdeveloped, as will be discussed in Section 10.5.

The approach used in *RECOMBINE* and *LAMARC* is computationally expensive for long sequences. A variant of the CS approach has been developed by Wang and Rannala (2008), (expanded in Wang and Rannala (2009)) and is available as the program *InferRho*. They consider only the genealogies of the variable sites (SNPs). When SNPs are well-spaced, as in the human genome, this simplification greatly reduces the search space. Some recombinations need not be tracked because they do not affect the genealogy of any SNP, and the locations of recombinations are abstracted to being between adjacent SNPs, reducing the number of potential recombinations that must be tried during the MCMC run. No measurements have yet been made of the information loss involved in this abstraction, but it should be modest for human-like data. *InferRho* is therefore more suitable than *LAMARC* for large-scale data with sparse SNPs.

Another attempt to speed up CS sampling has been developed by Brendan O'Fallon in the distributed, but as yet unpublished program *ACG* (O'Fallon,

2011). *ACG* differs from other recombinant genealogy samplers in that it does not restrict the ARG to ancestral lineages only, but incorporates all lineages which give rise to any part of any ancestral lineage. This leads to huge ARGs, but enormously simplifies the rearrangement algorithm, so that more rearrangements can be tried in the same amount of computer time. The ACG approach performs well when recombinations are rare, but may struggle with long sequences or high recombination rates.

Gene mapping. The other subfield in which recombinant inference has been of interest is gene mapping from population data samples with associated phenotypic data. Interval trees that are highly concordant with the phenotypic data are good candidates for possession of a trait-affecting locus. This possibility has been exploited in several ways. Smith and Kuhner (2009) used *LAMARC* to infer recombinant genealogies without use of the phenotypic data and then measure fit of the phenotypic data to each interval tree, leading to a relative score for presence of the trait locus in each interval tree. They also proposed an alternative method in which the location of the trait locus is initially assigned at random and then updated as a step of the Markov chain Monte Carlo procedure; however, this strategy was found to be much slower than the previous one and to offer no clear advantages.

To reduce the scope of the problem, other groups (Morris et al., 2002; Zollner and Pritchard, 2005) have explored the "shattered coalescent," in which only lineages ancestral to a specific short region of the chromosome are explored, so that as the coalescent genealogy is traced backward in time segments that drop out of consideration as they lose connection with the target region. This results in a score for presence of the trait locus in the target region, rather than scores globally across the entire sequence, but can be repeated with different hypotheses for the trait location. Shattered coalescent models focus on the target region and are thus not informative about the ARG as a whole, though information on the whole ARG could be built up via a sliding-window approach. The *GeneRecon* program (Mailund et al., 2006) implements shattered-coalescent fine-scale mapping.

Table 10.1 shows programs which, in some sense, infer ARGs from data.

Approximate methods. Searching the full space of possible ARGs is a challenging undertaking. An approximate approach to the recombinant ARG is the composite likelihood method of *LDhat* (McVean et al., 2002), based on work by Hudson (Hudson, 2001). This method considers adjacent pairs of sites and builds up an approximate likelihood based on treating successive pairs as independent. *LDhat* is widely used as an estimator of the local recombination rate because it can handle much larger datasets than any known full-ARG method. However, it can offer little information about the ARG itself.

Programs such as *PHASE* (Stephens et al., 2001) infer haplotypes by minimizing recombinations, and there is some implicit conception of an ARG in this task, but they do not even approximately infer the ARG.

Approximate Bayesian computation (ABC) methods. For many evolutionary models it is fairly easy to simulate trees (or ARGs) under a

Table 10.1: Software which infers or approximately infers a cloud of ARGs.

Program name	Citation	Sampler type
recom	Griffiths and Marjoram (1996)	IS likelihood
fins	Fearnhead and Donnelly (2001)	IS likelihood
RECOMBINE	Kuhner et al. (2000)	CS likelihood
LAMARC	Kuhner (2006)	CS likelihood or Bayesian
InferRho	Wang and Rannala (2008)	CS Bayesian
ACG	O'Fallon (2011)	CS Bayesian
LDHat	McVean et al. (2002)	Composite likelihood
GeneRecon	Mailund et al. (2006)	Shattered coalescent
LATAG	Zollner and Pritchard (2005)	Shattered coalescent
BEAST	Heled and Drummond (2010)	CS Bayesian

given set of parameters, but very difficult to infer the tree or ARG of an actual dataset. This advantage of simulation over inference gives rise to the ABC approach, in which many trees are simulated and summary statistics collected from them are compared to summary statistics on the actual data. ABC methods in population genetics are discussed by Beaumont et al. (2002).

I am not aware of ARG-based ABC algorithms, but the means for building them are certainly available. ARGs can be simulated by a variety of programs including the general-purpose simulators *ms* (Hudson, 2002) and *SPLATCHE2* (Ray et al., 2010) and several simulators which add some type of natural selection including *SelSim* (Spencer and Coop, 2004) for directional selection and the simulator of Nordborg and Innan (2003) for balancing selection. ARG simulation has been scaled up to whole-chromosome levels in the *GENOME* program of Liang et al. (2006).

The key to success with the ABC approach would be finding appropriate summary statistics. Existing summary-statistic estimators of the recombination rate are not very satisfactory as stand-alone estimators (Wall, 2000), and it is not clear that they would perform any better in an ABC.

10.4 Mechanics of sampling ARGs

This section delves into the technical issues of rearranging an ARG so as to write a Bayesian phylogeny or genealogy sampler. It is directed to readers who wish to implement such algorithms or understand them in detail, and may be of limited interest to the general reader.

Rearranging the ARG topology. The basic tradeoff of ARG rearrangement is which lineages to exclude from the ARG. The more lineages that can be excluded, the smaller and simpler the ARG will be, but the more complex the rearrangement algorithm will become. O'Fallon (2011) has experimented with a completely non-restrictive view of the ARG, permitting a very simple rearrangement algorithm. This approach is successful when recombination is rare, but seems likely to produce an intractably large ARG in cases with high recombination. In an application of his method (O'Fallon, personal communication), he breaks the human X chromosome into overlapping windows in order to keep the ARGs manageable, a strategy developed by Wang and Rannala (2009).

As soon as one tries to limit the lineages admitted into the ARG, rearrangement becomes complex: a tipward change may affect the validity of structures rootward of it. This can be visualized by imagining connection to the observed tip data as water flowing down from the tips toward the root and being shunted about by recombinations. Cutting a connection near the tips can cause rootward sections of the tree to "dry up" so that they no longer explain any of the observed data. Such lineages violate the standard ARG definition and should therefore not be present in the ARG.

It seems tempting to apply a rearrangement algorithm that ignores this issue and then reduce the result to a legal ARG by pruning unwanted lineages. However, the distribution of ARGs that results from beginning with a reduced ARG, applying a rearrangement that assumes non-reduction, and then reducing the result is not the same as the distribution that results from maintaining the reduction at all times. Unless this discrepancy could be quantified and compensated for by an acceptance/rejection term, the sampler will not recover the desired distribution.

This is one of countless examples of attractive-looking rearrangement schemes that violate essential rules of MCMC. In this case, the rule being violated is that every step must be *reversible:* if you can go from ARG A to ARG B in one step, you must be able to go back in one step. I will work through a sample rearrangement algorithm, taken from Kuhner et al. (2000), to show how challenging this can be in practice.

Kuhner et al. (2000) began with a non-recombinant proposal algorithm which could be termed "branch cutting and resimulation" (BCR), first proposed by Beerli and Felsenstein (1999). A branch of the tree is chosen at random and erased up to its tipward end. It is then simulated downward until it reconnects to the remaining structure of the tree (the "residual tree"). This simulation is guided by the current parameter values of the sampler: for example, the probability that the resimulated lineage will coalesce in a given time interval depends on the number of lineages available for it to coalesce with and the current value of the drift parameter $\Theta = 4N_e\mu$. Its coalescence partner, when it does coalesce, is chosen at random.

This algorithm can be applied almost unchanged to cases which model migration or similar forces. One complication arises when migration is allowed (and will also appear when recombination is allowed). Migration events appear as nodes in the tree with one ancestor and one descendant; they mark the point at which a lineage changes its population membership. Migration nodes are not placed on the root lineage, as such migrations could never be detected in the data. However, when another lineage is cut and resimulated downward past the original location of the root, part of the root branch will become an interior branch in the tree. This new interior branch must have a fair chance of containing migrations. Thus, when the old root position is passed, the root branch must now be treated as a branch undergoing resimulation (Beerli and Felsenstein, 1999). Failure to do this will result in trees with an asymmetrical distribution of migration events on the two bottommost branches, a subtle and difficult-to-detect MCMC failure.

To apply the BCR algorithm to an ARG (creating a recBCR algorithm) we must first decide which lineages will be included in the ARG: all lineages ancestral to any part of the sampled sequences, or only lineages tipward of the interval-tree roots (the FC optimization). I will consider the non-FC case for simplicity, but it is not simple enough to avoid many knotty details!

As in BCR, recBCR begins by choosing a branch at random to cut. A problem immediately arises: whereas in the non-recombinant case the number of branches in a tree is fixed, an ARG with more recombinations has more total branches. This means that if we choose a specific branch, cut it, and obtain a more recombinant ARG as our result, the reverse change will be less probable as it will require choice of that specific branch out of a larger number of branches. We need a Hastings term (Hastings, 1970) to compensate for this unfairness, or the sampler will move inexorably toward more and more recombinant genealogies. The Hastings term is simple: the acceptance probability is multiplied by the ratio of cut-eligible branches in the old and new ARGs.

Which branches are eligible? It is meaningless to cut the root branch, as this would only result in recombinations arising below the root, where they are not of interest. Any other branch can be cut: following Beerli and Felsenstein (1999) we ignore migration nodes in defining branches for this purpose.

Once the branch is chosen, it is erased downward. In BCR it is erased until it reaches its coalescence; in recBCR we must decide what to do if we encounter a recombination first. Such recombinations are meaningless as they no longer have any path to the sampled tips, so we will remove them, too. This means that we may remove multiple coalescences: at least one, plus one more for every recombination encountered. An example is shown in Figure 10.4.

We now have a residual tree and a dangling lineage. The dangling lineage is resimulated downward according to the coalescent with recombination (Hudson, 1983). It may coalesce with any existing lineage, or it may undergo a recombination (splitting) event to produce two dangling lineages which may now coalesce with existing lineages (including each other) or may split further. This resimulation process continues downward until there are no more dangling lineages. As with the BCR with migration, if this process passes the old location of the root, the root branch must be added to the list of dangling lineages (otherwise the old root branch, now an interior branch, would be unnaturally devoid of recombinations).

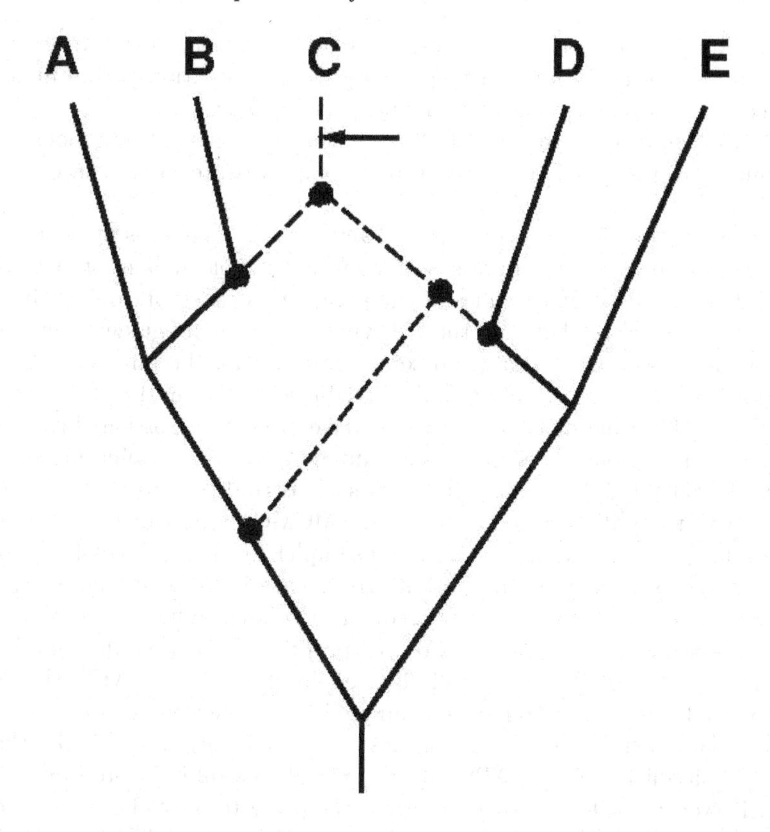

Figure 10.4: Lineages removed during branch cutting. The arrow shows the initial target branch. Dotted lines indicate branches to be removed and circles indicate nodes to be removed.

During the resimulation process, we never insert an event which is inappropriate under our model of an ARG which contains only branches ancestral to some part of a sampled tip. It may be surprising, therefore, to find that the resulting ARG sometimes contains inappropriate events. This happens because of the cutting process: when a branch is cut, branches rootward of it may lose their last connection to the tips and therefore their justification for inclusion in the ARG.

What should we do with these inappropriate events? The obvious solution is to remove them, leaving a fully legal ARG. The authors of RECOMBINE spent a painful summer wondering why this procedure did not create a correct Markov chain. The problem is violation of reversibility. We are allowing ourselves to remove branches which are not part of the resimulation path. Suppose that we have made a rearrangement which orphans a particular branch far from the resimulation path, and have removed it. We may now want to make a single step which will return us to the original ARG, as required by the MCMC reversibility property. But we cannot: our rearrangement process, as described so far, never inserts a branch except along the resimulation path, but it has allowed itself to *remove* a branch distant from that path.

A naive implementation of recBCR therefore removes recombinations which it is unable to restore. This causes a directional pressure toward fewer recombinations.

A powerful test for the correctness of any MCMC sampler is to see whether it can recover its own prior. In this case, we would sample millions of rearranged ARGs, holding the driving values of the parameters constant, and setting our data to "unknown" so that only these driving values should guide which ARGs appear in the sample. For a given set of parameters, the number of recombinations per ARG has a predictable distribution. We can directly calculate expectations for some small cases, such as the expected proportion of trees with 0 or 1 recombinations. Expectations for any number of recombinations can be estimated using an independent simulator such as Hudson's *ms* (Hudson, 2002). A simple χ^2 test will diagnose failure of the MCMC to meet its expectations. I venture to guess that no large genealogy sampler has been successfully coded without a test such as this (my working group calls it the "stationaries test" as it assesses the stationary distribution of the Markov chain).

It is tempting to forgo removal of inappropriate branches, but that would have to be done wholesale — including all branches in the ARG that give rise to any branch ancestral to the sampled data, whether they explain any sampled data or not. O'Fallon (2011) has tried this approach and found that it trades off simplicity against ARG size and may be useful in low-recombination cases. However, for the recBCR we are attempting to limit the ARG to only lineages carrying some ancestral material, as in the original ARG definition by Griffiths and Marjoram (1997).

The clear, if unpalatable, alternative is that if the sampler can remove branches that are not on the path of resimulation, it must insert such branches as well. This requires additional bookkeeping in the ARG. Before beginning the rearrangement, we must note for each branch, which positions on it are eligible for recombination. As we resimulate downward, we must track the effect of all our cuttings and rejoinings on the list of eligible positions per branch. When a range of sites becomes ineligible, recombinations within that range are fated to be removed as already described. When a range of sites that were previously ineligible becomes eligible during the rearrangement, we must treat the branch carrying them as susceptible to new recombinations, even though it is in the residual tree and would normally not be part of resimulation. When such a recombination occurs, one part of the sequence retains its current path through the residual tree; the other part becomes a new dangling lineage and joins the resimulation process.

This algorithm has been termed "hidden passages" by Joe Felsenstein. The concept is that complex recombinational relationships involving non-ancestral material exist at all times, but are not displayed in the ARG and thus are "hidden." When a rearrangement causes previously non-ancestral material to become ancestral, we must "reveal" this hidden complexity as it is now relevant to the ARG. Since we were not storing it, the only way to "reveal" it is to resimulate it.

The recBCR algorithm thus consists of three phases: (1) Cut a randomly chosen branch and dissolve unwanted material rootward of it. (2) Resimulate downward, being sure to track newly eligible site ranges and allow them to develop new recombinations. (3) When the ARG has completely coalesced, traverse it to remove any meaningless events: recombinations and their associated coalescences which no longer carry any ancestral material, and also recombinations and coalescences which are now below the ARG root.

It can be seen that the need to resimulate the root when the resimulation process passes the bottommost node in the tree is another example of ensuring reversibility. We allow ourselves to prune events which have been stranded by upward movements of the root. We must equally insert events when the root has moved downward, or reversibility is violated. (Again, a test of the stationary distribution will quickly reveal the problem.)

The recBCR is a sufficient proposal to eventually search all of the ARG-space. Its performance may not be satisfactory, however, as it makes relatively large changes in the ARG. In particular, suppose that we have proposed a recombination which is almost correct, but a few bases too far to the left. The recBCR can fix this problem only by removing this recombination and then fortuitously inserting a new one in a better chromosomal location but with the same parental lineages. We might get better breakpoint inference by supplementing the recBCR with a proposal that changes recombination points. (For reasons of reversibility, this will turn out to be a resimulation proposal: it must track hidden passages and will sometimes have to remove old recombinations or insert new ones as a result of changing the breakpoint of an existing recombination.)

This overview of one recombination proposal does not exhaust the possibilities by any means, but it does showcase how tricky manipulating the ARG can be. An alternative approach to ARG rearrangement is not to attempt to insert or remove recombinations in proportion to their probability, but instead to insert or remove them based on a simple algorithm, and to incorporate the probabilities into the acceptance/rejection step of the Metropolis-Hastings algorithm. This approach has been used in non-recombinant genealogy samplers such as *IM* (Hey and Nielsen, 2004) but to my knowledge, the only recombinant version is unpublished (Brendan O'Fallon, personal communication).

10.5 Hazards of Bayesian inference in the presence of recombination

Sticky problem. CS recombination samplers tend, in my experience, to mix poorly. The ARG cannot nimbly change from a low-recombination to a high-recombination state or vice versa without suffering a high chance of rejection. The sampler may therefore stick at a point where the recombinations in the current ARG are a good fit to the current value of the recombination rate parameter, and be unwilling to move away from that point. This is particularly pressing if updates of recombination rate and updates of the ARG are separate, as a change in the recombination rate which poorly fits the number of recombinations in the ARG, or vice versa, will tend to be rejected even if it is a move toward the truth. This has been named the "sticky problem" by Shili Lin (personal communication).

Metropolis-coupled MCMC or "heating" (Geyer, 1991) can improve mixing substantially. An alternative is to run several searches from different starting points ("replication") and combine the results. In a preliminary test based on migration-rate inference, heating and replication performed about equally well and much better than unaugmented searches (Peter Beerli, personal communication).

Much work remains to be done in developing additional techniques, such as alternative proposal mechanisms, to reduce stickiness of the search. In current samplers, this stickiness is quite dangerous. If the area of ARG-space searched by a given run is not adequate, not only will the most probable estimate (MPE) be potentially unreliable, but the support intervals will be systematically too narrow. All Bayesian phylogenetic methods share this unpleasant property: a too-short search, which should ideally be associated with a claim of low certainty about the results, is instead associated with a claim of high certainty as it misses potentially conflicting information. But the issue is particularly pressing for sticky-problem searches such as existing CS searches over the space of ARGs.

In my experience, runs suffering from stickiness often overestimate the recombination rate, probably due to the following scenario. The randomized starting ARG fits the data poorly. Its deficiencies are quickly plastered over with numerous recombinations that improve the fit. The sampler is then slow to find cleaner, less-recombinant solutions that may lie quite far away in ARG space.

IS samplers, which do not use the previous ARG to guide generation of the current one, avoid the sticky problem but encounter problems of their own. The space of possible ARGs is so large that an independent-sampler search may scarcely ever visit the regions containing valuable solutions. If a way to direct attention to the correct parts of ARG space were found, IS samplers could become the preferable approach to ARG inference. Fearnhead and Donnelly (2001) have explored improved sampling distributions based on approximation of the optimal sampling distribution, a promising approach that has not received much attention. The performance of IS samplers could also be improved by incorporation of the sequential Markovian coalescent approximation of McVean and Cardin (2005), which reduces the expected number of recombinations in the ARG substantially. It is much easier to see how to generate Markovian ARGs *de novo* in an IS sampler than it is to see how to rearrange them while maintaining the MCMC properties needed for a CS sampler.

Reliance on the prior. All Bayesian inference methods are vulnerable to bad priors. Methods which incorporate recombination share this vulnerability. If the true value of the recombination parameter is outside the bounds of the prior, it cannot be recovered. More subtly, if the prior includes the truth but is unreasonably narrow, the support intervals will be unreasonably narrow as well. It is tempting to therefore choose very wide "uninformative" priors. However, if the prior is too wide the MCMC search becomes extremely inefficient and may not recover a good estimate in a tolerable amount of time.

Many datasets have limited information about the recombination rate parameter. In such cases, the posterior is heavily influenced by the prior. This is not "wrong" but can be surprising and potentially misleading. In particular, for any given amount of data there is a minimum recombination rate which corresponds to no practical chance of recombination; it is effectively zero. A prior with much of its weight below this minimal value is, in effect, a prior with a high weight on zero, and will behave as such. In my experience, the researcher often did not mean to use a zero-weighted prior, but simply did not realize what the meaningful lower bound of the recombination rate was.

It is helpful to know that when $4N_ec$ (four times the effective population size times the recombination rate per inter-base link) is less than 1/(sequence length-1), the number of expected recombinations in the sequence is less than 1 (Robertson and Hill, 1983). (In *LAMARC* and similar programs, multiply estimated Θ by estimated recombination rate to obtain $4N_ec$.) Values of the recombination rate much below this are therefore practically equivalent to zero (and to each other). Log priors which extend several orders of magnitude below this effective zero lead to particularly misleading estimates which may reject the actual value generating the data. (This problem occurs with inference of other parameters as well, particularly migration rate.)

In general, Bayesian inference is vulnerable to poorly chosen priors. It is important to examine the prior and posterior of any Bayesian run and to regard with skepticism any posterior which appears similar to its prior, unless the prior is much better founded than is usual in phylogenetics.

Methods such as *LAMARC, InferRho,* and *recom*, which rely on explicitly representing recombination events in an ARG, show a practical vulnerability to too-high upper boundaries on the recombination rate prior. Recall that a large proportion of recombinations are cryptic or silent; the data can provide no basis for accepting or rejecting these. Hyper-recombinant ARGs which nonetheless fit the data very well are therefore readily available even when the data actually come from a scenario with little or no recombination. A sampler's demands for time and space rise with the number of recombinations in the ARG, and searching areas of ARG-space containing massively recombinant ARGs can cause the programs to bog down or even crash. Bayesian methods have an advantage over likelihood-based ones in this regard as the prior can be used as a practical tool to restrain the search. In my own research, I have seen cases that do not run successfully in maximum likelihood mode but can be coaxed into running with a carefully chosen prior. Of course, the prior that leads to optimal performance is not guaranteed to be a prior containing the truth!

10.6 Directions for future research

Missing tools. The field lacks several basic tools for analysis of ARGs. There is no agreed-upon format for communicating an ARG. Simulation programs such as *ms* (Hudson, 2002) output the interval trees in Newick format, but the ARG often cannot be uniquely reassembled from its interval trees, and even when it can, no efficient algorithm for doing this is known. It would be preferable to output the ARG in its entirety. Existing formats for representing directed acyclic graphs may be appropriate; for example, the Gene Ontology database (Gene Ontology Consortium, 2000) offers several DAG-output formats. A recent proposal (McGill et al., 2013) defines an ArgML standard based on GraphML (Brandes et al., 2002) but it is too early to tell if this will be widely accepted.

A more severe problem is that there is no way to compare ARG topologies analogous to distance metrics for ordinary trees. Lacking this, we can only say that one ARG-reconstruction algorithm is better than another at recovering the ARG topology when one succeeds completely and the other does not. It is well known in phylogenetics that asking for complete success restricts attention to trivial trees only, and the situation is no better with ARGs.

The computer-science literature contains many proposals for distance metrics between arbitrary directed acyclic graphs (DAGs). These are reviewed in Arenas et al. (2008), which concludes that all of them are unsatisfactory (most require unreasonable constraints on the structure of the DAG, such as no two hybridizations occurring without an intervening coalescence). For the special case of the ARG, it should be possible to do better by comparing interval trees. However, most extant studies of ARG reconstruction appear to focus on recovery of the breakpoints, rather than recovery of the ARG topology.

ARGs are also difficult to display. Many programs will display topological networks, but these lack information about the breakpoint position. Brendan O'Fallon (personal communication) has created an ARG-visualization tool that can show the ARG for a window onto the sequence, which helps to keep the image comprehensible. As an interactive tool, this is successful, but the question of how to display an ARG for publication remains. Example ARGs in books such as this one are engineered to be easily drawn, but real ARGs with more than a few recombinations generally require numerous crossing branches.

Finally, Bayesian phylogenetics makes heavy use of the construction of consensus trees, but no algorithm is known for making a consensus ARG. One can make a consensus of the interval trees for each interval, but it will generally require far more recombinations than the ARGs from which it is derived. A consensus of the breakpoints is easy to create, but for researchers interested in the actual ARG, it is frustrating not to be able to summarize features of the cloud of ARGs. If a consensus ARG is not feasible, it would at least be helpful to have software which can take a cloud of ARGs and answer questions such as "What is the posterior probability that this sequence is recombinant?" and "If it is recombinant, what are the most probable parents?"

Simulation of the ARG. Several programs create simulated ARGs, usually under a backward-time coalescent model. Among them are the *ms* program (Hudson, 2002) and its derivative *msHOT* (Hellenthal and Stephens, 2007) and *mbs* (Teshima and Innan, 2009); *SelSim* (Spencer and Coop, 2004); *fin* (McVean et al., 2002); and *Recodon* (Arenas and Posada, 2007). These tools have been used to test success of recombination rate and breakpoint location inference, and could also be used to test the success of ARG inference if the problems described above were solved. They could also form the basis for an ABC algorithm.

Priors for inference beyond the population level. An obvious question is raised by the presence of this chapter in a book on Bayesian phylogeny reconstruction: Is a Bayesian phylogeny-inference program incorporating recombination feasible?

LAMARC, InferRho, and *ACG* are coalescent-based ARG inference programs (or at least would be if they output ARGs along with their parameter estimates). Their only substantive difference from a standard Bayesian inference program such as MRBAYES (Huelsenbeck and Ronquist, 2001; Ronquist et al., 2012b) is use of the recombination-enhanced coalescent as a prior distribution on ARGs. This is appropriate for a within-population or recent-speciation case where the coalescent prior can be defended, but becomes increasingly questionable as the problem moves away from the population level. I have seen unpublished results of an attempted use of LAMARC to infer rates of horizontal gene transfer between plants and bacteria. The coalescent prior seems poorly justified here, but what could be substituted?

Realism of the recombination model. Modeling the recombination process in terms of single crossovers is clearly an oversimplification. Practically all organisms which have recombination also have gene conversion, and therefore apparent "double recombinations" may actually be single conversions. However, on a practical level, the difference in kinetics may be unimportant, given the low statistical power of current recombination estimators.

A more worrisome situation arises in organisms with a circular genome, where only double recombinations or conversions are viable. The sampler needs to search the space containing only double recombinations. While, in the limit, a single-recombination sampler should be able to do so, it may be hopelessly inefficient. For this reason, an explicit gene conversion model would be a good addition to the software repertoire.

Gene conversion is a minor violation of the simplifying rule that recombinations occur one at a time. Organisms such as HIV-1 can violate that rule much more dramatically: circulating recombinant forms are usually a patchwork suggestive of three or more simultaneous recombination events (for example, Gao et al. (1996)). No existing methods model this well.

Inference of recombinational hotspots. Both population-level analyses and sperm typing confirm that recombination is not uniform across the chromosome in many species, but instead clusters in recombinational hotspots. Many existing methods ignore this, assuming a constant recombination rate across the sequence. The LDHat composite likelihood method of McVean et al. (2002) is a notable exception.

If information about hotspots is externally available, say from sperm typing, it could be incorporated into existing genealogy samplers fairly readily. Trying to co-infer hotspot structure and the ARG is more challenging. Such a sampler, based on *RECOMBINE*, was developed by Kang (2008) in his thesis work. He used a hidden Markov model of the underlying hot/cold structure: the program used a Markov chain Monte Carlo approach to sum over many possible hotspot structures while inferring parameters such as hot recombination rate, cold recombination rate, and the transition rates between hot and cold along the chromosome (which determine the mean lengths of hot and cold regions). This model is promising in theory but showed severe stickiness as well as issues with identifiability of its parameters. If the hot and cold rates become too similar, the transition rates can no longer be inferred; there are also configurations of the transition rates at which the whole sequence is effectively either hot or cold, preventing any inference about the other state. A Bayesian approach might outperform Kang's maximum likelihood approach as priors could be used to restrain the parameters to regions where they are identifiable.

10.7 Open questions

A number of existing questions in the literature would be amenable to recombination-aware Bayesian methods. A sampling is given here.

HIV-1 history. Researchers have attempted to trace the early history of the HIV-1 epidemic via phylogenies of HIV-1 subtypes. This attempt is complicated by the high recombination rate of HIV-1. Early in the epidemic there was probably limited opportunity for between-subtype recombination (within-subtype recombination is ubiquitous but does not disrupt phylogeny inference as much), but as the virus became more widespread, mixed infections allowed between-subtype recombinations, and the resulting circulating recombinant forms are now major drivers of the epidemic in many countries. Methods such as bootscanning and use of a Bayesian genealogy sampler in a sliding window were used by Abecasis et al. (2007) to reach the conclusion that "subtype G" is actually a recombinant and that one of the currently classified recombinants is its parent, not its offspring. However, this study, while persuasive, lacks a strong statistical basis. The question "Is subtype G a recombinant?" is tailor-made for a full recombination-aware Bayesian analysis, but would require facilities for testing a large collection of ARGs to see how many contained this feature. Such an analysis might need to take special account of the high probability that multiple recombinations occur in a single burst during a rare co-infection, rather than occurring at a uniform rate over time (e.g., Jung et al., 2002).

Coadaptation in HLA genes. It has been suggested that the high linkage disequilibrium seen across the human leukocyte antigen (HLA) gene cluster is due to natural selection favoring coadapted sets of alleles. If this is true, observed recombinations in modern sequences should tend to be recent rather than ancient, by the reasoning of O'Fallon et al. (2010): ancient unfavorable recombinations will have been eliminated by selection, whereas recent ones may still persist. This could be tested via a Bayesian skyline plot of recombinations analogous to the skyline plots of migrations offered by *MIGRATE-N* (Beerli and Felsenstein, 2001a).

Does mtDNA undergo recombination? There is an ongoing controversy (reviewed by Eyre-Walker and Awadalla, 2001) over whether the rate of recombination in human mtDNA is actually zero, as is assumed by many studies. A Bayesian analysis with a carefully constructed mutational model might be able to put bounds on the possible values of the recombination rate, though it is troublesome that the value of interest is at the extreme edge of the prior. Such an analysis would require a good model for the double recombinations characteristic of circular genomes; development of such a model would also permit inference on a variety of bacterial and viral species with circular genomes or chromosomes.

10.8 Conclusions

The ancestral recombination graph is a recalcitrant creature which no method can hope to recover with high accuracy. ARG inference is tailor-made for Bayesian phylogenetic methods, as a cloud of inferred ARGs is a much better match to what can be known about the history of the data than any single ARG can ever be.

Good progress has been made in simulating and sampling ARGs, but almost all programs doing so are directed toward parameter analysis or gene mapping, treating the ARG as a nuisance parameter. Actual inference of the ARG is greatly hampered by the absence of basic statistical tools such as ARG distance measures and ARG consensus graphs.

Genealogy-sampler methods, many of them Bayesian, have been fairly powerful for parameter inference (Wall, 2000) and may also be powerful for inference of the ARG itself. These computationally intensive algorithms generally struggle with long sequences. Progress in this area often involves finding ways to reduce the search space: simplifying the mutational model (Griffiths and Marjoram, 1996), considering only recombinations with potential (Hudson, 2002) or actual (Wang and Rannala, 2008) detectable effects, approximating the ARG with a Markovian process (McVean and Cardin, 2005), or tracing only lineages ancestral to an area of special interest (Morris et al., 2002). Less success has been reported for attempts to improve mixing by varying the proposal distributions. It is quite difficult to create even a single correct proposal algorithm for recombination, let alone a collection of them.

As an alternative to genealogy samplers, approximate Bayesian computation (ABC) approaches seem plausible but have not been explored in the recombinant context despite the ready availability of simulation programs. The key issue for ABC design will likely be the choice of appropriate summary statistics.

Use of recombination-aware phylogenetic methods beyond the population level is an almost untouched field. In general, inference and analysis of ARGs are wide open for future innovation, and are likely to become increasingly important as the significance of recombination, reassortment, gene transfer, and other reticulate processes in evolution is increasingly appreciated.

Acknowledgments

I thank Peter Beerli, Joe Felsenstein, Jim McGill, Bob Giansiracusa, and Jon Yamato for helpful comments on the manuscript, Jim McGill and Joe Felsenstein for assistance with figure preparation, and Joe Felsenstein for identifying sources in the literature. This work was supported by NIH grant R01HG004839 and NSF grant 0814322 (to MKK).

11

Bayesian nonparametric phylodynamics

Julia A. Palacios
Department of Statistics, University of Washington, Seattle, Seattle, Washington, USA

Mandev S. Gill
Department of Biostatistics, University of California, Los Angeles, Los Angeles, California, USA

Marc A. Suchard
Departments of Biomathematics, Human Genetics, and Biostatistics, University of California, Los Angeles, Los Angeles, California, USA

Vladimir N. Minin
Department of Statistics, University of Washington, Seattle, Seattle, Washington, USA

CONTENTS

11.1 Introduction

Changes in population size affect the variability of population gene frequencies in natural populations, allowing genetic variation in present-day and recent-past molecular sequence data to help recover the more distant past demographic history of the population. This variability also enables researchers to examine the factors driving past population dynamics and to establish molecular surveillance of emerging infectious diseases. For example, Campos et al. (2010) analyze ancient and modern musk ox mtDNA samples dated from 56,900 radiocarbon years old to present and recover the population dynamics throughout the late Pleistocene to the present; 63 RNA sequences of hepatitis C virus (HCV) obtained in 1993 effectively reveal the dynamics of HCV infections in Egypt over the past century (Pybus et al., 2003); and human influenza A/H3N2 subtype sequences sampled over a 12-year period in New York state return estimates of the seasonal population dynamics of human influenza A/H3N2 (Rambaut et al., 2008).

In 1931, Sewall Wright introduced the concept of the *effective population size* of a population (Wright, 1931). The study of effective population sizes has grown into a central theme in population genetics ever since. Molecular epidemiologists often employ estimates of effective population size to approximate census population size by incorporating knowledge about the expected number of molecular sequence substitutions (e.g., DNA substitutions) per calendar time unit along the inferred genealogy, generation time in calendar units, and the population variance in number of offspring (Wakeley and Sargsyan, 2008). However, even when such prior information is available, interpreting estimates of the effective population size remains challenging, especially in studies of infectious diseases (Frost and Volz, 2010).

Initially, most studies focused on two summary statistics of a multiple alignment of molecular sequences — the number of segregating sites (Watterson, 1975) and the mean number of nucleotide differences between two sequences in a sample (Tajima, 1983) — to quantify the effective population size. However, with the introduction of coalescent theory (Kingman, 1982a) and the coalescent with variable population size (Slatkin and Hudson, 1991; Griffiths and Tavaré, 1994), the genealogical relationships among the sequences in a sample have begun to inform estimates of effective population size and its dynamics over time.

The coalescent provides a probability model that describes the relationship between the coalescent times in a gene genealogy and the effective population size (Nordborg, 2001; Hein et al., 2005). Some coalescent-based methods assume that a single genealogy is available (Fu, 1994; Pybus et al., 2000b) and others produce estimates of effective population size trajectories directly from molecular sequence data through an unknown genealogy (Kuhner et al., 1995; Drummond et al., 2002, 2005; Minin et al., 2008). Methods that take into account the genealogical uncertainty more efficiently use the information present in the data (Felsenstein, 1992b); however, the achieved efficiency comes at substantial computational cost of Monte Carlo methods needed to integrate over the space of genealogies. When molecular sequences contain sufficient phylogenetic information, the computationally expensive Monte Carlo can be omitted in favor of inferring the population size trajectory from a single estimated genealogy (Pybus et al., 2000b; Minin et al., 2008). However, such a two-step estimation procedure can lead to substantial underestimation of uncertainty in population size estimates and, therefore, should be used with caution (Minin et al., 2008).

When all molecular sequence data are sampled at the same time under an *isochronous sampling* scheme, it is possible to estimate $\theta = 4N_e\mu$, where the effective population size N_e is measured in units of generations in a diploid population and μ represents the substitution rate per site per generation. If one has access to an independent estimate of μ via previous studies or from other sources, one can estimate N_e directly, otherwise N_e and μ remain confounded. Felsenstein and Rodrigo (1999) extend the coalescent model to incorporate genealogies with sequence data sampled at different times under a *heterochronous sampling* scheme. Here, the sampling times of noncontemporary sequences can provide information about μ and help to identify N_e and μ separately (Rambaut, 2000; Pybus et al., 2003). Such serially sampled data are common in studies of analyses of ancient DNA and rapidly evolving viruses.

Scientific interest often lies in the changes of the effective population size over time or, in other words, in the effective population size trajectory, $N_e(t)$. Most inferential tools for estimating such a trajectory assume a simple *parametric* form of $N_e(t)$, such as exponential or logistic growth. Maximum likelihood (Griffiths and Tavaré, 1994; Kuhner et al., 1998) and full Bayesian (Drummond et al., 2002) approaches provide estimates of the parameters that characterize these functional forms. However, for poorly studied populations, a simple parametric form remains difficult to justify and more flexible *nonparametric* methods are preferable.

Over the last 15 years, the development of nonparametric methods to infer $N_e(t)$ has blossomed. A common characteristic of most of these methods is an underlying assumption of a piecewise constant or linear trajectory describing $N_e(t)$. Early methods, such as the *skyline plot* (Pybus et al., 2000b) and its regularized version, the *generalized skyline plot* (Strimmer and Pybus, 2001), provide fast but noisy estimates of $N_e(t)$ from a fixed genealogy. Drummond et al. (2005), who call their method the *Bayesian skyline plot*, and Opgen-Rhein et al. (2005) introduced more sophisticated multiple change-point models to estimate population trajectories in a Bayesian framework. The most popular implementation of the Bayesian skyline plot, available in the Bayesian Evolutionary Analysis by Sampling Trees (BEAST) software package (Drummond et al., 2012), starts from the sequence data, models $N_e(t)$ as a piecewise constant function, and assumes *a priori* a fixed number of changes in height (change-points). Opgen-Rhein et al. (2005) propose a similar model, but these authors infer the number of change-points simultaneously with other model parameters. However, in contrast to Drummond et al. (2005), Opgen-Rhein et al. (2005) condition on a single genealogy.

Recently, Minin et al. (2008) and Palacios and Minin (2013) proposed and implemented Bayesian nonparametric approaches that rely on Gaussian processes (GPs) for prior specification of $N_e(t)$. GP-based nonparametric methods enjoy a long and successful history in spatial statistics (Cressie, 1993) and machine learning literature (Rasmussen and Williams, 2006), but GP-based inference started to appear in the evolutionary genetics literature only recently. In the context of effective population size estimation, GP-based methods allow for more flexible prior specification than previous approaches based on piecewise continuous prior formulations. The Bayesian skyride model of Minin et al. (2008) *a priori* assumes that the population size trajectory follows a discretized log-Gaussian process, while the continuous time GP-based method of Palacios and Minin (2013) puts a continuously defined exponential Gaussian process prior on the population size trajectory.

The main goal of this chapter is to provide a general overview of modern Bayesian nonparametric methods for inference of effective population size trajectories. First, we formulate the problem of effective population size estimation in general terms within a Bayesian framework. We then develop a notation that allows us to work with both isochronous and heterochronous sampling, and unifies the presentation of multiple change-point and Gaussian process models. Next, we show that coalescent-based estimation of population size trajectories can be thought of as estimation of an intensity function of a temporal point process. This point process representation allows us to make explicit connections between Bayesian nonparametric phylodynamics methods and Bayesian nonparametric estimation of an intensity of an inhomogenous Poisson process. This connection is important, because borrowing statistical and computational techniques from the point process literature should be fruitful for extending Bayesian nonparametric phylodynamics models in the future. To illustrate the performance of Bayesian nonparametric phylodynamics, we analyze simulated and real data. We conclude by describing ongoing and future work on extensions of Bayesian nonparametric methods for estimation of population size trajectories.

11.2 General model formulation

11.2.1 Likelihood for sequence alignment

Let $\mathbf{Y}$ denote an $n \times L$ sequence alignment matrix, where n represents the number of individuals randomly sampled from the population of interest and L refers to the length of the sequence alignment. The sequence alignment usually represents DNA or RNA sequences from a protein coding region or a "gene." We assume that sites within the alignment are fully linked, meaning that there is no recombination possible between the sequences. This last assumption implies that we can postulate an existence of a genealogy/phylogeny $\mathbf{g}$, in the form of a rooted bifurcating tree, that represents ancestral relationships among the sampled individuals. We assume that sequence alignment $\mathbf{Y}$ is generated by a substitution process, defined by a parameter vector $\mathbf{m}$, acting on the genealogy $\mathbf{g}$. This construction yields the following likelihood function:

$$P(\mathbf{Y} \mid \mathbf{g}, \mathbf{m}). \tag{11.1}$$

Although specifics of the substitution process are not important for the general model formulation, it is commonly assumed that at each alignment site, substitutions occur according to a continuous-time Markov chain. In such cases, the likelihood function (11.1) is referred to as the Felsenstein likelihood (Felsenstein, 1981).

11.2.2 Coalescent prior

We proceed in a hierarchical Bayesian framework by putting a coalescent prior distribution on the genealogy $\mathbf{g}$. When all sequences are sampled at effectively the same time (*isochronous sampling*), this prior becomes

$$P(\mathbf{g} \mid N_e(t)) \propto \prod_{k=2}^{n} \frac{C_k}{N_e(t_{k-1})} \exp\left[-C_k \int_{t_k}^{t_{k-1}} \frac{1}{N_e(t)} dt\right], \tag{11.2}$$

where $t_n = 0$ denotes the time of sampling, $0 < t_{n-1} < ... < t_1$ are *coalescent times*, times at which two lineages meet their most recent common ancestor, and $C_k = \binom{k}{2}$ is the coalescent factor that depends on the number of lineages $k = 2, ..., n$. The prior (11.2) is a product of $(n-1)$ conditional densities of coalescent times, where each density is quadratic in the number of lineages and inversely proportional to the effective population size. Figure 11.1 shows an example of a population that experiences growth and then decay in population size. To appreciate the effect of the effective population size on the distribution of the coalescent times, consider coalescent times t_5, t_4, t_3, and t_2 in Figure 11.1. The elapsed times since the last coalescent event are long for t_5, t_4, owing to relatively large population sizes in the vicinity of these times. In contrast, coalescences occur vary fast at times t_3 and t_2 when the population size becomes small.

The *heterochronous coalescent* or *serially sampled coalescent* arises when not all sequences are sampled at the same time (Figure 11.2). In this case, the coalescent prior is

$$P(\mathbf{g} \mid N_e(t)) \propto \prod_{k=2}^{n} \frac{C_{0,k}}{N_e(t_{k-1})} \exp\left[-\int_{I_{0,k}} \frac{C_{0,k}}{N_e(t)} dt - \sum_{i=1}^{m} \int_{I_{i,k}} \frac{C_{i,k}}{N_e(t)} dt\right], \tag{11.3}$$

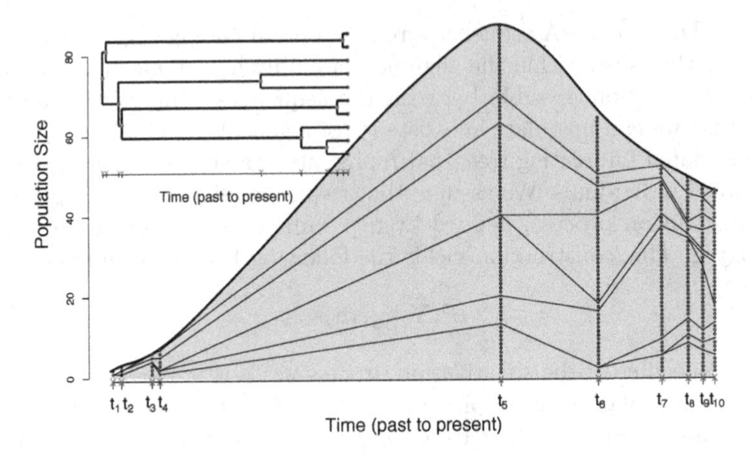

Figure 11.1: Example of a genealogy of 10 individuals (depicted as black circles in the inset) randomly sampled from the population (circles at time t_{10}). When we follow their ancestry back in time, two of the lineages coalesce at time t_9; the rest of the lineages continue to coalesce until the time to the most recent common ancestor of the sample at time t_1. The population size trajectory is shown as the solid black curve. When the population size is large (around t_5), any pair of lineages coming from time t_5 (circles at t_5) take longer to meet a common ancestor at time t_4. The figure in the top left corner shows the genealogy reconstructed by following the ancestry of the 10 individuals. It is an aligned representation of the genealogy depicted in the main plot.

where $t_n = 0 < t_{n-1} < \cdots < t_1$ denote the coalescent times as before, but the coalescent factor $C_{i,k} = \binom{n_{i,k}}{2}$ depends on the number of lineages $n_{i,k}$ in the interval $I_{i,k}$ defined by coalescent times and sampling times $s_m = 0 < s_{m-1} < ... < s_1 < s_0 = t_1$ of $n_m, ..., n_1$ sequences, respectively, $\sum_{j=1}^{m} n_j = n$. We denote intervals that end with a coalescent event by

$$I_{0,k} = (\max\{t_k, s_j\}, t_{k-1}], \text{ for } s_j < t_{k-1} \text{ and } k = 2, \ldots, n, \tag{11.4}$$

and intervals that end with a sampling event by

$$I_{i,k} = (\max\{t_k, s_{j+i}\}, s_{j+i-1}], \text{ for } s_{j+i-1} > t_k \text{ and } s_j < t_{k-1}, k = 2, \ldots, n.$$

The main difference between (11.2) and (11.3) is that the conditional density for the next coalescent time t_{k-1} is the product of the density of the coalescent time $t_{k-1} \in I_{0,k}$ and the probability of not having a coalescent event during the period of time spanned by intervals $I_{1,k}, ..., I_{m_k,k}$, where m_k is the number of intervals that end with a sampling event in $(t_k, t_{k-1}]$ (Felsenstein and Rodrigo, 1999). Lastly, we point out that the densities on the right-hand sides of equations (11.2) and (11.3) are in fact densities of the coalescent times. The corresponding genealogical densities are obtained by dropping the factors

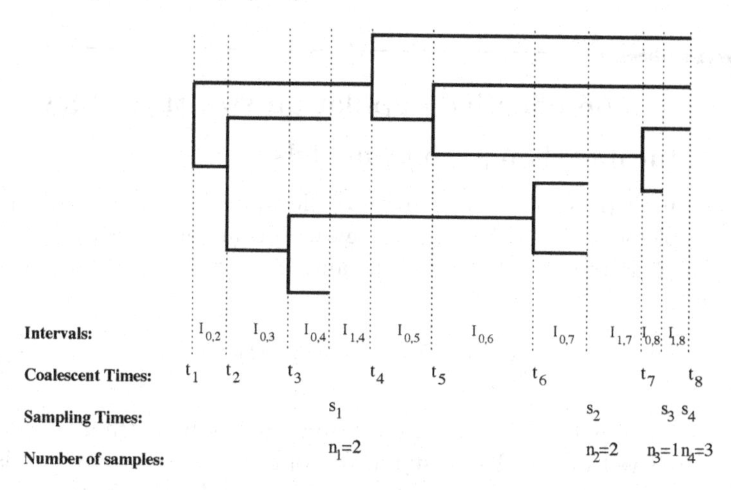

Figure 11.2: Example of a genealogy relating serially sampled sequences (heterochronous sampling). The number of lineages changes every time we move between intervals ($I_{i,k}$). Each endpoint of an interval is a coalescent time ($\{t_k\}_{k=1}^n$) or a sampling time ($\{s_j\}_{j=1}^m$). The number of sequences sampled at time s_j is denoted by n_j.

C_k and $C_{0,k}$, because for a given genealogy, we do not need to enumerate all possible orders in which lineages coalesce one pair at a time.

11.2.3 Posterior inference

We have now defined the likelihood function (11.1) and two priors for genealogies (11.2) and (11.3), corresponding to contemporaneously and serially sampled data. The next step is to define a hyper-prior $P(N_e(t) \mid \boldsymbol{\theta})$ for the effective population size trajectory with hyper-parameters $\boldsymbol{\theta}$, accompanied by their own prior $P(\boldsymbol{\theta})$. Various approaches to this prior specification will be discussed in the next section. We also need a prior for the substitution process parameters, $P(\mathbf{m})$. Such a prior heavily depends on the choice of the substitution model and usually follows standard practice in Bayesian phylogenetics (Suchard et al., 2001; Ronquist et al., 2012b).

We are now ready to define the posterior distribution of all model parameters:

$$P(\mathbf{g}, \mathbf{m}, N_e(t), \boldsymbol{\theta} \mid \mathbf{Y}) \propto P(\mathbf{Y} \mid \mathbf{g}, \mathbf{m}) P(\mathbf{m}) P(\mathbf{g} \mid N_e(t)) P(N_e(t) \mid \boldsymbol{\theta}) P(\boldsymbol{\theta}),$$

where we assume that the substitution process parameters $\mathbf{m}$ and genealogy $\mathbf{g}$ are *a priori* independent. If the genealogy is assumed to be fixed, the posterior distribution simplifies as follows:

$$P(N_e(t), \boldsymbol{\theta} \mid \mathbf{g}) \propto P(\mathbf{g} \mid N_e(t)) P(N_e(t) | \boldsymbol{\theta}) P(\boldsymbol{\theta}). \tag{11.5}$$

11.3 Priors on effective population size trajectory

11.3.1 Multiple change-point models

A Bayesian multiple change-point method, developed by Opgen-Rhein et al. (2005), is implemented for a fixed genealogy with contemporaneous data in the R package APE (Paradis et al., 2004). The authors assume a piecewise constant trajectory

$$N_e(t) = \sum_{j=1}^{k+1} \gamma_j 1_{(a_{j-1},a_j]}(t), \tag{11.6}$$

where the change-points $a_1, ..., a_k$ are *a priori* uniformly distributed in $[0, t_1]$, $a_{k+1} = t_1$ is the time to the most recent common ancestor, $a_0 = 0$ is the sampling time of the contemporaneous sequences, and for an interval A

$$1_A(t) = \begin{cases} 1 & \text{if } t \in A, \\ 0 & \text{otherwise.} \end{cases}$$

The number of change-points, k, has a truncated Poisson prior with hyperparameter $\lambda \sim \text{Gamma}(a, b)$. *A priori*

$$\gamma_j \sim \text{Gamma}(\alpha_j, \beta_j) \text{ for } j = 1, \ldots, k+1. \tag{11.7}$$

The authors approximate the posterior distribution of $N_e(t)$ and other model parameters by Markov chain Monte Carlo (MCMC) sampling.

The *Bayesian skyline plot* (Drummond et al., 2005) is implemented in BEAST (Drummond et al., 2012) to estimate population size trajectories directly from serially sampled sequence data. This method considers the following piecewise constant prior

$$N_e(t) = \sum_{j=1}^{k+1} \gamma_j 1_{(a_{j-1},a_j]}(t), \tag{11.8}$$

with $\gamma_k > 0$ for $k \leq n-2$ change-points $a_1, ..., a_k$. These change-points are an ordered subset of the coalescent times $\{t_{n-1}, ..., t_2\}$, and $a_0 = t_n = 0$ and $a_{k+1} = t_1$. That is, the effective population size changes only at some coalescent events. The number of change-points is fixed by the user. The heights of step function (11.8) follow an auto-exponential prior:

$$\gamma_j \sim \text{Exponential}(\gamma_{j-1}), \text{ for } j = 1, \ldots, k+1. \tag{11.9}$$

The first step function height, γ_1, receives an improper scale-invariant prior with density $f(x) \propto 1/x$. Drummond et al. (2005) approximate the posterior distribution of $N_e(t)$ and other model parameters, including the genealogy, using MCMC.

11.3.2 Coalescent as a point process

Before proceeding to the GP-based priors on the effective population size trajectory, we take a moment to view the coalescent as a point process. First, we notice that (11.2) and (11.3) suggest that once we extract the coalescent times from a genealogy, the genealogy does not provide any further information about the effective population size dynamics. Next, recall that the coalescent and its companion ancestral process, which tracks the number of genealogical lineages, are continuous-time Markov chains and the coalescent times are transition times of these two chains (Tavaré, 2004). Therefore, the coalescent times form a Markov point process with a conditional intensity inversely proportional to $N_e(t)$ (Andersen et al., 1995). See (Palacios and Minin, 2013) for a more detailed exposition.

Benefits of the above reflection may not be immediately obvious, but viewing the coalescent as a point process allows us to recognize that the problem of reconstructing population size trajectory from coalescent times closely resembles the problem of estimating an intensity function of the inhomogeneous Poisson process. The latter task is well studied in the statistical literature, providing opportunities to adapt point process estimation tools to the coalescent framework. In fact, the multiple change-points discussed above can be viewed as modifications of the change-point approach to Poisson process intensity estimation (Raftery and Akman, 1986; Green, 1995).

The aforementioned change-point modeling falls into the area of nonparametric statistics, in which the number of parameters can grow indefinitely with dimensionality of the data. One popular alternative to the change-point approaches is GP-based nonparametric inference. A GP is a stochastic process such that any finite sample taken from the realization of the process has a multivariate normal distribution (Rasmussen and Williams, 2006). For example, Brownian motion and Ornstein-Uhlenbeck processes are GPs. GP-based models are very flexible and amenable to multivariate extensions, making these models dominant players in the point process and spatial statistics literature (Cressie, 1993). We now turn to describing the methods that use GP-based approaches to nonparametric inference of population dynamics.

11.3.3 Gaussian process-based nonparametrics

There are two GP-based approaches to estimation of effective population size trajectories. The first approach — the *Bayesian skyride* (Minin et al., 2008) — assumes that given a genealogy, the effective population size trajectory is a piecewise constant function with change-points coinciding with the coalescent times:

$$N_e(t) = \sum_{k=2}^{n} \exp(\gamma_k) 1_{(t_k, t_{k-1}]}(t).$$

In contrast to the *Bayesian skyline*, all elements in $\boldsymbol{\gamma} = (\gamma_2, \ldots, \gamma_n)$ are allowed to be distinct. Instead, Minin et al. (2008) place a potentially strong smoothing prior on γ:

$$P(\gamma \mid \tau) \propto \tau^{(n-2)/2} \exp\left[-\frac{\tau}{2}\sum_{k=2}^{n}\frac{(\gamma_k - \gamma_{k-1})^2}{\delta_k}\right], \tag{11.10}$$

where τ is a precision parameter that determines how much differences between adjacent γs are penalized, and δs are chosen either to be all one or are set to midpoint distances between inter-coalescent intervals. The precision hyperparameter τ receives a Gamma prior distribution. The prior specified by (11.10) can be thought as a first-order random walk with normal increments and with initial distribution unspecified. Alternatively, we can view γ *a priori* as a discretized GP, discretely observed Brownian motion to be specific, on an irregular grid. In light of our discussion of point processes, it is not surprising that a very similar random walk prior was used for Bayesian nonparametric estimation of Poisson process intensity (Arjas and Heikkinen, 1997). Minin et al. (2008) approximate the posterior distribution of γ, τ, and other model parameters, including the genealogy, using MCMC, implemented in BEAST (Drummond et al., 2012).

To avoid artificial discretization of $N_e(t)$ and potential statistical problems associated with such discretization, Palacios and Minin (2013) propose a more flexible GP-based prior for $N_e(t)$. They model the effective population size trajectory $N_e(t)$ as a linear transformation of an exponential Brownian motion:

$$N_e(t) = \{1 + \exp[-\gamma(t)]\}/\lambda, \tag{11.11}$$

where $\gamma(t) \sim \mathcal{BM}(\tau)$ and $\mathcal{BM}(\tau)$ denotes a Brownian motion process with mean function $\mathbf{0}$ and precision parameter τ. Parameter λ is necessary for the technical reasons we outline below.

Using a continuous stochastic process prior for $N_e(t)$ has its price: the densities (11.2) and (11.3) become intractable due to the stochastic integration involved in computing these densities. A similar problem occurs during estimation of the Poisson process intensity. Until recently, the only available solution involved discretization of the Poisson intensity (Møller et al., 1998), which is not too different from the *Bayesian skyride* solution above. For estimation of the Poisson process intensity, Adams et al. (2009) propose a data augmentation that bypasses stochastic integration and avoids discretization of the intensity. Palacios and Minin (2013) develop a similar solution for the effective population size estimation. Their data augmentation procedure requires that $N_e(t)$ has a lower bound denoted by λ^{-1} in (11.11). Palacios and Minin (2013) place a Gamma prior distribution on the precision hyperparameter τ and a mixture of uniform and exponential distributions on λ as follows:

$$P(\lambda) = \epsilon\frac{1}{\hat{\lambda}}1_{\{\lambda<\hat{\lambda}\}} + (1-\epsilon)\frac{1}{\hat{\lambda}}e^{-\frac{1}{\hat{\lambda}}(\lambda-\hat{\lambda})}1_{\{\lambda\geq\hat{\lambda}\}}, \tag{11.12}$$

where $\epsilon > 0$ is a mixing proportion and $\hat{\lambda}$ is our best guess of the upper bound, possibly obtained from previous studies.

The discretization-free GP-based estimation of $N_e(t)$ is currently available only for a fixed genealogy. Palacios and Minin (2013) approximate the posterior distribution of $N_e(t)$, τ, λ, and latent variables, introduced by the data augmentation, by MCMC sampling. Of course, one cannot keep track of the infinite dimensional object $N_e(t)$ without discretization. During MCMC, $N_e(t)$ is sampled at a finite number of time points at each iteration. *After MCMC is finished*, a grid of points $\{s_1, ..., s_B\}$ is formed and for each $g = 1, \ldots, B$, the posterior distribution of $N_e(s_g)$ is obtained from the MCMC samples by drawing from the posterior predictive distribution of $N_e(s_g)$. We emphasize that the coarseness of the grid has no bearing on statistical inference, because the grid is not used during the MCMC sampling. The grid can be made as fine as necessary after the MCMC is finished.

11.4 Examples

11.4.1 Fixed genealogy

Consider a population that experienced an expansion followed by a contraction under the following demographic scenario:

$$N_e(t) = \begin{cases} e^{4t} & t \in [0, 0.5], \\ e^{-2t+3} & t \in (0.5, \infty). \end{cases} \tag{11.13}$$

Starting with 100 contemporaneous samples from this population, we simulated coalescent time points according to the specified demographic scenario. These coalescent times are shown as crosses at the bottom of each plot in Figure 11.3. The first plot in Figure 11.3 shows the piecewise constant nature of the reconstructed trajectory using the Bayesian skyride method for a fixed genealogy and the second plot shows a smoother reconstructed trajectory using the continuous-time GP-based method. In both plots, the truth (11.13) is represented by a dashed line, posterior medians by solid lines, and 95% Bayesian credible intervals (BCIs) by shaded areas. Although both methods recover a trajectory of a population that experiences growth and then a decay, the methods have difficulty timing the peak of the population size. We will see further in the chapter how increasing the number of independent loci under analysis improves precision of the phylodynamic reconstruction.

We now consider real data examples. There are two major areas of application of phylodynamic methods. The first area corresponds to evolutionary studies of rapidly evolving infectious agents, such as RNA viruses. The second application area seeks to uncover past population size dynamics from ancient DNA. We showcase the usefulness of phylodynamic methods in these two areas by re-analyzing HCV in Egypt and bison across Beringia.

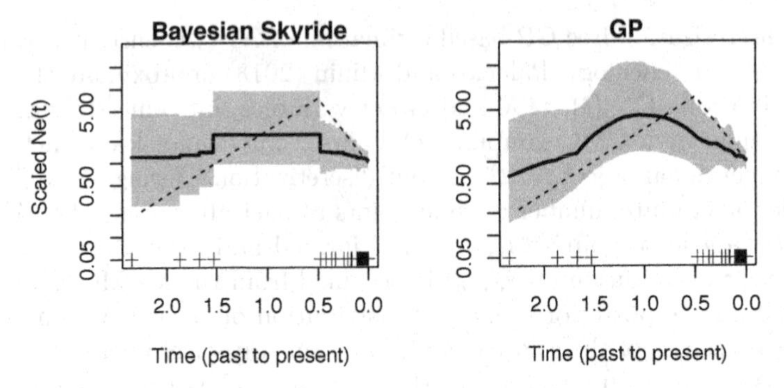

Figure 11.3: Example of a population that experienced expansion followed by a crash in population size. The true population size trajectory is depicted as dashed lines in the two plots. The simulated coalescent times are represented by the points at the bottom of each plot. We show the log of the effective population size trajectory estimated under the Bayesian skyride (left plot) and continuous time GP inference of effective population size (right plot). Posterior medians are shown as solid black lines and 95% Bayesian credible intervals (BCIs) are represented by gray shaded areas.

We first consider an estimated gene genealogy from 63 HCV E1 sequences sampled in 1993 in Egypt (Pybus et al., 2003). This genealogy is depicted by the 62 coalescent time points at the bottom of the plots in Figure 11.4. We apply the Bayesian skyride and GP-based phylodynamic reconstruction to this genealogy. Both methods recover a population size trajectory that increases exponentially after the 1920s and decreases after the 1970s. This reconstruction is consistent with a hypothesized role of parenteral antischistosomal therapy

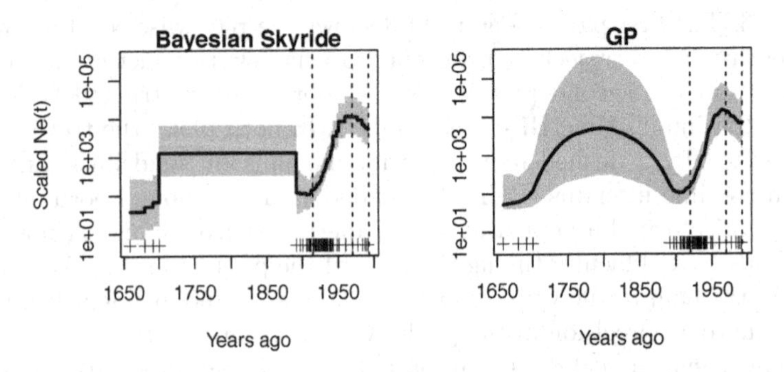

Figure 11.4: Egyptian HCV. The plots show the log of the scaled effective population size estimated using the Bayesian skyride (left) and the GP-based method (right) from the majority clade support genealogy with median node heights. Vertical dashed lines mark the years 1920, 1970, and 1993 from left to right. See the caption of Figure 11.3 for the rest of the legend.

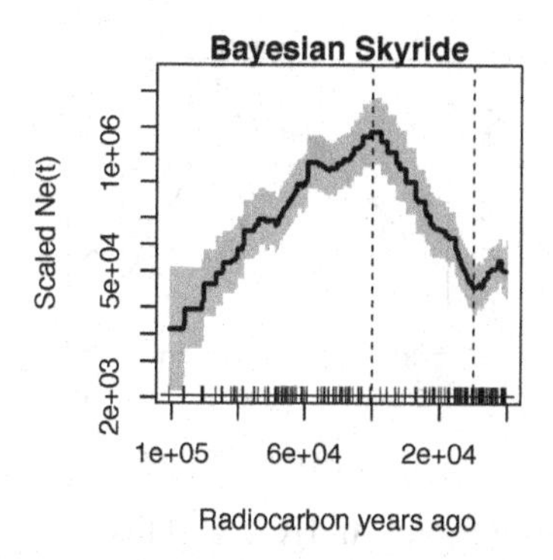

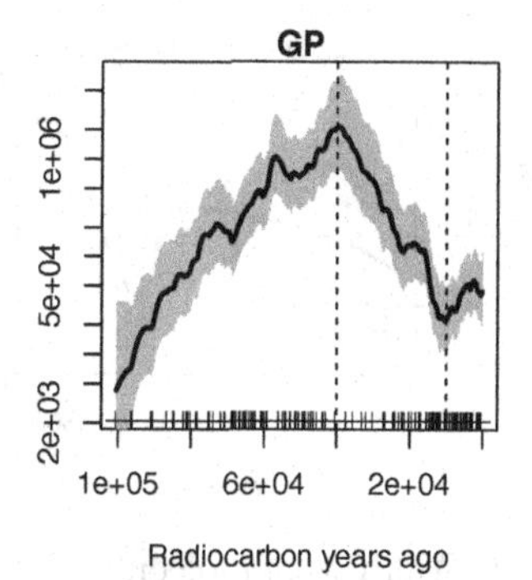

Figure 11.5: Berigian bison. The plots show the log of the scaled effective population size estimated using the Bayesian skyride (left) and the GP-based method (right) from the majority clade support genealogy with median node heights. Vertical dashed lines mark 40 ka B.P. and 10 ka B.P. See the caption of Figure 11.3 for the rest of the legend.

(PAT) in HCV spread in Egypt (Frank et al., 2000). The PAT campaign started in the 1920s, with the treatment administered intravenously, which together with a lack of sanitary practices, is believed to have led to a rapid increase of HCV infections in Egypt. The intravenous administration of the PAT was gradually replaced by oral administration in the 1970s, a change that is reflected in the decay in the effective number of HCV infections in our phylodynamic reconstructions.

To investigate the evolution and demographic history of Pleistocene bison, Shapiro et al. (2004) collected 152 mtDNA samples from bison fossils found in Alaska, Canada, Siberia, China, and the lower 48 United States with dates that spanned a period of more than 80,000 radiocarbon years before present. As with the HCV example, we used these DNA samples to reconstruct a genealogy of these samples under a molecular clock assumption. In Figure 11.5, we show bison population size trajectories reconstructed using the Bayesian skyride and the GP-based method from the estimated genealogy. Here, both methods agree in a recovered population size trajectory that reaches its maximum around 40 ka B.P., followed by a decay until it reaches a bottleneck around 10 ka B.P. This bottleneck occurs at the time of human settlement in Alaska, agreeing with previous analyses (Drummond et al., 2005).

11.4.2 Accounting for genealogical uncertainty

We now consider the same molecular data of HCV in Egypt and bison described earlier, but instead of conditioning on a reconstructed genealogy, proceed with the estimation of population size trajectories from the molecular data directly using the Bayesian skyride method. Figure 11.6 shows the recovered trajectories. In both cases, the key aspects of the population sizes are recovered from a

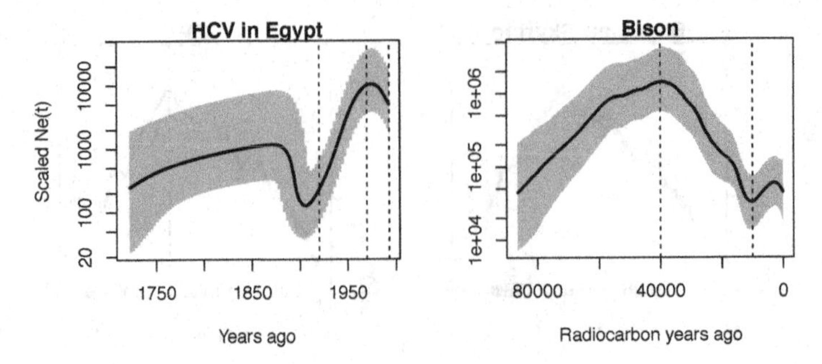

Figure 11.6: Analysis of molecular data from Egyptian HCV and bison examples. The left plot shows the log of the scaled effective population size of the Egyptian HCV and the right plot shows the log of the bison scaled effective population size, with both trajectories recovered by the Bayesian skyride method from the molecular data directly. See the captions of Figures 11.3, 11.4, and 11.5 for the rest of the legend.

fixed genealogy and from the sequence data directly, however, assessment of uncertainty and the degree of smoothness of the recovered trajectories differ substantially. These results reiterate that despite the attractive simplicity of the fixed-genealogy methods, methods that properly account for genealogical uncertainty should be preferred to the fixed-genealogy approaches.

11.5 Extensions and future directions

The coalescent prior on genealogies used in all the models described here assumes a random sample of orthologous, nonrecombining, and neutrally evolving sequences from a panmictic population. Any violation of these assumptions will result in an estimated effective population size trajectory that is not directly comparable to census population size. *Selection* effectively shifts the distribution of mutations on the genealogy; therefore, interpretation of effective population size estimation under selection needs to be done with caution (Pybus and Rambaut, 2009; Ho and Shapiro, 2011). Recombination is clearly a problem for all the methods described so far, since these methods assume a single genealogy. In principle, it should be possible to extend Bayesian nonparametric phylodynamic methods to include a possibility of recombination similar to the work of Kuhner and Smith (2007), but software development and computational costs will be significant. Alternatively, one can use a sequentially Markov approximation of the coalescent process (McVean and Cardin, 2005), as was done in a recent attempt to leverage whole-genome sequence data of a single individual to estimate population size dynamics (Li and Durbin, 2011).

Although the inferential framework in this case is very different from the one described in this chapter, Bayesian nonparametric approaches similar to those detailed here, can also be applied in this setting. In the rest of this section, we discuss further possible extensions of the currently available phylodynamic methods.

11.5.1 Multiple loci

Data from multiple effectively unlinked genetic loci are rapidly becoming the norm in the era of next-generation sequencing. Evolutionary dynamics of such independently evolving loci are governed by the same demographic history of the population under study, enabling straightforward estimation of effective population size trajectories based on multilocus genetic data. Increasing the number of loci improves precision of the phylodynamic estimation, which is critical for these nonparametric procedures that often suffer from large BCIs. One of the primary difficulties in estimating population dynamics is that most of the coalescent events in the reconstructed genealogy usually occur in a short time span. During the long periods of time in which few coalescent events occur, there is not much data to infer the population dynamics. Increasing the sample size mitigates this problem to a certain extent, but the additional coalescent events also tend to occur in a small stretch of time. It is more advantageous to increase the number of loci since this provides extra information during the long time frames with few coalescent events (Heled and Drummond, 2008).

To allow for inference of effective population size trajectories from multiple loci, Heled and Drummond (2008) implemented the *extended Bayesian skyline plot* in BEAST (Drummond et al., 2012). This new version of the Bayesian skyline places a different prior on $N_e(t)$. Instead of a piecewise constant function with jumps at some coalescent points, the estimated trajectory is piecewise linear with straight lines connecting "heights" γ_j at change-points $a_1, ..., a_k$, that is,

$$N_e(t) = \sum_{j=1}^{k+1} \left(\gamma_{j-1} + (\gamma_j - \gamma_{j-1}) \frac{t - a_{j-1}}{a_j - a_{j-1}} \right) 1_{(a_{j-1}, a_j]}(t). \tag{11.14}$$

The change-points are an ordered subset of the coalescent times $\{t_n, \dots, t_1\}$, which now include coalescent times from genealogies at all loci, and $a_0 = t_n = 0$ and $a_{k+1} = t_1$. Here, *a priori*

$$\gamma_j \sim \text{Exponential}(\theta), \text{ for } j = 1, \dots, k+1, \tag{11.15}$$

where θ may be either fixed or have a prior $P(\theta)$. The number of change-points k has a truncated Poisson distribution with mean $\log(2)$.

Gill et al. (2012) have developed a model, called the *Bayesian skygrid*, that generalizes and improves the *Bayesian skyride*. It differs from the *Bayesian skyride* not only in its ability to incorporate data from several loci, but also in that the estimated piecewise constant trajectory has change-points at fixed user-specified times rather than coalescent times. Given ordered times $s_B < \cdots < s_1$,

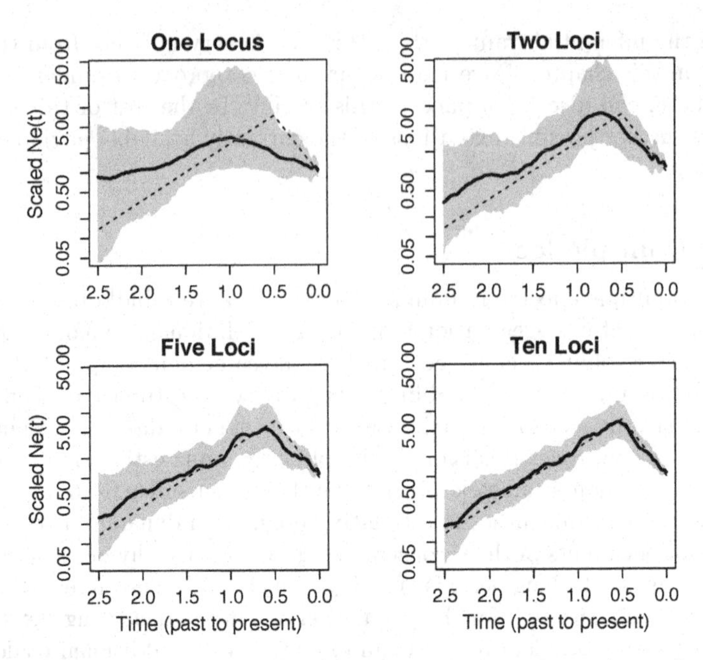

Figure 11.7: The log effective population size trajectories estimated under the *Bayesian skygrid* from 1, 2, 5, and 10 simulated genealogies. The legend is the same as in Figure 11.3.

with $s_B = 0$ and $s_0 = \infty$, we have

$$N_e(t) = \sum_{k=1}^{B} \exp(\gamma_k)\, 1_{[s_k, s_{k-1})}(t), \tag{11.16}$$

where $\boldsymbol{\gamma} = (\gamma_1, \ldots, \gamma_B)$ is *a priori* a Gaussian Markov random field.

To illustrate the benefits of estimating population dynamics from multilocus data, we simulate genealogies under the same demographic scenario as in Example 1.4. We estimate the effective population size from 1, 2, 5, and 10 genealogies using the *Bayesian skygrid* with $n = 99$ and $s_{100}, \ldots, s_1$ equally spaced times between 0 and 2.5. Figure 11.7 demonstrates that increasing the number of loci even modestly leads to appreciable gain of estimation precision.

11.5.2 Effect of population structure

The coalescent with variable population size assumes that there is random mixing in the population and that the samples are taken randomly from the population. This former assumption is clearly violated for many real populations, because individuals tend to mate with other individuals in geographic or social proximity. In the presence of a well-defined population structure, a simple random sample of the whole population will not efficiently capture the diversity of each subpopulation and the coalescent with variable population size applied directly to the whole sample will not account for the "blocking/clustering" of sampled lineages in the genealogy. Instead, a better strategy would be to consider gathering a stratified sample by subpopulation and estimate the total population size trajectory under the structured coalescent prior on genealogies (Hudson, 1990; Beerli and Felsenstein, 2001b). The structured coalescent accounts for subpopulations with different population sizes and allows for migration between subpopulations. The structured coalescent has been used for parametric estimation of population size trajectories and migration parameters from isochronous and heterochronous data in Bayesian and maximum likelihood frameworks (Ewing and Rodrigo, 2006a,b; Beerli and Felsenstein, 2001b). To our knowledge, no implementation of the structured coalescent equipped with Bayesian nonparametric estimation of subpopulation size trajectories is available.

A more complex violation of random mixing occurs when individuals in the population are connected by a social or contact network (Welch et al., 2011). The standard structured coalescent assumes random mating within subpopulations and constant variability in number of offspring, making this modeling framework incapable of handling network-based population structure. In order to account for contact heterogeneity, and hence a variable reproductive variance, attempts to equip the coalescent with social network modeling are emerging. If we are interested in a population of infected individuals with a certain rapidly evolving infectious disease, knowing the social or contact network of the sampled individuals should help in reducing uncertainty in the genealogical reconstruction (transmission network) among the samples. However, knowledge about the social network and, moreover, knowledge of the dynamics of the social network, are not readily available to us in practice. Network-based coalescent presents further challenges in formulating a model for the social network of the sample and the population size affect the shape of the genealogy in a statistical (likelihood-based) framework. In light of these difficulties, it is not surprising that so far progress on merging the coalescent and network-based approaches advanced mainly through simulations with an aim to measure the impact of network structure on the estimation of effective population sizes (Goodreau, 2006; O'Dea and Wilke, 2010; Leventhal et al., 2012).

11.5.3 Coalescent and infectious disease dynamics

Phylodynamic methods have been widely applied to study the evolution of rapidly evolving diseases. Here, inference is based on sampled disease agent molecular sequences from infected hosts. If superinfection is rare and mutations accumulate fast relative to epidemic growth, each coalescent time in a genealogy of sampled consensus viral isolates from infected individuals corresponds to a transmission event (Volz et al., 2009). Estimation of effective population size under the coalescent with variable population size prior on genealogies assumes that generation length and variability in number of transmissions are constant through time. For some pathogens, this may be unrealistic and interpretability of the effective population size as the number of infections becomes imprecise. In order to gain interpretability of epidemiologically relevant parameters, there has been a growing interest in formalizing the integration of phylodynamic methods and standard epidemiological and ecological models (Grenfell et al., 2004; Pybus and Rambaut, 2009; Volz et al., 2009; Bennett et al., 2010; Frost and Volz, 2010; Kühnert et al., 2011). In the most recent effort on this front, Volz (2012) considered a population under a continuous time birth–death process with varying birth–death rates and expressed the coalescence rate in terms of the birth rate (incidence) and the number of infected individuals (prevalence). Further, the dynamic population model was extended to include migration and two stages of infection to accommodate cases where the transmission probability per contact changes over the course of infection. In addition to incorporating disease dynamics into the coalescent, there is a growing need to integrate molecular data with clinical, socio-demographic, and other relevant data in order to measure the correlation between the population size and the environment (Rasmussen et al., 2011). We hope that more sophisticated Bayesian modeling will be able to solve these challenging problems.

12

Sampling and summary statistics of endpoint-conditioned paths in DNA sequence evolution

Asger Hobolth

Bioinformatics Research Center, Aarhus University, Aarhus C, Denmark

Jeffrey L. Thorne

Bioinformatics Research Center, North Carolina State University, Raleigh, North Carolina, USA

CONTENTS

12.1 Introduction

While some probabilistic models of DNA or protein sequence change are not based on an instantaneous rate matrix (e.g., Barry and Hartigan (1987)), most are. With an instantaneous rate matrix, there is an opportunity to go beyond the sequences that begin and end a branch on a phylogenetic tree — inferences can be made about the sequence changes that happened between these endpoints. At the most detailed level, inferences would be about which changes transformed the beginning sequence into the ending one and about exactly when these changes occurred. At a less detailed level, various summary statistics about the evolutionary trajectory from the beginning to ending of the branch might be of interest. A variety of techniques are available for making inferences about evolutionary trajectories conditional upon the endpoints of a branch and one objective of this chapter is to introduce them. To parallel the "Brownian bridge" that results when the endpoints of a Brownian motion process are conditioned upon, an endpoint-conditioned Markov process is known as a Markov bridge (Al-Hussaini and Elliot, 1989). This chapter is not intended to be comprehensive regarding inference techniques for Markov bridges. Instead, the focus is on endpoint-conditioning with Markov models for molecular sequence evolution.

The other purpose of this chapter is to emphasize how endpoint-conditioned evolutionary inference promotes an attractive alternative to conventional approaches for studying molecular evolution in a likelihood-based framework. As will be described below, the attraction is particularly strong when the evolutionary models are too biologically rich for Felsenstein's pruning algorithm (Felsenstein, 1981) to be computationally feasible.

For interspecific genetic data, the pruning algorithm is widely employed as the workhorse of likelihood-based evolutionary inference. As input, it requires a tree topology of interest, the states of homologous characters at the tips of the tree, and a set of transition probabilities for each branch on the phylogeny. In addition to any calculations needed to obtain transition probabilities, the algorithm relies on the assumption that character change over time is Markovian. It determines the likelihood with an amount of computation that is proportional to the number of internal nodes on the tree multiplied by the square of the number of possible character states. Because the number of possible DNA sequences of length L is 4^L, considering an entire DNA sequence as a character state and then applying the pruning algorithm will be prohibitive unless the sequence length is so short that the number of rate matrix entries $4^L \times 4^L = 4^{2L}$ is not overly large. The usual practice is to salvage the computational feasibility of the pruning algorithm by imposing severe restrictions on the evolutionary model that governs sequence change.

Most often, evolutionary models are simplified by adding the stringent constraint that individual sequence positions (or sometimes individual codons) change independently of one another but evolve according to a shared phylogeny. In this case, the likelihood of a set of aligned sequences of length L can be expressed as the product over sequence positions of individual site likelihoods. This converts the amount of computation needed for pruning with DNA data to a linear rather than exponential function of L. In addition, models with independent change among sites can have a much lower computational burden associated with determining transition probabilities, but the degree to which the burden is alleviated will depend on the parametric form of the model of sequence change.

The tremendous gain in computational feasibility that is achieved by assuming that sequence positions change independently is countered by the loss in biological plausibility. The origin of differences between homologous DNA sequences is mutation, and mutation processes violate the independent change among sites assumption in a variety of ways. It is well established that mutations at different genomic positions do not occur independently of one another. Schrider et al. (2011) estimate that multiple nearby sequence positions are simultaneously affected in about 3% of eukaryotic mutation events. In addition, rates of point mutation at a genomic position can be highly dependent on the local sequence context of the position being mutated. The most prominent example of context-dependent mutation is the elevated rate of change from C to T in mammalian genomes when the C is immediately followed in the sequence by a G. Other cases of context-dependent mutation have also been characterized to varying degrees (reviewed in Hodgkinson and Eyre-Walker (2011)).

Natural selection can be another important source of dependent change among sequence positions. Selection acts on phenotype and phenotype does not emerge from environment and genotype (i.e., DNA sequence) via independent contributions of each position in the genome. For instance, protein sequences can be specified as a one-dimensional string of amino acids or as a one-dimensional string of the codons that specify these amino acids, but the biological function of a protein and the natural selection that acts upon it are contingent upon the three-dimensional shape of the protein just as the folding of a protein into its tertiary structure is influenced by the interactions between the amino acids that comprise the proteins. Interactions between amino acids that are spatially close to one another can be especially strong. Because protein structure is phenotype at a fundamental molecular level, natural selection on proteins cannot and does not result in exclusively independent change among codon positions. There is, therefore, a strong scientific incentive for relaxing the assumption of independent change among codons.

Natural selection to maintain protein tertiary structure is far from being the only way that natural selection can induce dependent change among sequence positions. Natural selection on gene expression and RNA secondary structure are two other important sources of dependent change among positions. More generally, the fitness of a genome is not simply the sum of the contributions of individual genomic positions. Because limitations of the pruning algorithm prevent it from being employed in many situations where sequence positions change in a dependent fashion, alternative inference strategies are needed. Promising alternatives include the strategies discussed here that augment the observed sequence data with possible histories that generated these data. Strategies that do not involve augmentation are also available for making evolutionary inferences with dependent change among sites (e.g., Arndt et al. (2003); Siepel and Haussler (2004); Baele et al. (2008)) and are nicely summarized in Baele et al. (2008).

12.2 Independent sites models and summary statistics

We begin by considering a model of sequence evolution that can be represented as a continuous time Markov chain (CTMC) with a discrete and finite state space. Such a model is described by a rate matrix Q with non-negative off-diagonal entries $Q_{\alpha\beta} \geq 0$ and negative diagonal entries $Q_{\alpha\alpha} = -Q_{\alpha} = -\sum_{\beta\neq\alpha} Q_{\alpha\beta} < 0$. We make the further assumption that the CTMC is time-homogeneous (i.e., Q does not change over time) and positive recurrent (i.e., the Markov chain has strictly positive probability to reach any state from any other state in a finite number of jumps). These conditions ensure that a stationary distribution π exists. We let $X(t)$ be the state of the CTMC at time t and we use $X = \{X(t) : 0 \leq t \leq T\}$ to represent all values of the CTMC between time 0, when the CTMC is initially observed, and time T, when the process stops being observed.

12.2.1 Likelihood inference

A CTMC that conforms to these conditions can be thought of as a Poisson process where the intensity is Q_{α} when the CTMC occupies state α and where the intensity can change after each event. Specifically, when an event occurs, the intensity immediately changes from Q_{α} to Q_{β} with probability $Q_{\alpha\beta}/Q_{\alpha}$. A fully observed sample $x = \{x(t) : 0 \leq t \leq T\}$ can therefore be divided into intervals according to the intensity of the Poisson process (see Figure 12.1). The first interval begins at time 0 and ends when the first change occurs. The second interval represents the time between the first and second changes and so on. The final interval begins when the final change occurs and it ends at time T. Conditional upon the initial state of the Markov chain, the probability density of the fully observed sample is a product of factors corresponding to each interval with some factors representing the time duration of the intervals and with other factors representing the changes that end the intervals. Specifically, if the Markov chain occupies state α at the beginning of an interval, the amount of time until the interval ends because the next change occurs will have an exponential distribution with parameter Q_{α}. Given that a change

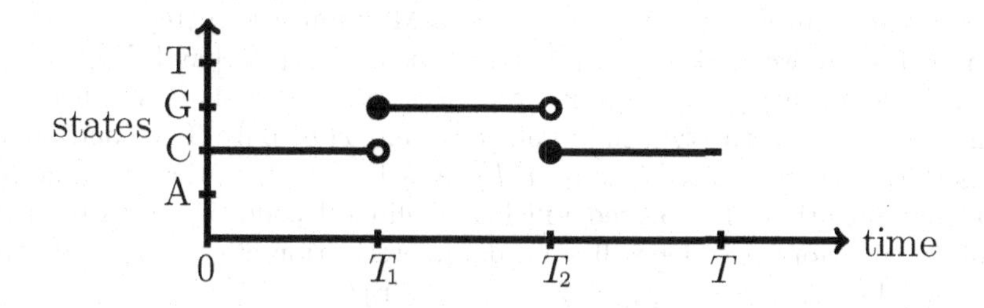

Figure 12.1: History of a sequence site. The substitutions to G at time T_1 and back to C at time T_2 divide the time from 0 to T into three intervals. Conditional upon starting with a C at time 0, the probability density of the history is $Q_C e^{-Q_C(T_1-0)} \times (Q_{CG}/Q_C) \times Q_G e^{-Q_G(T_2-T_1)} \times (Q_{GC}/Q_G) \times e^{-Q_C(T-T_2)}$.

will occur at some instant, the probability that the change is to state β is $Q_{\alpha\beta}/Q_\alpha$. When the change to β is made, the waiting time until the next event is again an exponential distribution, but the parameter of this exponential distribution is Q_β. The single factor representing the final interval corresponds to the probability of no event occurring during the time from the final event to the end of the branch. If we condition upon the initial state of the Markov chain, the likelihood of the fully observed data is

$$P(X = x|Q, X(0) = x(0)) = \Big(\prod_\alpha e^{-Q_\alpha D_\alpha}\Big)\Big(\prod_\alpha \prod_{\beta\neq\alpha} Q_{\alpha\beta}^{N_{\alpha\beta}}\Big), \tag{12.1}$$

where $N_{\alpha\beta} = N_{\alpha\beta}(x)$ is the number of substitutions from α to β, and $D_\alpha = D_\alpha(x)$ is the total time spent in state α. For example, the fully observed history of Figure 12.1 would yield an endpoint-conditioned likelihood that can be obtained by applying (12.1) with $D_C = (T_1 - 0) + (T - T_2)$, $D_G = (T_2 - T_1)$, $N_{CG} = N_{GC} = 1$, and with all other time durations and substitution counts being set to 0. In general, the full log likelihood is given by

$$\log P(X = x|Q, X(0) = x(0)) = \Big(\sum_\alpha -Q_\alpha D_\alpha\Big) + \Big(\sum_\alpha \sum_{\beta\neq\alpha} N_{\alpha\beta} \log Q_{\alpha\beta}\Big),$$

and in the unrestricted case we obtain the maximum likelihood estimates $\hat{Q}_{\alpha\beta} = N_{\alpha\beta}/D_\alpha$.

12.2.2 The EM algorithm

Suppose the chain is discretely observed in a set of points $\{t_1, \ldots, t_n\}$ and let $y = \{x(t_1), \ldots, x(t_n)\}$ denote the corresponding observations. The EM algorithm is an efficient tool for estimating the rates in this case (Holmes and

Rubin (2002); see also Hobolth and Jensen (2005)). In the algorithm, we must iterate between an E(xpectation) step and a M(aximization) step.

In the **E-step** we need the conditional means $\mathrm{E}\left[N_{\alpha\beta}|Q,y\right]$ and $\mathrm{E}\left[D_{\alpha}|Q,y\right]$, and because of the Markov property it suffices to consider the case where the chain is observed in the two endpoints $x(0)$ and $x(T)$ only. The conditional means then become $\mathrm{E}\left[N_{\alpha\beta}|Q,x(0),x(T)\right]$ and $\mathrm{E}\left[D_{\alpha}|Q,x(0),x(T)\right]$. The **M-step**, maximization of expected full log–likelihood conditional on data, is usually straightforward (depending on parameterization of $Q = Q(\theta)$). In the unrestricted case, we have $Q_{\alpha\beta} = \mathrm{E}[N_{\alpha\beta}|Q,y]/\mathrm{E}[D_{\alpha}|Q,y]$.

12.2.3 Bayesian inference

A natural prior for the model is a gamma distribution with shape parameter $\kappa_{\alpha\beta} > 0$ and scale parameter $\lambda_{\alpha} > 0$ on the rates $Q_{\alpha\beta}$ (e.g., Bladt and Sørensen, 2005),

$$P(Q) \propto \prod_{\alpha}\prod_{\beta\neq\alpha}\left(Q_{\alpha\beta}^{\kappa_{\alpha\beta}-1}e^{-\lambda_{\alpha}Q_{\alpha\beta}}\right) = \left(\prod_{\alpha} e^{-Q_{\alpha}\lambda_{\alpha}}\right)\left(\prod_{\alpha}\prod_{\beta\neq\alpha} Q_{\alpha\beta}^{\kappa_{\alpha\beta}-1}\right).$$

The posterior is proportional to $P(X = x|Q, X(0) = x(0))P(Q)$. The maximum is given by

$$Q_{\alpha\beta} = \frac{N_{\alpha\beta} + \kappa_{\alpha\beta} - 1}{D_{\alpha} + \lambda_{\alpha}}. \tag{12.2}$$

Bladt and Sørensen (2005) choose $\kappa_{\alpha\beta} = \lambda_{\alpha} = 1$; choosing $\kappa_{\alpha\beta} \geq 1$ avoids negative offdiagonal entries in the rate matrix. In case the CTMC is discretely observed, the posterior mode can be found by iterating between an E-step and an M-step where in (12.2) $N_{\alpha\beta}$ and D_{α} are substituted by their means.

12.2.4 Conditional means on a phylogeny

Conditional means on a phylogeny are obtained from Felsenstein's pruning algorithm. Consider the situation in Figure 12.2. By conditioning on the inner nodes, the total number of expected changes

$$\mathrm{E}[N_{\alpha\beta}|Q,y] = \sum_{z_1,z_2,z_3} \mathrm{E}[N_{\alpha\beta}|Q,y,z]P(z|Q,y)$$

reduces to a sum of expected changes on each branch. The expected number of changes $N_{\alpha\beta}^{z_1 y_1}$ from α to β on the branch from internal node z_1 to tip y_1 becomes

$$\mathrm{E}[N_{\alpha\beta}^{z_1 y_1}|Q,y,z] = \sum_{z_1} \mathrm{E}[N_{\alpha\beta}^{z_1 y_1}|Q,z_1,y_1]P(z_1|Q,y),$$

where $y = (y_1, y_2, y_3, y_4, y_5)$ and $P(z_1|Q,y)$ is obtained from Felsenstein's pruning algorithm. Similarly for the branch from internal node z_1 to internal

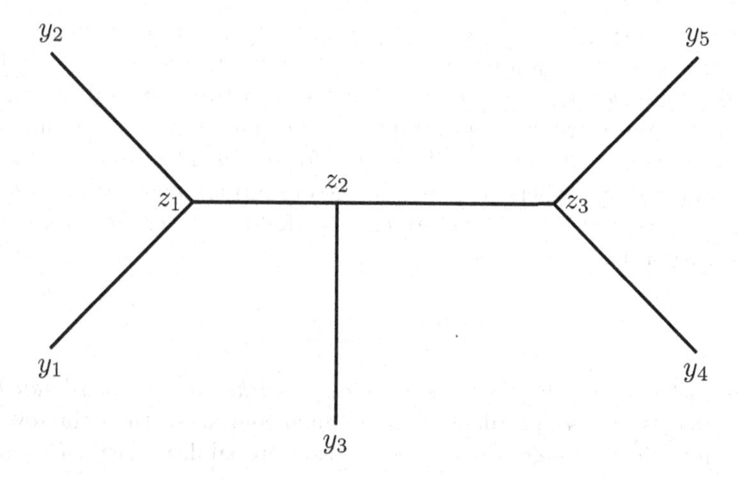

Figure 12.2: Phylogenetic tree. The values y in the leaf nodes are known. By conditioning on the possible values $z = (z_1, z_2, z_3)$ at the inner nodes, we obtain an endpoint-conditioned inference problem.

node z_2,

$$E[N_{\alpha\beta}^{z_1 z_2}|Q, y, z] = \sum_{z_1, z_2} E[N_{\alpha\beta}^{z_1 z_2}|Q, z_1, z_2]P(z_1, z_2|Q, y).$$

The conditional means for the time spent in a state $E[D_\alpha|Q, y]$ are obtained in a similar way.

12.2.5 Endpoint-conditioned summary statistics from uniformization

At least four methods exist for calculating the conditional means for the time spent in a state and for the number of substitutions between any two states. These methods are based upon (i) uniformization (Hobolth and Jensen, 2011), (ii) an eigenvalue decomposition of the rate matrix (Holmes and Rubin, 2002), (iii) results for integrals of matrix exponentials (Van Loan, 1978), and (iv) Laplace transforms and continued fractions (Crawford and Suchard, 2012); see Tataru and Hobolth (2011) for a comparison of methods (i)–(iii). We present the uniformization method because this procedure allows us to solve the more general problem of determining the distributions (not only the means) for the two types of statistics.

Uniformization is an alternative way of defining a CTMC (Jensen, 1953; Fearnhead and Sherlock, 2006; Lartillot, 2006; Mateiu and Rannala, 2006; Rodrigue et al., 2008). In the conventional representation of a CTMC, the rate Q_α away from a state α depends on the state, and state changes are always between different states. With uniformization, the rate away from a state is the same for all states and is equal to $\mu = \max_\alpha Q_\alpha$. The consequence is that waiting times between jumps are exponentially distributed with a uniform rate μ or, equivalently, the number of jumps in a time interval of length T follows a Poisson distribution with rate μT, and the jump times are uniformly distributed in the time interval. The price for uniform waiting times is that virtual jumps (jumps where no state change occurs) are allowed and occur for state α with rate $\mu - Q_\alpha$. The state changes follow a discrete Markov chain with transition matrix

$$R = I + \frac{1}{\mu}Q.$$

The stochastic process that results is called a *Markov chain subordinated to a Poisson process* and is equivalent to the original continuous time Markov chain $X(t)$, as invoking the definition of a transition probability matrix $P(t)$ shows

$$\begin{aligned} P(t) = \exp(Qt) = \exp(\mu(R-I)t) &= \exp(-\mu t)\sum_{n=0}^{\infty}\frac{(\mu Rt)^n}{n!} \\ = \sum_{n=0}^{\infty} R^n \exp(-\mu t)\frac{(\mu t)^n}{n!} &= \sum_{n=0}^{\infty} R^n \mathrm{Pois}(n;\mu t). \end{aligned} \tag{12.3}$$

Let $0 = T_0 < T_1 < T_2 < \cdots < T_J < T$ denote the J jump times for the Poisson process with rate μ, and let $z_0, z_1, z_2, \ldots, z_J$ be the states for the discrete time Markov chain with transition matrix R. Then the CTMC $\{X(t) : 0 \le t \le T\}$ is given by $X(t) = z_k$ for $T_k \le t < T_{k+1}$, $0 \le k \le J$, where $T_{J+1} = T$.

Uniformization can be used to calculate various types of endpoint-conditioned summary statistics. Note that for any statistic H, we have

$$\mathrm{E}[H|X(0) = a, X(T) = b] = \frac{\mathrm{E}[H \cdot 1(X(T) = b)|X(0) = a]}{P_{ab}(T)},$$

where we suppress the conditioning on Q and the denominator is the entry in the transition probability matrix corresponding to state b at time T given state a at time 0. We formulate our results through terms of the form $\mathrm{E}[H \cdot 1(X(T) = b)|X(0) = a]$. We are mainly interested in the number of state changes from α to β,

$$N_{\alpha\beta} = \sum_{i=1}^{J} 1((z_{i-1}, z_i) = (\alpha, \beta)),$$

and the time spent in a state α,

$$D_\alpha = \sum_{i=0}^{J} 1(z_i = \alpha)(T_{i+1} - T_i).$$

Theorem 1. *The endpoint-conditioned mean value for $N_{\alpha\beta}$ is given by*

$$E[N_{\alpha\beta} \cdot 1(X(T) = b)|X(0) = a] = \sum_{n=1}^{\infty} M_{\alpha\beta}(n, a, b)\text{Pois}(n; \mu T),$$

where

$$M_{\alpha\beta}(n, a, b) = \sum_{i=1}^{n} (R^{i-1})_{a\alpha} R_{\alpha\beta} (R^{n-i})_{\beta b},$$

and $R^0 = I$.

Proof. We find that

$$\begin{aligned}
&E[N_{\alpha\beta} \cdot 1(X(T) = b)|X(0) = a] \\
=&\sum_{n=0}^{\infty} E[N_{\alpha\beta} 1(X(T) = b) 1(n \text{ jumps in } [0, T])|X(0) = a] \\
=&\sum_{n=0}^{\infty} E[N_{\alpha\beta} 1(X(T) = b)|n \text{ jumps in } [0, T], X(0) = a]\text{Pois}(n; \mu T) \\
=&\sum_{n=1}^{\infty} E[\sum_{i=1}^{n} 1(z_{i-1}, z_i) = (\alpha, \beta) 1(z_n = b)|z_0 = a]\text{Pois}(n; \mu T) \\
=&\sum_{n=1}^{\infty} \sum_{i=1}^{n} P(z_{i-1} = \alpha, z_i = \beta, z_n = b|z_0 = a)\text{Pois}(n; \mu T) \\
=&\sum_{n=1}^{\infty} \sum_{i=1}^{n} (R^{i-1})_{a\alpha} R_{\alpha\beta} (R^{n-i})_{\beta b}\text{Pois}(n; \mu T).
\end{aligned}$$

Qed

Theorem 2. *The endpoint-conditioned mean value for D_α is given by*

$$E[D_\alpha \cdot 1(X(T) = b)|X(0) = a] = \sum_{n=0}^{\infty} \frac{T}{n+1} M_\alpha(n, a, b)\text{Pois}(n; \mu T),$$

where

$$M_\alpha(n, a, b) = \sum_{i=0}^{n} (R^i)_{a\alpha} (R^{n-i})_{\alpha b},$$

and with $R^0 = I$.

Proof. A calculation similar to the above gives

$$
\begin{aligned}
&E[D_\alpha \cdot 1(X(T) = b)|X(0) = a] \\
&= \sum_{n=0}^{\infty}\sum_{i=0}^{n} P(z_i = \alpha, z_n = b|z_0 = a)\times \\
&\qquad E(T_{i+1} - T_i|n \text{ jumps in } [0,T])\text{Pois}(n;\mu T) \\
&= \sum_{n=0}^{\infty}\sum_{i=0}^{n} (R^i)_{a\alpha}(R^{n-i})_{\alpha b}\frac{T}{n+1}\text{Pois}(n;\mu T).
\end{aligned}
$$

Qed

Hobolth and Jensen (2011) consider a more general statistic

$$
H = \psi(z_0)f(T_1) + \sum_{i=1}^{J}\phi(z_{i-1}, z_i)f(T_{i+1} - T_i), \tag{12.4}
$$

and calculate means, variances, and covariances for such variables. If $\psi(X) \equiv 0$, $\phi(X,Y) = 1((X,Y) = (\alpha,\beta))$, and $f \equiv 1$, we get $H = N_{\alpha\beta}$, and if $\psi(X) = 1(X = \alpha)$, $\phi(X,Y) = 1(Y = \alpha)$ and $f(t) = t$, we get $H = D_\alpha$.

We now move on to the endpoint-conditioned distributions of the number of jumps and time spent in a state.

Theorem 3. *The endpoint-conditioned distribution for $N_{\alpha\beta}$ is given by*

$$
P(N_{\alpha\beta} = k, X(T) = b|X(0) = a) = \sum_{n=0}^{\infty} P_{\alpha\beta}(k,n,a,b)\text{Pois}(n;\mu T),
$$

where $P_{\alpha\beta}(k,n,a,b) = 0$ for $k > n$. For $n = 0$, $P_{\alpha\beta}(0,0,a,b) = 1(a = b)$. For $n = 1$,

$$
P_{\alpha\beta}(0,1,a,b) = R_{ab}(1 - 1(a = \alpha)1(b = \beta)), \tag{12.5}
$$

$$
P_{\alpha\beta}(1,1,a,b) = R_{ab}1(a = \alpha)1(b = \beta). \tag{12.6}
$$

For $n \geq 2$ ($k = 0,\ldots,n$), we have the recursion

$$
\begin{aligned}
&P_{\alpha\beta}(k,n,a,b) \\
&= \begin{cases} P_{\alpha\beta}(k-1,n-1,a,\alpha)R_{\alpha b} + \sum_{c\neq\alpha} P_{\alpha\beta}(k,n-1,a,c)R_{cb} & \text{if } b = \beta, \\ \sum_c P_{\alpha\beta}(k,n-1,a,c)R_{cb} & \text{if } b \neq \beta, \end{cases}
\end{aligned} \tag{12.7}
$$

and we define $P_{\alpha\beta}(-1,\cdot,\cdot,\cdot) = 0$.

Proof.

$$\begin{aligned}
&P(N_{\alpha\beta} = k, X(T) = b|X(0) = a) \\
&= \sum_{n=0}^{\infty} P(N_{\alpha\beta} = k, X(T) = b, n \text{ jumps in } [0,T]|X(0) = a) \\
&= 1(k = 0, a = b)\text{Pois}(0; \mu T) \\
&\quad + \sum_{n=1}^{\infty} P(\sum_{i=1}^{n} 1((z_{i-1}, z_i) = (\alpha, \beta)) = k, z_n = b|z_0 = a)\text{Pois}(n; \mu T).
\end{aligned}$$

Let

$$P_{\alpha\beta}(k, n, a, b) = P(\sum_{i=1}^{n} 1((z_{i-1}, z_i) = (\alpha, \beta)) = k, z_n = b|z_0 = a).$$

We immediately find (12.5) and (12.6), and the recursion (12.7) for $n \geq 2$ is derived by conditioning on the state c just before the last jump and noting if the state change from c to b is of type α to β or not:

$$P_{\alpha\beta}(k, n, a, b) = \sum_{c} P(\sum_{i=1}^{n} 1((z_{i-1}, z_i) = (\alpha, \beta)) = k, z_{n-1} = c, z_n = b|z_0 = a).$$

Qed

Implementation is easier when the initialization and recursion are written in matrix form. Let $U(\alpha, \beta)$ be the matrix with 1 in entry (α, β) and 0 in all other entries, and let $*$ denote elementwise matrix multiplication. We get the initialization

$$\begin{aligned}
P_{\alpha\beta}(0, 0) &= I, \\
P_{\alpha\beta}(0, 1) &= R - R * U(\alpha, \beta), \\
P_{\alpha\beta}(1, 1) &= R * U(\alpha, \beta).
\end{aligned}$$

For $n \geq 2$ and $k = 0$, we have

$$P_{\alpha\beta}(0, n) = P_{\alpha\beta}(0, n-1)(R - R * U(\alpha, \beta)),$$

and when $k = 1, \ldots, n$, we get

$$\begin{aligned}
&P_{\alpha\beta}(k, n) \\
&= P_{\alpha\beta}(k-1, n-1)(R * U(\alpha, \beta)) + P_{\alpha\beta}(k, n-1)(R - R * U(\alpha, \beta)).
\end{aligned}$$

Theorem 4. *The endpoint-conditioned distribution for D_α has a mixed distribution with point probabilities at 0 and T and a continuous distribution between these endpoints. The distribution of D_α is given as follows. Avoid state α:*

$$P(D_\alpha = 0, X(T) = b|X(0) = a) = \sum_{n=0}^{\infty} P_\alpha(0, n, a, b)\text{Pois}(n; \mu T).$$

Never leave state α:

$$P(D_\alpha = T, X(T) = b|X(0) = a)$$
$$= \sum_{n=0}^{\infty} P_\alpha(n+1, n, a, b)\text{Pois}(n; \mu T) = e^{-Q_a T} 1(a = b = \alpha).$$

Continuous distribution for $0 < t < T$:

$$f(t; a, b) = \sum_{n=1}^{\infty}\sum_{k=1}^{n} \frac{1}{T} f_B(t/T; k, n-k+1) P_\alpha(k, n, a, b)\text{Pois}(n; \mu T),$$

where $f_B(u; \lambda_1, \lambda_2)$ *is the Beta density. For* $n = 0$,

$$P_\alpha(0, 0, a, b) = 1(a = b \neq \alpha), \quad \textit{and also} \quad P_\alpha(1, 0, a, b) = 1(a = b = \alpha).$$

For $n \geq 1$ *and* $k = 0, \ldots, n+1$, *we have the recursion*

$$P_\alpha(k, n, a, b) = \begin{cases} \sum_c P_\alpha(k-1, n-1, a, c) R_{cb} & \textit{if } b = \alpha, \\ \sum_c P_\alpha(k, n-1, a, c) R_{cb} & \textit{if } b \neq \alpha, \end{cases} \tag{12.8}$$

where we define $P_\alpha(-1, \cdot, \cdot, \cdot) = 0$.

Proof. Let

$$P_\alpha(k, n, a, b) = P(\sum_{i=0}^{n} 1(z_i = \alpha) = k, z_n = b|z_0 = a).$$

Suppose the chain makes n jumps and let $\sum_{i=0}^{n} 1(z_i = \alpha) = k$. The chain never visits state α when $k = 0$, and never leaves α when $k = n+1$.

Now consider the case $1 \leq k \leq n$. The interarrival times are $W_i = T_i - T_{i-1}$ for $i = 1, \ldots, n+1$, and where $T_0 = 0$ and $T_{n+1} = T$. The $W_1, \ldots, W_{n+1}$ are independent and exponentially distributed variables with rate μ. In particular, the vector $(W_1, W_2, \ldots, W_{n+1})/T$ conditional on the sum of the interarrival times being T, $\sum_{i=1}^{n+1} W_i = T$, follows a Dirichlet distribution with parameter $(1, \ldots, 1)$. Consequently, the marginal distribution of any k interarrival times, $\sum_{i=1}^{k} W_i/T$, follows a Beta distribution with parameter $(k, n+1-k)$.

The recursion (12.8) for $n \geq 1$ is derived by conditioning on the state c just before the last jump and noting if the last state is of type α or not:

$$P_\alpha(k, n, a, b) = \sum_c P\Big(\sum_{i=0}^{n} 1(z_i = \alpha) = k, z_{n-1} = c, z_n = b \Big| z_0 = a\Big).$$

Qed

In matrix notation, the recursion is given by

$$P_\alpha(0, 0) = I - U(\alpha, \alpha), \quad P_\alpha(1, 0) = U(\alpha, \alpha),$$

and for $n \geq 1$

$$P_\alpha(k, n) = P_\alpha(k-1, n-1)(R * V(\alpha)) + P_\alpha(k, n-1)(R - R * V(\alpha)),$$

where $V(\alpha)$ is the matrix with 1 in all entries in column α and 0 otherwise.

Example

As an illustration of densities for endpoint-conditioned numbers of substitutions and time spent in a state, consider the HKY model with rate matrix

$$Q = Q(\kappa, \pi) = \begin{bmatrix} \cdot & \kappa\pi_{\mathtt{G}} & \pi_{\mathtt{C}} & \pi_{\mathtt{T}} \\ \kappa\pi_{\mathtt{A}} & \cdot & \pi_{\mathtt{C}} & \pi_{\mathtt{T}} \\ \pi_{\mathtt{A}} & \pi_{\mathtt{G}} & \cdot & \kappa\pi_{\mathtt{T}} \\ \pi_{\mathtt{A}} & \pi_{\mathtt{G}} & \kappa\pi_{\mathtt{C}} & \cdot \end{bmatrix}.$$

The parameter κ distinguishes between *transitions* (substitutions within purines (A,G) or pyrimidines (C,T)) and *transversions* (substitutions between purines and pyrimidines), and is reversible with stationary distribution $\pi = (\pi_{\mathtt{A}}, \pi_{\mathtt{G}}, \pi_{\mathtt{C}}, \pi_{\mathtt{T}})$.

We let $\kappa = 2$, $\pi = (0.2, 0.3, 0.3, 0.2)$ and scale the matrix such that one substitution is expected per time unit $\sum_i Q_i \pi_i = 1$. In Figure 12.3, we show the distribution of the number of substitutions from G to T when $Time = 1$.

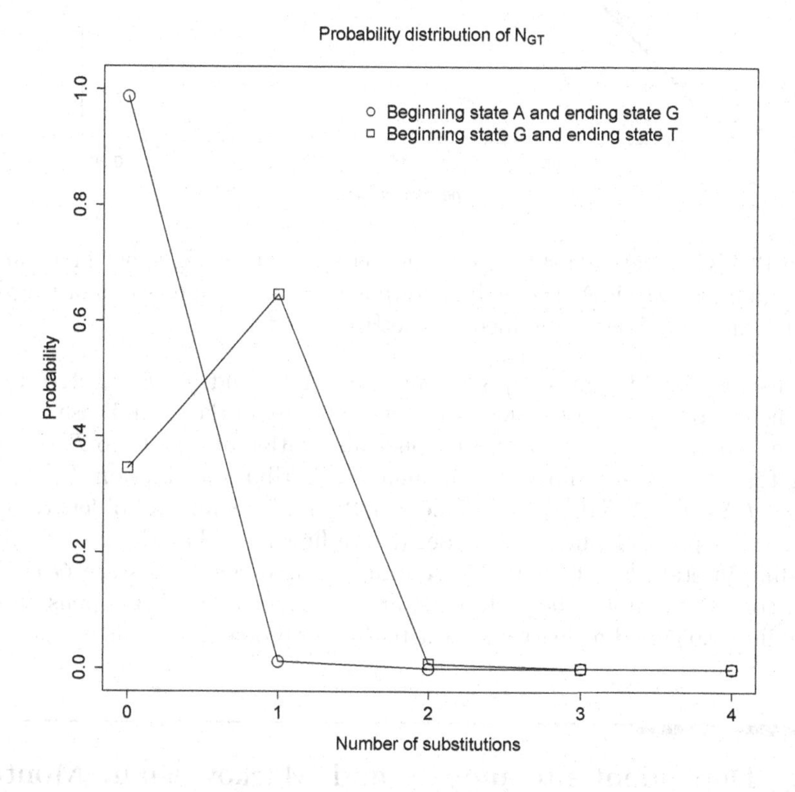

Figure 12.3: Probability distribution of the number of substitutions N_{GT} from state G to state T with two sets of endpoints: Beginning state A and ending state G, and beginning state G and ending state T. The CTMC is determined by the HKY model, and the time interval is of length 1.

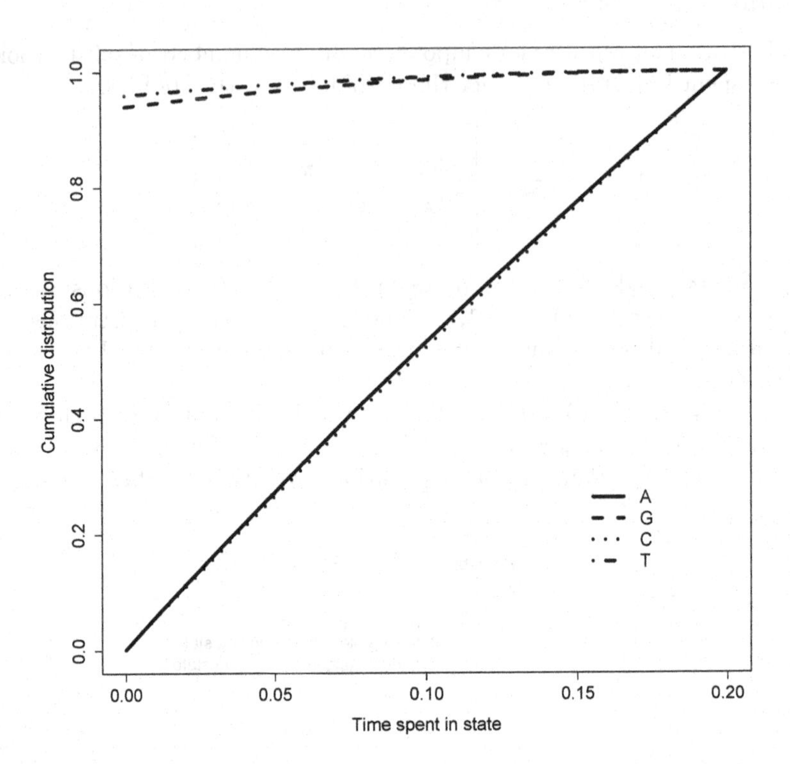

Figure 12.4: Cumulative distribution function for the time D_α spent in a state. The beginning state is A, the ending state is G, the time interval is of length 0.2, and the CTMC is determined by the HKY model.

We consider two different endpoints, namely (A, G) and (G, T). In the first case, there is rarely a substitution of the desired type. In the second case, there is more than a 60% chance of exactly one substitution of type G to T.

In Figure 12.4, we show the cumulative distribution function $F_\alpha(t) = P(D_\alpha \leq t | X(0) = A, X(0.2) = C, Time = 0.2)$ for a time interval of length 0.2 and for all 4 possible nucleotide types α. We first note that the probability of ending in state $b = C$ is 0.055. A sample might not visit state G or T at all; conditional upon the endpoints, at least one visit to G happens with probability 0.055 and at least one visit to T with probability 0.036.

12.3 Dependent-site models and Markov chain Monte Carlo

As noted earlier, Felsenstein's pruning algorithm can become computationally infeasible when nucleotide substitution rates at a sequence position are affected by the types of nucleotides at other sites. The problem is that the number of possible DNA sequences of length L is 4^L and L need not be very large for it to be computationally onerous to determine and handle the $4^L \times 4^L$ transition probability matrix that the pruning algorithm requires. Rather than computing a likelihood via the pruning algorithm, an alternative is to exploit the simple form for the likelihood of a fully observed Markov process (see (12.1)).

Markov chain Monte Carlo facilitates such exploitation. The Markov process that generates sequence data at the endpoints (i.e., tips) of the phylogenetic tree is not fully observed, but the observed sequence data at the tips can be augmented with a possible history that generated them. Markov chain Monte Carlo can then be used to sample sequence histories according to their endpoint-conditioned probability densities. This was the idea of the pioneering work by Jensen and Pedersen (Jensen and Pedersen, 2000; Pedersen and Jensen, 2001). Although their approach was originally applied only to sequence pairs, the extension of their technique to a larger number of sequences that are related by a phylogeny is relatively straightforward.

To apply (12.1) and to successfully implement the Jensen-Pedersen technique, calculation of Q_α (i.e., the rate at which the Markov process exits state α) must be computationally tractable for each possible state α. This is a potentially important limitation of the Jensen-Pedersen technique. We have

$$Q_\alpha = \sum_{\beta \neq \alpha} Q_{\alpha\beta}.$$

If α represents a specific sequence of length L, then Q_α is a sum of $4^L - 1$ rates. Unless L is small, determining this sum can be prohibitive. However, the conventional assumption with models of molecular evolution is that only a single sequence position can change at a time (but see Whelan and Goldman (2004)). With this conventional assumption, Q_α can be determined by summing the $3L$ rates $Q_{\alpha\beta}$ that may be non-zero because they represent single nucleotide changes from α. As the number of terms to be summed is linear rather than exponential in L, determining this sum can be accomplished relatively quickly even when the parameterization underlying the rates $Q_{\alpha\beta}$ is relatively involved.

The Jensen-Pedersen approach can be applied to evolutionary models that have a relatively simple form of dependence among sites such as that which may be due to neighbor-dependent mutation, but it can also be successfully applied to evolutionary models that involve more complicated forms of dependence among sites such as that which arises from natural selection on protein tertiary structure or on some other phenotypic characteristic that is partially or completely encoded by genotype. Whether the dependence arises from mutation or natural selection, some way must be developed for proposing modifications to endpoint-conditioned histories so that the Markov chain Monte Carlo procedure can sample histories according to the appropriate probability density. Next, we overview a Gibbs sampling procedure that seems well-suited for local dependence due to context-dependent mutation. A crucial element of the Gibbs sampling procedure is the ability to sample endpoint-conditioned histories from a CTMC. We provide a detailed summary of the various path sampling algorithms, and then we follow with a brief discussion of Metropolis-Hastings proposals of endpoint-conditioned histories.

12.3.1 Gibbs sampling with context dependence

Hwang and Green (2004) published one of the most important works on molecular evolution in mammals. These authors developed a careful nucleotide substitution model of a 1.7 megabase genomic region in a dataset representing 19 species of mammals. One important finding was that different substitution types violate, to different degrees, the molecular clock hypothesis of a constant rate of sequence change over time. Specifically, Hwang and Green (2004) concluded that the high rates of substitution from CpG to TpG in mammals exhibit relatively little variation over time when compared to other substitution types.

Because Hwang and Green analyzed such a large genomic region, they could consider a relatively parameter-rich and a relatively biologically plausible treatment of the substitution process. The process allowed for different substitution rates and patterns in transcribed and nontranscribed regions. It also permitted distinct substitution parameters in annotated repeats. Most importantly, the substitution rate at a particular position was context-dependent in that it could be influenced by the 5' and 3' flanking nucleotides. For nontranscribed regions, the parameter space was somewhat reduced by forcing the rate at which one sequence changes to another to be equal to the value when the sequence being changed and the sequence that results are replaced by their reverse complements. This constraint on the substitution rates is sometimes referred to as strand symmetry (e.g., Yap and Speed (2004)).

The inference procedure adopted by Hwang and Green (2004) was innovative. One clever decision made by Hwang and Green (2004) was to discretize time along branches of the evolutionary tree. Special care was taken to make the time intervals so small that each particular sequence position and its local context were unlikely to change during a given interval. This means that the possibility of multiple sequence changes affecting a position and its context during some time interval could be neglected. It also means that the probability of a specific change can be well approximated as the product of its rate and the length of the time interval. This can eliminate the need for computationally expensive calculations that convert a matrix of substitution rates into a transition probability matrix.

Hwang and Green (2004) adopted a Bayesian perspective and employed Markov chain Monte Carlo techniques. They augmented the observed sequences with sequence paths along the branches of the phylogeny that were constrained to end at the observed sequences. The Markov chain Monte Carlo procedure was employed to approximate the joint posterior density of sequence paths and parameters. Changes to the sequence path were made by randomly selecting a particular sequence site and then using Gibbs sampling to generate a new

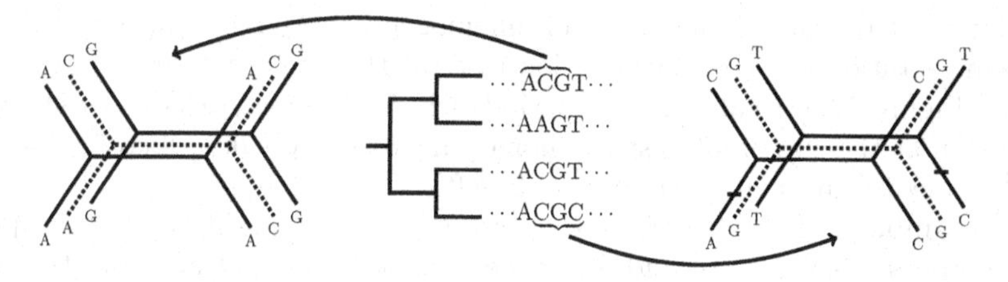

Figure 12.5: Gibbs sampling. Left: A history (represented as the dotted line) at a particular site in the alignment (shown in the middle) is updated, conditional on the histories of the left and right sites. The left and right histories are constant (the left nucleotide is A and the right nucleotide is G), and the history is sampled using the corresponding neighbor-dependent rate matrix. Right: Updating the history at this site must take into account that the flanking nucleotides are not constant. Three rate matrices are needed corresponding to the (left, right) nucleotides being (A,T), (C,T), or (C,C).

endpoint conditioned history at the site. Therefore, consecutive sequence histories in the Markov chain formed by Gibbs sampling only differ by the path at the randomly selected site (see Figure 12.5).

Whereas Hwang and Green (2004) present their Gibbs sampling algorithm when time is discretized, Hobolth (2008) introduces a Gibbs sampling algorithm for continuous time. Rather than detailing the Gibbs sampler of Hwang and Green (2004) here, we instead describe the version of Hobolth (2008) because it uses the continuous time framework that is employed throughout this chapter.

We consider a context-dependent model where the substitution rate from state a to state b at a site is potentially affected by the nucleotide type l to the left (i.e., 5') of the site and by the nucleotide type r to the right (i.e., 3') of the site. We denote this substitution rate as $Q(b; l, a, r)$. In total we have 4^3 possibilities for (l, a, r) and 3 possibilities for b; the rate matrix has $4^3 \times 3 = 192$ free parameters. The parameters can be estimated using the iterative EM algorithm described earlier for the independent sites model. In iteration $i + 1$, the new value $Q^{(i+1)}(b; l, a, r)$ of the neighbor-dependent rates are updated from the conditional means $E[N(b; l, a, r)|Q^{(i)}, y]$ of the number of substitutions $N(b; l, a, r)$ from a to b with left l and right r neighbors, and the conditional time $E[D(l, a, r)|Q^{(i)}, y]$ spent in a state (l, a, r), conditional on the data y of sequences observed in the leaves of the tree and the present value of the rate parameters $Q^{(i)}$. However, for neighbor-dependent models, we cannot find the conditional means analytically, but must resort to approximative or sampling-based procedures. Gibbs sampling seems an attractive choice for estimating the conditional means.

Gibbs sampling of an endpoint-conditioned path with context-dependence is more complicated than doing so for a model with independently evolving sequence positions because the rate $Q(b; l, a, r)$ can be affected by substitutions that alter the nucleotide type to either the left or the right of the site of interest.

Evolution at the site of interest is an inhomogeneous Markov process. When a flanking position changes, substitution rates at the site of interest can change (recall Figure 12.5, right). During periods where flanking sites do not change, evolution at the site of interest is a homogeneous process described by a 4×4 instantaneous rate matrix (Figure 12.5, left).

To sample a history at the site of interest that is conditioned both upon the endpoints for the site and upon the nucleotide types that flank the site of interest at each instant during the history, Hobolth (2008) modifies the sampling techniques as described below. The key modification is to divide the time duration represented by a branch into intervals. Each substitution at a flanking site ends one time interval and starts another. Within each interval, a single 4×4 instantaneous rate matrix applies and the transition probability matrix for the interval is obtained as usual by exponentiating the rate matrix. With the transition probability matrices as well as the states that begin and end the branch, a straightforward modification of Felsenstein's pruning algorithm can jointly sample the states of the site of interest at the breakpoints that separate time intervals. Sampling the path of the site conditional upon the states that begin and end the time intervals can be performed with any of the path sampling techniques described below. Hobolth (2008) employed direct sampling.

Gibbs sampling has the advantage that Markov chain Monte Carlo proposals are never rejected. However, models with widespread dependencies among sites may be poorly suited to Gibbs sampling. A potential obstacle with Gibbs sampling is that the substitution process at a site can be altered each time a substitution occurs at some sequence position that influences it. This means that the time intervals during which the site evolves according to a homogeneous Markov process are likely to be large in number and short in duration. Because the continuous-time Gibbs sampling procedure must carefully account for the 4×4 instantaneous rate matrix that operates during each time interval, it may not be worthwhile to use Gibbs sampling when the 4×4 instantaneous rate matrix is apt to change many times in the history of the sequence position.

12.3.2 Path sampling

In this section, we review four strategies for simulating a realization of a finite-state CTMC $\{X(t) : 0 \leq t \leq T\}$ conditional on its beginning state $X(0) = a$ and ending state $X(T) = b$. See Hobolth and Stone (2009b) and Asmussen and Hobolth (2012) for a more detailed comparison of the path sampling algorithms.

To understand the sampling difficulties associated with conditioning a CTMC on its endpoints, it is useful to first review how one proceeds when the ending state $X(T)$ is unobserved. Simulating a sample path of $\{X(t) : 0 \leq t \leq T\}$ that begins at $X(0) = a$ can be accomplished by a simple iterative procedure. The key observation is that the waiting time τ to the first state change is exponentially distributed with mean $1/Q_a$. If $\tau > T$, there is no state change in the interval $[0, T]$, and the corresponding sample path is constant; otherwise, a new state c is drawn from the discrete probability distribution with probability masses Q_{ac}/Q_a and the procedure is iterated for the shorter time interval $[\tau, T]$ (or equivalently, for $[0, T-\tau]$). For reference, we present this forward sampling algorithm below:

Algorithm 1. *Forward Sampling*

1. Sample $\tau \sim \text{Exponential}(Q_a)$. If $\tau \geq T$, we are done: $X(t) = a$ for all $t \in [0, T]$.
2. If $\tau < T$, choose a new state $c \neq a$ from a discrete probability distribution with probability masses Q_{ac}/Q_a. Repeat the procedure with new beginning state c and new time interval $[\tau, T]$.

Under the assumption that the ending state $X(T) = b$ is observed, conditioning excludes all paths sampled from the preceding algorithm that fail to end in state b. This is the essence of the rejection sampling approach, whose modification by Nielsen (2002) we discuss in the next subsection.

12.3.2.1 Rejection sampling

Naive rejection sampling uses forward sampling to generate candidate sample paths of an endpoint-conditioned CTMC. From these, the acceptable candidates are those for which the simulated ending state and the observed ending state are the same. In particular, when sampling forward, the probability of hitting the observed ending state b is $P_{ab}(T) = \exp(Qt)_{ab}$. Thus, if T is large, this probability approximately equals the stationary probability π_b of b. Conversely, if T is small and $a \neq b$, the probability is approximately $Q_{ab}T$. It follows that in case of (i) large time T and small stationary probability π_b, or (ii) different states $a \neq b$ and small time T, naive rejection sampling requires excessive computation because so many sample paths are rejected.

Sometimes, the goal is to generate sample paths for a large number of combinations of starting and ending points. This can occur if Q governs evolution at a large number of independently evolving sequence positions and an endpoint-conditioned path is desired for each site. In this situation, computational feasibility of naive rejection sampling can be partially salvaged by caching sample paths that do not end with the desired state. These sample paths can then be employed later when the state at which they end is desired (Yu and Thorne, 2006).

Nielsen's modification improves naive rejection sampling when the time T is small and the beginning and ending states differ (Nielsen, 2002). By a conditioning argument, the time τ to the first state change, given that at least one state change occurs before T and $X(0) = a$ has density according to a truncated exponential distribution

$$f(\tau) = \frac{Q_a e^{-\tau Q_a}}{1 - e^{-TQ_a}}, \quad 0 \leq \tau \leq T. \tag{12.9}$$

The corresponding cumulative distribution function is

$$F(\tau) = \frac{1 - e^{-\tau Q_a}}{1 - e^{-TQ_a}}, \quad 0 \leq \tau \leq T, \tag{12.10}$$

with explicit inverse

$$F^{-1}(u) = -\log\left[1 - u(1 - e^{-TQ_a})\right]/Q_a. \tag{12.11}$$

Thus, upon sampling u from a Uniform$(0,1)$ distribution, transformation yields the sample waiting time $F^{-1}(u)$ to the first state change of the CTMC.

Algorithm 2. *Modified Rejection Sampling*

If $a = b$:

1. Simulate from $\{X(t) : 0 \leq t \leq T\}$ using the forward sampling algorithm.
2. Accept the simulated path if $X(T) = a$; otherwise, return to step 1 and begin anew.

If $a \neq b$:

1. Sample τ from the density (12.9) using the inverse transformation method, and choose a new state $c \neq a$ from a discrete probability distribution with probability masses Q_{ac}/Q_a.
2. Simulate the remainder $\{X(t) : \tau \leq t \leq T\}$ using the forward sampling algorithm from the beginning state $X(\tau) = c$.
3. Accept the simulated path if $X(T) = b$; otherwise, return to step 1 and begin anew.

In short, modified rejection sampling explicitly avoids simulating constant sample paths when it is known that at least one state change must take place. This is particularly beneficial when T is small, as the naive approach will be dominated by wasted constant sample paths whose ending state remains a (which occurs with probability approximately $(1 - Q_a T)$). Nevertheless, if the transition from a to b is unlikely, so that Q_{ab}/Q_a is small, then essentially every sample path will still be rejected. In such a setting, either direct sampling, uniformization, or bisectioning is required.

12.3.2.2 Direct sampling

The direct sampling procedure of Hobolth (2008) requires that the instantaneous rate matrix Q admits an eigenvalue decomposition. Under that assumption, let U be a matrix with orthogonal eigenvectors as columns and let D_λ be the diagonal matrix of corresponding eigenvalues such that $Q = UD_\lambda U^{-1}$.

Then, for any time t, the transition probability matrix of the CTMC can be calculated as

$$P(t) = e^{Qt} = Ue^{tD_\lambda}U^{-1} \text{ and } P_{ab}(t) = \sum_j U_{aj}U_{jb}^{-1}e^{t\lambda_j}. \qquad (12.12)$$

Consider first the case where the endpoints of the CTMC are identical so that $X(0) = X(T) = a$. The probability that there are no state changes in the time interval $[0, T]$ conditional on $X(0) = a$ and $X(T) = a$ is given by

$$p_a = \frac{e^{-Q_a T}}{P_{aa}(T)}. \qquad (12.13)$$

Furthermore, with probability $(1 - p_a)$, at least one state change occurs. Thus, when $X(0) = X(T) = a$, the sample path is constant with probability p_a, and has at least one change with probability $(1 - p_a)$.

Next consider the case where $X(0) = a$ and $X(T) = b$, with $a \neq b$. Let τ denote the waiting time until the first state change. The conditional probability that the first state change is to i at a time smaller than t is

$$\begin{aligned}
&P(\tau \leq t, X(\tau) = i | X(0) = a, X(T) = b) \\
&= P(\tau \leq t, X(\tau) = i, X(T) = b | X(0) = a)/P(X(T) = b | X(0) = a) \\
&= \int_0^t Q_a e^{-Q_a s} \frac{Q_{ai}}{Q_a} \frac{P_{ib}(T - s)}{P_{ab}(T)} ds \\
&= \int_0^t f_i(s) ds,
\end{aligned}$$

where $f_i(s)$ is the integrand. Specifically, conditional on the endpoints $X(0) = a$ and $X(0) = b$, the probability p_i that the first state change is to i is

$$p_i = \int_0^T f_i(t)dt, \quad i \neq a, \quad a \neq b. \qquad (12.14)$$

Using (12.12) we can rewrite the integrand as

$$f_i(t) = Q_{ai}e^{-Q_a t}\frac{P_{ib}(T - t)}{P_{ab}(T)} = \frac{Q_{ai}}{P_{ab}(T)} \sum_j U_{ij}U_{jb}^{-1}e^{T\lambda_j}e^{-t(\lambda_j + Q_a)}, \qquad (12.15)$$

which renders the integral in (12.14) straightforward. We get

$$p_i = \frac{Q_{ai}}{P_{ab}(T)} \sum_j U_{ij}U_{jb}^{-1}J_{aj}, \qquad (12.16)$$

where

$$J_{aj} = \begin{cases} Te^{T\lambda_j} & \text{if } \lambda_j + Q_a = 0, \\ \frac{e^{T\lambda_j} - e^{-Q_a T}}{\lambda_j + Q_a} & \text{if } \lambda_j + Q_a \neq 0. \end{cases}$$

We now have a procedure for simulating the next state and the waiting time before the state change occurs. Iterating the procedure allows us to simulate a sample path $\{X(t) : 0 \leq t \leq T\}$ that begins in $X(0) = a$ and ends in $X(T) = b$.

Algorithm 3. *Direct Sampling*

1. If $a = b$, sample $Z \sim \text{Bernoulli}(p_a)$ where p_a is given by (12.13). If $Z = 1$ we are done: $X(t) = a,\ 0 \leq t \leq T$.
2. If $a \neq b$ or $Z = 0$, then at least one state change occurs. Calculate p_i for all $i \neq a$ from (12.16). Sample $i \neq a$ from the discrete probability distribution with probability masses $p_i/p_{-a},\ i \neq a$, where $p_{-a} = \sum_{j \neq a} p_j$. (Note that $p_{-a} = 1$ when $a = b$ and $p_{-a} = (1 - p_a)$ otherwise.)
3. Sample the waiting time τ in state a according to the continuous density $f_i(t)/p_i,\ 0 \leq t \leq T$, where $f_i(t)$ is given by (12.15). Set $X(t) = a,\ 0 \leq t < \tau$.
4. Repeat procedure with new starting value i and new time interval of length $T - \tau$.

Remark 1. *In step 3 above, we simulate from the scaled density (12.15) by finding the cumulative distribution function and then use the inverse transformation method. To calculate the cumulative distribution function, note that*

$$\int_0^t e^{T\lambda_j} e^{-s(\lambda_j + Q_a)} ds = \begin{cases} te^{T\lambda_j} & \textit{if } \lambda_j + Q_a = 0, \\ e^{T\lambda_j} \frac{1}{\lambda_j + Q_a}(1 - e^{-t(\lambda_j + Q_a)}) & \textit{if } \lambda_j + Q_a \neq 0. \end{cases}$$

To use the inverse transformation method, we must find the time t such that $F(t) - u = 0$ where $F(t)$ is the cumulative distribution function and $0 < u < 1$. We have used a numerical root finder for this purpose.

12.3.2.3 Uniformization

Here, we describe how uniformization can be used to construct an algorithm for exact sampling from $X(t)$, conditional on the beginning and ending states. It follows directly from (12.3) that the transition function of the Markov chain subordinated to a Poisson process is given by

$$P_{ab}(t) = P(X(t) = b | X(0) = a) = e^{-\mu t} 1_{(a=b)} + \sum_{n=1}^{\infty} e^{-\mu t} \frac{(\mu t)^n}{n!} R^n_{ab}.$$

Thus, the number of state changes N (including the virtual) for the conditional process that starts in $X(0) = a$ and ends in $X(T) = b$ is given by

$$P(N = n | X(0) = a, X(T) = b) = e^{-\mu T} \frac{(\mu T)^n}{n!} R^n_{ab} / P_{ab}(T). \tag{12.17}$$

Putting these things together, we have the following algorithm for simulating a continuous time Markov chain $\{X(t) : 0 \leq t \leq T\}$ conditional on the starting state $X(0) = a$ and ending state $X(T) = b$.

Algorithm 4. *Uniformization*

1. Simulate the number of state changes n from the distribution (12.17).
2. If the number of state changes is 0 we are done: $X(t) = a,\ 0 \leq t \leq T$.
3. If the number of state changes is 1 and $a = b$ we are done: $X(t) = a,\ 0 \leq t \leq T$.
4. If the number of state changes is 1 and $a \neq b$ simulate t_1 uniformly random in $[0, T]$, we are done: $X(t) = a,\ t < t_1$, and $X(t) = b,\ t_1 \leq t \leq T$.
5. When the number of state changes n is at least 2, simulate n independent uniform random numbers in $[0, T]$ and sort the numbers in increasing order to obtain the times of state changes $0 < t_1 < \cdots < t_n < T$. Simulate $X(t_1), \ldots, X(t_{n-1})$ from a discrete-time Markov chain with transition matrix R and conditional on starting state $X(0) = a$ and ending state $X(t_n) = b$. Determine which state changes are virtual and return the remaining changes and corresponding times of change.

Remark 2. *In Step 1 above, we find the number of state changes n by simulating u from a Uniform(0,1) distribution and letting n be the first time the cumulative sum of (12.17) exceeds u. When calculating the cumulative sum, we need to raise R to powers 1 through n. These powers of R are stored because they are required in Step 5 of the algorithm. The matrix exponential $P_{ab}(t) = (\exp(Qt))_{ab}$ can be calculated in many different ways and is available in many software programs; see Moler and Loan (2003) for a review and Al-Mohy and Higham (2009) for recent progress.*

Remark 3. *Rao and Teh (2012) have combined uniformization with Gibbs sampling in order to generate paths $\{X(t) : 0 \leq t \leq T\}$ conditional upon a beginning state $X(0) = a$ and an ending state $X(T) = b$. The Rao-Teh Gibbs sampler alternates between sampling new state changes (real and virtual) conditional upon jump times and new virtual jump times conditional upon the real events (real state changes and corresponding times). Their algorithm starts with a possible path of n events with times $0 < t_1 < \ldots < t_n < T$. Some of these events may be virtual and the others are real. The Rao-Teh Gibbs sampler iterates between two steps. The first step is to update the states conditional upon the times $t_1, \ldots, t_n$. In the new path, an event occurring at a certain time might be virtual whereas the previous path had the event being real (or vice versa). This step amounts to sampling from a discrete Markov chain with known endpoints. The second step is to update the virtual jump times. Rao and Teh note that the distribution of virtual events during a time interval where $X(t) = \alpha$ is determined by a Poisson process with intensity $\mu - \alpha$. This means that a new set of virtual events between 0 and T can be sampled conditional upon the times of the real events. The new set of virtual events together with the real events constitute a new path conditional upon* *beginning state $X(0) = a$ and ending state $X(T) = b$. It is essential that $\mu > \max_i Q_i$ in order to introduce new virtual jumps. Rao and Teh (2012) set $\mu = 2\max_i Q_i$ and find that the Gibbs sampler typically converges after less than five iterations. A computational advantage of this Gibbs sampler is that paths can be generated without the need to determine the transition probability $P_{ab}(T)$ that is in the denominator of Equation 12.17. This is especially likely to be advantageous when the instantaneous rate matrix is sparse but with a high dimension.*

12.3.2.4 Bisectioning

The bisection algorithm (Asmussen and Hobolth, 2008) is based on two fundamental observations:

1. If $X(0) = X(T) = a$ and there are no jumps, we are done: $X(t) = a,\ 0 \le t \le T$.
2. If $X(0) = a$ and $X(T) = b \ne a$ and there is precisely one jump, we are basically done: $X(t) = a,\ 0 \le t < \tau$, and $X(t) = b,\ \tau \le t \le T$.

Regarding the second observation, the jump time τ is determined by the lemma and corresponding remark below.

Algorithm 5. *Bisection Sampling*
The bisection algorithm is a recursive procedure where we finish off intervals with zero or one jumps according to the two fundamental observations above, and keep bisecting intervals with two or more jumps. The recursion ends when no intervals with two or more jumps are present.

The following lemma and remark show that intervals with precisely one jump are easy to handle.

Lemma 1. *Consider an interval of length T with $X(0) = a$, and let $b \ne a$. The probability that $X(t) = b$ and there is only one single jump (necessarily from a to b) in the interval is given by*

$$R_{ab}(T) = Q_{ab}\begin{cases} \dfrac{e^{-Q_aT} - e^{-Q_bT}}{Q_b - Q_a} & Q_a \ne Q_b, \\[2ex] Te^{-Q_aT} & Q_a = Q_b. \end{cases} \tag{12.18}$$

The density of the time of state change is

$$f_{ab}(t;T) = \frac{Q_{ab}e^{-Q_bT}}{R_{ab}(T)}e^{-(Q_a-Q_b)t}, \quad 0 \le t \le T.$$

Furthermore, the probability that $X(T) = b$ and there are at least two jumps in the interval is $P_{ab}(T) - R_{ab}(T)$.

Proof. Let $N(T)$ denote the number of jumps in the interval $[0,T]$. The first two parts of the lemma follow from

$$\begin{aligned} R_{ab}(T) &= P\big(X(T)=j, N(T)=1 \,\big|\, X(0)=a\big) \\ &= \int_0^T Q_a e^{-Q_a t} \frac{Q_{ab}}{Q_a} e^{-Q_b(T-t)}\, dt \\ &= Q_{ab} e^{-Q_b T} \int_0^T e^{-(Q_a-Q_b)t}\, dt, \quad a \neq b. \end{aligned}$$

The last part is clear since the case of zero jumps is excluded by $a \neq b$. **Qed**

Remark 4. *If $Q_a = Q_b$, the single state change is uniformly distributed in the interval $[0,T]$. If $Q_a > Q_b$, the time of the state change is an exponentially distributed random variable truncated to $[0,T]$. Such a random variable V is easily simulated by inversion (recall (12.9)–(12.11)). If $Q_a < Q_b$, we have by symmetry that $f_{ab}(t)$ is the density of the random variable $T - V$, where V is an exponentially distributed random variable with rate $Q_b - Q_a$ truncated to $[0,T]$.* **Qed**

Thus far, we have outlined four competing strategies for simulating sample paths from an endpoint-conditioned CTMC. Though our discussion has been agnostic to the number of desired sample paths, this quantity has a direct and varied impact on the computational efficiency of each sampler. For example, while direct sampling requires a possibly time-consuming eigendecomposition of Q, and uniformization requires calculation of the matrix exponential $\exp(Qt)$, it is clear that one such computation will suffice even when multiple sample paths are desired. The number of sample paths desired from an endpoint-conditioned CTMC is application driven: estimation of some quantity, such as the expected number of visits to a given state, may require many sample paths, whereas the updating step in a Bayesian computation may require as few as one.

Updates of sample paths in Markov chain Monte Carlo approximations of posterior distributions warrant some additional discussion.

12.3.3 Metropolis-Hastings algorithm with dependence

The Metropolis-Hastings algorithm can also be used when augmenting observed sequence data with their detailed histories. As with the Gibbs samplers of Hwang and Green (2004) and Hobolth (2008), the Markov chain of sequence paths can be constructed with the Metropolis-Hastings algorithm by randomly selecting a site and then proposing a new sequence path via the sampling of a new endpoint-conditioned path for the selected site. Because they are generated according to the dependence model of interest, the proposed site paths are always accepted with the Gibbs sampling techniques. In contrast, the Metropolis-Hastings algorithm can have an endpoint-conditioned site path generated according to a simple independent-sites model and can then have the proposed sequence path be accepted or rejected according to the Metropolis-Hastings ratio.

Robinson et al. (2003) considered an evolutionary model that had dependent change among codons due to protein tertiary structure. They employed a Metropolis-Hastings algorithm to approximate the posterior distribution of (augmented) sequence histories. Changes were made to the sequence histories by modifying the endpoint-conditioned site path at a randomly selected sequence position. Site paths were proposed with a strategy closely resembling uniformization that was designed for a nucleotide substitution model that has independent evolution among sites and that is commonly referred to as the Felsenstein 1984 model (Felsenstein, 1989).

The dependence model of Robinson et al. (2003) was motivated by the observation that protein sequences seem to change more quickly during evolution than protein structures. The model was designed for datasets of aligned homologous protein-coding DNA sequences. In addition, the encoded proteins were assumed to share a known and experimentally determined tertiary structure. Rates of nonsynonymous substitution were parameterized so that changes which improve the fit of protein sequence to the known tertiary structure could have high rates whereas changes that worsen the fit of protein sequence to structure could have low rates. The model possessed dependent change among codons because compatibility between protein sequence and structure was partially determined by the pairwise interactions between amino acids that would be induced when the protein sequence was folded into the known structure.

Rodrigue and collaborators (Rodrigue et al., 2005, 2006) improved upon the protein-coding DNA substitution model of Robinson et al. (2003) in diverse ways that include using better measures of sequence–structure compatibility, being able to analyze more sequences, having a better and faster software implementation, and developing better ways to assess the fits of the resulting models. These authors randomly selected sequence sites and proposed endpoint-conditioned site paths with independent-sites models. They adopted the Metropolis-Hastings algorithm to accept or reject the proposed sequence paths that resulted from changing the site paths. In some of their implementations, Rodrigue and collaborators (e.g., see Rodrigue (2007)) proposed sequence paths that differ from the current one because of path changes at multiple sequence positions. They find that simultaneously changing multiple site paths in each Metropolis-Hastings proposal can yield Markov chains that mix well and that yield good and computationally tractable approximations to posterior distributions.

Yu and Thorne (2006) adopted much the same inference strategy as Robinson et al. (2003) but they investigated an evolutionary model where substitution rates could be influenced by the effect of the substitution on RNA secondary structure. Substitutions that were predicted to make the secondary structure more stable could have higher rates than substitutions that disrupted the secondary structure. Proposed changes to the sequence path in the

Metropolis-Hastings algorithm of Yu and Thorne (2006) were made in a similar fashion to the procedure of Robinson et al. (2003) except that Yu and Thorne (2006) employed naive forward sampling with an independent-sites model along with a system for caching simulated site paths that were not immediately needed.

Often, the times at which sequence changes occur can be considered nuisance parameters. Miklós et al. (2004) refer to the ordered set of sequences visited from the beginning to ending of a time interval as a trajectory and they refer to the trajectory together with the time of each sequence change as a history. Rather than augmenting observed data with a history (path), Miklós et al. (2004) consider augmentation with only the trajectory (see also Robinson (2003)). They present a recursive algorithm to calculate the likelihood of a trajectory by marginalizing over all histories that are consistent with the trajectory. Miklós et al. (2004) developed their innovative approach for analyzing non-aligned homologous sequence data, but the idea is also promising for the structurally constrained dependence-among-sites models described above. Potentially, marginalizing over times would make Markov chain Monte Carlo techniques more computationally tractable. However, the gains obtained via marginalization would be offset by computational expenses associated with the recursive algorithm of Miklós et al. (2004). A compromise would be to ignore trajectories that specify an improbably large number of sequence changes and that therefore have extremely low probability densities. Dividing time intervals into short stretches and then ignoring low-probability trajectories was one of the strategies underlying the extremely fast likelihood-based inference strategy of de Koning et al. (2010).

12.4 Future directions for sequence paths with dependence models

Widely used models of sequence change tend to be phenomenological (for a nice discussion, see Rodrigue (2007) and Rodrigue and Philippe (2010)). Their parameters may improve statistical fits of the models to data, but it can be challenging to assign a biological interpretation to the parameters. Because a major reason for developing and employing models of sequence change is to study natural selection, this phenomenological flavor of evolutionary models is likely to lessen over time.

Models of sequence evolution that provide the basis for characterizing mutation–selection balance are increasingly available. The mutation–selection approach was pioneered by Halpern and Bruno (1998). The key idea is that genetic differences between species are the result of mutations that fixed. Halpern and Bruno (1998) framed their model with some parameters that reflected mutation and others that represented natural selection. With their approach, Halpern and Bruno (1998) could use interspecific sequence data to infer population genetic parameters that Nielsen and Yang (2003) termed "scaled selection coefficients" and that are the product of twice the effective population size and the fitness difference caused by a mutation.

A variety of subsequent studies have adopted the Halpern-Bruno perspective (e.g., Nielsen and Yang (2003); Choi et al. (2007); Thorne et al. (2007); Choi et al. (2008); Yang and Nielsen (2008); Rodrigue et al. (2010)). The approach facilitates inferences about how changes in phenotype due to a mutation can be translated into scaled selection coefficients. The approach greatly benefits from the ability to predict the phenotypic effect of a mutation. Genotype (i.e., DNA) to phenotype mapping is a daunting prediction problem that is made even more difficult because environment can combine with genotype to influence phenotype. Fortunately, the mapping of genotype (and environment) to phenotype is arguably the dominant pursuit of modern biological research. As genotype–phenotype mapping improves due to this major investment of research effort, the potential for quantifying the influence of natural selection on phenotype will strengthen. Careful incorporation of natural selection into models of sequence evolution is bound to induce dependent change among sequence positions because it is the collective action of genomes that produces phenotype.

Augmentation of observed sequence data with possible sequence histories is not the only way to handle dependent change among positions due to natural selection, but it is a promising one. To fully realize the promise of augmentation for inference via Markov chain Monte Carlo, better ways of proposing changes to possible sequence histories are needed. This will be especially true for cases where dependence among sites is strong. Inferential challenges caused by dependence will be closely tied to the fitness landscape. If many of the most direct mutational paths between observed sequences are implausible because they pass through fitness valleys, only a small proportion of possible trajectories will have high posterior probability. Path proposal schemes that ignore the fitness landscape will then be apt to result in a high proportion of rejected path proposals and mixing may be poor for Markov chain Monte Carlo analyses. Rather than proposing a change to a sequence history by changing a single sequence position or an arbitrarily selected subset of positions, a better approach might be to simultaneously propose histories or portions of histories at a large number of interdependent sites. This would be especially attractive if the dependence structure among sites could be reflected in the scheme for simultaneously proposing the site histories. Ideas from other fields may be relevant for generating path proposals that are "aware" of the fitness landscape. For example, techniques exist for studying the path from one conformation of a protein (or other system of atoms) to another (Bolhuis et al., 2002) and maybe they can be adapted to endpoint-conditioned evolutionary inference.

Acknowledgments

We thank Alex Griffing for helpful comments. J.L. Thorne was supported by N.I.H. grants GM090201 and GM070806 and by NSF grant MCB-1021883.

13

Bayesian inference of species divergence times

Tracy A. Heath
Department of Integrative Biology, University of California, Berkeley, California, USA

Brian R. Moore
Department of Evolution and Ecology, University of California, Davis, California, USA

CONTENTS

13.1 Introduction

Phylogenies provide an explicit historical perspective that critically informs biological research in a vast and growing number of scientific disciplines. For many inference problems, the tree topology and/or branch lengths—rendered as the expected amount of character change—provide sufficient information. Many other inference problems, however, require an ultrametric tree that confers temporal information: i.e., where branch durations and node heights are proportional to absolute or relative time. The study of continuous and discrete traits, for example, relies on stochastic models in which the probability of change is proportional to relative time; the study of lineage diversification relies upon stochastic branching process models that leverage information on the relative waiting times between events, etc. For such inference problems, a relative time scale is often sufficient. Other evolutionary questions require an absolute time scale. The study of biogeographic history, for instance, may use models that incorporate information on the changing proximity of areas through time, studies that seek to assess the correlation between shifts in rates of lineage diversification and events in earth history require branch lengths proportional to geological time, etc.

The potential to inform numerous biological questions motivates the development of methods to estimate divergence times. Unfortunately, estimating divergence times under the most common (and most general) phylogenetic model—the "unconstrained" model—is problematic. Under this model, each branch length is an independent parameter that reflects the amount of character divergence (e.g., the expected number of substitutions per site) rather than time. Specifically, the branch-length parameter, ν, is the product of the rate of substitution along a branch, r, and the time duration of that branch, t; hence $\nu = rt$. Rate and time are therefore confounded under the unconstrained model: a given branch length may reflect a relatively slow substitution rate over a long interval of time, or a relatively fast substitution rate over a shorter interval, etc. (Thorne and Kishino, 2005). Either of these two combinations of rate and time (and an infinite number of other combinations) provide an equally likely explanation of the data under the unconstrained model.

Because rate and time are conflated, we cannot infer one without making an assumption about the other. The molecular clock hypothesis (Zuckerkandl and Pauling, 1962, 1965), for example, makes the bold assumption that all branches share a common substitution rate. Accordingly, trees inferred from data with clock-like rates of molecular evolution will have branch lengths (ν) that are proportional to relative time. Moreover, it is straightforward to provide an absolute time scale for these trees if a fossil of known age can be assigned to one of its nodes. Alas, reality has proven to be far more complex: variation in the rate of molecular evolution is a pervasive feature of empirical datasets. This observation is perhaps not surprising, in light of the many factors that might contribute to departures from constant rates of substitution, such as historical changes in generation time, effective population size, selection intensity, etc.

The desire to estimate divergence times from molecular sequence datasets that depart from a *strict* molecular clock has motivated the development of

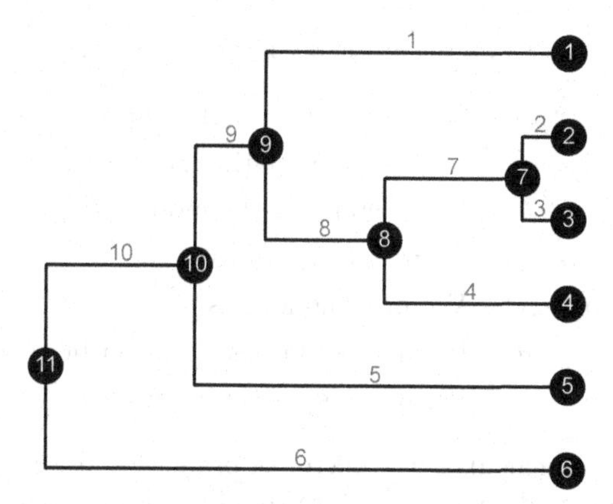

Figure 13.1: Notation for labeling branches and nodes. The tip nodes are labeled $1, \ldots, N$ and internal nodes labeled $N+1, \ldots, 2N-1$ (circles). Branch indices correspond to the index of their terminal node. Bayesian divergence-time methods focus on the ages of the internal nodes $(a_7, \ldots, a_{11})$ and the rates associated with the branch$(r_1, \ldots, r_{10})$. Some models describe the rates associated with the nodes instead of branches, for such models we use the notation r'_i to indicate the rate at node i.

so-called "*relaxed* molecular clock" models. These relaxed-clock models have mainly been pursued in a Bayesian statistical framework because it provides a natural means for us to impose assumptions about rates and times as prior probability densities and accommodate uncertainty in divergence-time estimates. Additionally, numerical methods used in Bayesian inference permit estimation under more complex models than would be possible using alternative approaches.

Bayesian divergence-time methods seek to compute the posterior probability density of substitution rates, node ages, and other phylogenetic model parameters given the sequence data and tree topology. Although some methods simultaneously estimate the tree topology and divergence times, we treat the topology as fixed (i.e., known) in the interest of simplifying our model notation. Figure 13.1 illustrates a fixed, rooted tree topology (τ) with $N = 6$ taxa. The tips are labeled $1, 2, \ldots, N$ and the internal nodes are labeled in postorder sequence, such that the root label is $2N - 1$. For example in Figure 13.1, the species are labeled $1, \ldots, 6$ and the internal nodes are labeled $7, \ldots, 11$. Branches are labeled $1, 2, \ldots, 2N - 2$, such that the index of the branch corresponds to the index of its terminal node. For many divergence-time models, it is convenient to identify the ancestor of a given node. We indicate the ancestor of node i using the notation $\sigma(i)$; e.g., the ancestor of node 5 is $\sigma(5) = 10$ (Figure 13.1).

We use the following notation for model parameters:

X	Sequence data (alignment assumed known)
τ	Rooted tree relating N species (assumed known)
θ_s	Parameters for the model of sequence evolution
$\boldsymbol{r} = (r_1, \ldots, r_{2N-2})$	Vector of branch rates
$\boldsymbol{a} = (a_{N+1}, \ldots, a_{2N-1})$	Vector of node ages
θ_r	Parameters for the prior on branch rates
θ_a	Parameters for the prior on node times

Additionally, we can calculate the vector of branch times, $\boldsymbol{t} = (t_1, \ldots, t_{2N-2})$, from the vector of node ages, $\boldsymbol{a}$, where the duration of the ith branch, t_i, is simply equal to the difference between the ages of its two nodes, $t_i = a_{\sigma(i)} - a_i$. The branch-length vector, $\boldsymbol{\nu} = (\nu_1, \ldots, \nu_{2N-2})$, is determined by the branch rates and times. Specifically, the length of the ith branch (in units of the expected number of substitutions per site) is $\nu_i = r_i t_i$. Note that although the branch lengths are used in the likelihood function, these parameters are deterministic variables—they are calculated using deterministic functions on other stochastic variables—and thus we do not include them in the equations below in order to simplify the notation.

Bayesian inference of divergence times is focused on estimating the joint conditional posterior probability distribution of branch rates and node ages:

$$f(\boldsymbol{r}, \boldsymbol{a}, \theta_r, \theta_a, \theta_s \mid X, \tau) = \frac{f(X \mid \boldsymbol{r}, \boldsymbol{a}, \theta_r, \theta_a, \theta_s, \tau) f(\boldsymbol{r}, \boldsymbol{a}, \theta_r, \theta_a, \theta_s)}{f(X|\tau)}, \tag{13.1}$$

where the joint posterior probability density of the parameters $[f(\cdot \mid X, \tau)]$ is equal to the likelihood $[f(X \mid \cdot)]$ times the joint prior probability density of the parameters $[f(\cdot)]$ divided by the marginal likelihood $[f(X|\tau)]$.

The likelihood function, $f(X \mid \boldsymbol{r}, \boldsymbol{a}, \theta_r, \theta_a, \theta_s, \tau)$, depends on the node ages (specifically the branch times, $\boldsymbol{t}$) and branch rates ($\boldsymbol{r}$). The likelihood can be simplified because the probability of the data relies only on the rate and time parameters (as well as the parameters of the model of sequence evolution, θ_s, and the tree topology, τ), but is independent of the processes generating the rates and ages:

$$f(X \mid \boldsymbol{r}, \boldsymbol{a}, \theta_r, \theta_a, \theta_s, \tau) = f(X \mid \boldsymbol{r}, \boldsymbol{a}, \theta_s, \tau).$$

Furthermore, the joint prior density on the model parameters, $f(\boldsymbol{r}, \boldsymbol{a}, \theta_r, \theta_a, \theta_s)$, can be separated into distinct components because we assume that the processes generating the rates and times are independent of the parameters of the model of sequence evolution:

$$f(\boldsymbol{r}, \boldsymbol{a}, \theta_r, \theta_a, \theta_s) = f(\boldsymbol{r} \mid \theta_r) f(\boldsymbol{a} \mid \theta_a) f(\theta_r) f(\theta_a) f(\theta_s).$$

Thus, for Bayesian divergence-time estimation, we are primarily interested in

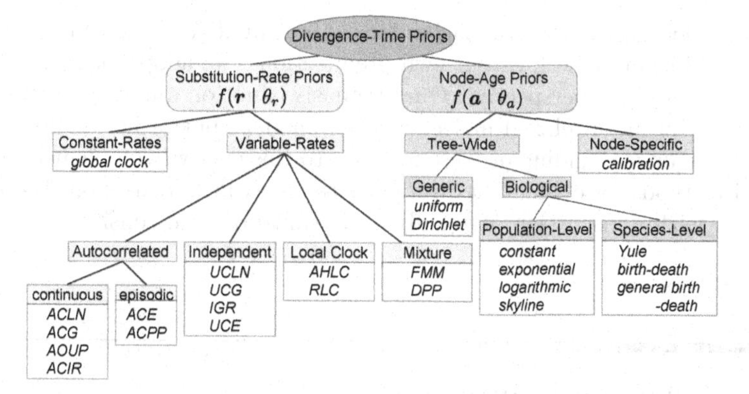

Figure 13.2: Bayesian approaches for estimating divergence times are all characterized by a prior describing the distribution of substitution rates across branches and a tree-wide prior on node ages. Substitution-rate priors span a range of variation from gradualistic to episodic; these models are arranged (to a very rough approximation) along this spectrum from left to right in the figure, and from top to bottom within a model category. Tree-wide priors on node ages have been proposed that are specific to population-level datasets, species-level datasets, or both (generic). In order to render times/rates on an absolute time scale, the model must also include one or more node-specific age priors. Calibrations are most commonly based (directly or indirectly) on information from the fossil record. Abbreviations are defined throughout the text.

specifying the prior model describing the vector of branch-specific substitution rates, $f(\boldsymbol{r} \mid \theta_r)$, and the prior distribution of node ages, $f(\boldsymbol{a} \mid \theta_a)$ (Figure 13.2). Furthermore, inference relies on numerical methods—Markov chain Monte Carlo (MCMC)—that conveniently eliminate the need to compute the marginal probability of the data, $f(X|\tau)$. This Bayesian framework allows us to calculate the joint conditional probability density on rates and times:

$$f(\boldsymbol{r}, \boldsymbol{a}, \theta_r, \theta_a, \theta_s \mid X, \tau) \propto \\ f(X \mid \boldsymbol{r}, \boldsymbol{a}, \theta_s, \tau) f(\boldsymbol{r} \mid \theta_r) f(\boldsymbol{a} \mid \theta_a) f(\theta_r) f(\theta_a) f(\theta_s). \quad (13.2)$$

Priors on rates of molecular evolution, $f(\boldsymbol{r} \mid \theta_r)$, are derived from a stochastic model that explicitly describes how rates vary across branches of the tree, and priors for times, $f(\boldsymbol{a} \mid \theta_a)$, are typically based on a stochastic branching process model (given our focus on estimating divergence times between species, we do not discuss priors on times within species). Additionally, if we wish to estimate absolute rather than relative times/rates, we must incorporate some external information to calibrate the tree. The structure of this chapter follows that of Bayes' theorem (13.2): we begin by reviewing recent advances in the development of models that describe substitution-rate variation and then discuss alternative approaches for specifying priors on node times (Figure 13.2). We then consider alternative approaches for calibrating divergence-time estimates. Given the forum, as well as our own predilections, we adopt an unabashedly Bayesian statistical perspective. Our emphasis is not on a mathematically rigorous description of relaxed molecular clock models, but rather on developing an intuitive understanding of these models. To this end, we complement our review of methods for estimating divergence times—which focuses on theoretical aspects—with a discussion of equally important practical considerations.

13.2 Priors on branch rates

A wide range of models have been proposed to describe how substitution rates vary across lineages (Figure 13.2). Although some models allow the substitution rate to vary continuously along a branch, most models assume that the rate is constant on each branch and treat these branch-specific rates as random variables.

The simplest model—the global molecular clock model—assumes that the substitution rate is constant through time (Zuckerkandl and Pauling, 1962, 1965). In this case, a single substitution rate, r, is applied uniformly to all branches of the tree. The Bayesian perspective views parameters as random variables, which requires that we specify a prior probability density that describes the precise nature of that random variation. Describing the molecular clock model in a Bayesian framework therefore requires that we specify a prior probability density for the clock rate. If we assume, for example, that the clock rate is an exponentially distributed random variable, the parameters of the rate prior, θ_r, would simply be the rate parameter of the exponential distribution, λ_r, and the prior density on the rate would be $f(r \mid \lambda_r)$. We can extend this framework to incorporate additional complexity into the prior on the vector containing all branch rates, $\boldsymbol{r}$, and the parameters associated with the prior, θ_r, to specify various models that can accommodate diverse patterns of substitution-rate variation.

13.2.1 Autocorrelated-rate models

Many of the factors thought to impact substitution rates—metabolic rate, generation time, effective population size, etc.—are heritable, suggesting that closely related lineages should share *similar* (but not necessarily *identical*) substitution rates. The evolution of substitution rate is therefore modeled as a stochastic process in which the rate of a branch is correlated with that of its immediate ancestor. More specifically, such autocorrelated models describe the rates at the endpoints of each branch, such that the substitution rate at the end of the ith branch, r'_i (where r'_i is the rate at *node i*), is a random variable drawn from a probability distribution that is centered on the rate of its ancestor, $r'_{\sigma(i)}$. Thus, the substitution rate of the ith branch, r_i, is the average of the rates at either end of that branch:

$$r_i = \frac{r'_i + r'_{\sigma(i)}}{2}.$$

The rate at the root of the tree—for which the ancestor has not been sampled—is drawn from a separate prior probability distribution (Kishino et al., 2001; Thorne and Kishino, 2005).

Models have been proposed that describe the autocorrelated evolution of substitution rates as both continuous and episodic processes. Generally speaking, continuous models assume that the scope for change in substitution rate is directly proportional to time, whereas episodic autocorrelated models assume that substitution rates evolve by shifts.

These models are quite modular: theorists are free to develop different relaxed-clock models by plugging in their favorite parametric distribution to describe the evolution of substitution rates. In some sense, the technical details of the various autocorrelated models are less important than their biological implications. Unfortunately, the biological motivations for (and implications of) the chosen parametric distributions are currently not well understood. Nevertheless, it appears that these models span a continuum of rate autocorrelation, ranging from a very gradual process of rate evolution at one extreme, to one in which rates evolve episodically at the other. We present them in this order below.

Continuous autocorrelated models. Under continuous autocorrelated models, the rate of a node follows a prior probability distribution that is centered on the rate of its ancestor, where the variance of the distribution, s^2, is directly proportional to the duration of the branch, $s^2 = \eta t$. The η parameter specifies the degree of substitution-rate variation: small η values correspond to more clock-like data; larger η values to more rate-variable data.

Thorne et al. (1998) pioneered the development of this modeling framework, proposing an autocorrelated log-normal (ACLN) model in which substitution rates evolve according to a geometric Brownian-motion process (also see Kishino et al., 2001; Thorne and Kishino, 2002a). Under this model, the rate at the end of the ith branch, r'_i, is drawn from a log-normal distribution that is centered on the rate of its ancestor, $r'_{\sigma(i)}$. The ACLN model has a location parameter, μ_i, and a variance equal to the product of the duration of the branch, t_i, and the Brownian-motion (BM) rate parameter, η. Therefore, $r'_i \sim \text{LN}(\mu_i, \eta t_i)$. The expectation of the log-normal distribution on r'_i is equal to the rate at the beginning of the branch:

$$\mathbb{E}[r'_i \mid r'_{\sigma(i)}] = r'_{\sigma(i)} = e^{\mu_i + \frac{\eta t_i}{2}},$$

where $r'_{\sigma(i)}$ is the rate of the ancestor of node i. The location parameter, μ_i, of the log-normal distribution is:

$$\mu_i = \log(r'_{\sigma(i)}) - \frac{\eta t_i}{2}.$$

Accordingly, the ACLN describes the evolution of substitution rate as a gradual process, where the variance in rate scales with the opportunity for change (branch duration) and the BM-rate parameter (η) controls the degree of rate autocorrelation. Specifically, if $\eta = 0$, the model collapses to the global molecular clock model with a single rate for every branch. Conversely, very large values of η reduce the correlation between ancestor-descendant branches (Kishino et al., 2001).

Aris-Brosou and Yang (2002) proposed a similar autocorrelated relaxed-clock model in which rates follow a gamma distribution (the ACG model; see also Aris-Brosou and Yang, 2003). Under the ACG, the rate of a node, r'_i, is gamma distributed with a mean equal to the rate of the ancestral node and a variance proportional to its length: $r'_i \sim \text{Gamma}(\alpha, \lambda)$. The mean and variance of rates under the ACG are:

$$\mathbb{E}[r'_i \mid r'_{\sigma(i)}] = r'_{\sigma(i)} = \frac{\alpha}{\lambda},$$
$$\text{Var}[r'_i] = \eta t_i = \frac{\alpha}{\lambda^2},$$

where the shape, α, and rate, λ, parameters of the gamma distribution are defined as:

$$\alpha = \frac{(r'_{\sigma(i)})^2}{t_i \eta},$$
$$\lambda = \frac{r'_{\sigma(i)}}{t_i \eta}.$$

As (geometric) Brownian-motion processes, the ACLN and ACG models have infinite variance—the scope for variation in substitution rate increases in time without bound. Consequently, these models are not ergodic: the substitution rate at time $t > 0$, denoted r_t, does not converge to a unique distribution (the stationary distribution) as t goes to infinity. Biologically, it seems reasonable to suppose that, during periods of extremely high rates, changes to lower rates are more likely than changes to even higher rates (Lepage et al., 2006). These considerations motivated the development of two (putatively) stationary autocorrelated models: the autocorrelated Ornstein-Uhlenbeck process (AOUP; Aris-Brosou and Yang, 2002, 2003) and the autocorrelated Cox-Ingersoll-Ross process (ACIR; Lepage et al., 2006, 2007) models. These stationary models are similar to their Brownian-motion counterparts but include a "rubber-band" parameter, ε, that acts to pull the rate of the process toward the mean. The force exerted is proportional to the deviation of the rate from the mean.

Under the autocorrelated Ornstein-Uhlenbeck process (AOUP; Aris-Brosou and Yang, 2002) model, the rate of the ith node, r'_i, follows a normal (Gaussian) distribution centered on the rate of the ancestral node, $r'_{\sigma(i)}$. Although OU processes typically have a stationary distribution (they are ergodic), the AOUP relaxed-clock model in fact is not stationary (Welch et al., 2005; Lepage et al., 2007). This is evident from the expected value of the rate of node i under the AOUP model:

$$\mathbb{E}[r'_i \mid r'_{\sigma(i)}] = r'_{\sigma(i)} e^{-\varepsilon t_i},$$

where t_i is the branch duration, and ε is a (positive) friction term. Accordingly, each node is expected to have a lower substitution rate than its immediate ancestor, such that the AOUP model implies a decreasing substitution rate over time (Welch et al., 2005). Moreover, the AOUP describes the change in substitution rate over branches as a Gaussian density, which is problematic. The rate of substitution must, of course, have a positive real value; however, the Gaussian probability density places non-zero prior mass on negative rates, and so is not appropriate for modeling substitution rates (Welch et al., 2005).

The autocorrelated Cox-Ingersoll-Ross (ACIR; Lepage et al., 2006, 2007) model is a mixture of *squared* OU processes, which thus preserves the positivity of the rate process and resolves the systematic trend of the AOUP model. Under this model, the expected value of the rate of the ith node is:

$$\mathbb{E}[r'_i \mid r'_{\sigma(i)}] = r'_{\sigma(i)} e^{-\varepsilon t_i} + (1 - e^{-\varepsilon t_i}),$$

where r'_i is the rate of node i, $r'_{\sigma(i)}$ is the rate of its ancestral node, t_i is the duration of the branch spanning $r'_{\sigma(i)}$ and r'_i, and ε is the (positive) friction term. The variance of the ACIR model is:

$$\mathrm{Var}[r'_i] = r'_{\sigma(i)} \frac{\sigma^2}{\varepsilon}(e^{-\varepsilon t_i} - e^{-2\varepsilon t_i}) + \frac{\sigma^2}{2\varepsilon}(1 - e^{-\varepsilon t_i})^2.$$

Accordingly, ε defines the degree of substitution-rate autocorrelation, and σ defines the "short-term" variance of the ACIR process.

Episodic autocorrelated models. Episodic autocorrelated models assume that substitution rates evolve by shifts. For example, Aris-Brosou and Yang (2002) proposed the autocorrelated exponential (ACE) model, which assumes that changes in substitution rate can only occur during speciation events. Under this model, the rate of the ith node, r'_i, is an exponentially distributed random variable with a mean centered on the rate of its ancestor, $r'_{\sigma(i)}$. The expected value of the rate of node i is equal to that of its ancestor, $r'_{\sigma(i)}$:

$$\mathbb{E}[r'_i \mid r'_{\sigma(i)}] = r'_{\sigma(i)},$$

and so the ACE model describes a non-directional process (but see below). Note that this model has no hyperparameters: the mean and variance for a given branch are determined by the rate of its ancestor (the variance of the exponential is equal to the square of the mean), $r'_i \sim \mathrm{Exp}[(r'_{\sigma(i)})^{-1}]$. Consequently, the variance in substitution rate under the ACE is not proportional to branch duration; instead, it simply scales with the mean rate. Welch et al. (2005) made the somewhat counterintuitive observation that, despite the stationarity of the ACE model, it nevertheless favors rate decreases. This aspect of the ACE model stems from the fact that the exponential density has a mode at zero.

An alternative episodic autocorrelated relaxed-clock model assumes that changes in substitution rate occur as discrete events in continuous time along the tree. Under the compound Poisson process model (ACPP; Huelsenbeck et al., 2000b), for example, the substitution rate changes at events placed on the tree under a Poisson process. The waiting times between rate-shift events are therefore exponentially distributed, with a frequency that is controlled by the rate hyperparameter, λ. At a given rate-shift event, the current substitution rate is multiplied by a gamma-distributed random variable, m_i, for i in K rate-change events. It might seem reasonable to use a mean-one gamma distribution for the rate multipliers: i.e., where $m_i \sim \text{Gamma}(\alpha, \alpha)$. It turns out, however, that the product of mean-one gamma-distributed multipliers converges to zero as K approaches infinity. Accordingly, Huelsenbeck et al. (2000b) parameterized the gamma prior on rate multipliers so that the expectation of the logarithm of m_i is equal to zero: $\mathbb{E}[\log(m_i)] = 0$. The rate multipliers are therefore approximately mean-one gamma distributed random variables:

$$m_i \sim \text{Gamma}(\alpha, e^{\psi(\alpha)}),$$

where $\psi(\cdot)$ is the derivative of the log of the gamma function. This relaxed-clock model has also been implemented for amino-acid sequence data (Blanquart and Lartillot, 2006, 2008) and for morphological data (Ronquist et al., 2012a,b).

The ACPP relaxed-clock model is known to be non-identifiable (e.g., Rannala, 2002; Ronquist et al., 2012a). That is, the model can explain the data equally well either by specifying relatively frequent rate-change events of small magnitude, or by specifying less frequent rate-change events of greater magnitude. In fact, there is an uncountably infinite number of model parameterizations for which the data have an identical likelihood (i.e., for which the model is nonidentifiable). One possible solution might be to impose strongly informative (possibly empirical) priors on the event rate and/or magnitude. For example, we might have prior evidence (or obtain estimates) that allow us to specify a density that places most of the prior mass on infrequent substitution-rate shifts. The use of such strong priors, however, is difficult to justify, as there is (typically) little empirical information on the expected frequency or magnitude of substitution-rate shifts.

13.2.2 Independent-rate models

Episodic substitution-rate change is the primary assumption of independent rate models. Interestingly, the biological motivation for these models is similar to that of the autocorrelated-rate models; however, the rationale leads to different expectations. Specifically, rate autocorrelation will occur when the largest component of rate variation is due to heritable factors. At very small time scales "autocorrelation is so strong that very little of the variation in rate can be attributed to inherited factors" (Drummond et al., 2006, p. 700). Conversely, at very large temporal scales "we might expect so much variation in the inherited determinants of rate that the autocorrelation from lineage to lineage begins to break down" (Drummond et al., 2006, p. 700).

The evolution of substitution rate is therefore modeled as a stochastic process in which the substitution rate of a branch is independent of the rate of its immediate ancestor. Specifically, these models assume that substitution rates on adjacent branches are independent draws from an underlying parametric distribution. Again, many different independent relaxed-clock models might be specified by selecting your favorite parametric distribution with which to sample branch-specific substitution rates.

Uncorrelated log-normal model (UCLN). Substitution rates on branches under the UCLN model are sampled independently from an underlying log-normal distribution (Drummond et al., 2006; Rannala and Yang, 2007). Thus, if $r_i \sim \text{LN}(\mu, \sigma)$, then $\log(r_i) \sim \text{Normal}(\mu, \sigma)$. Similar to the ACLN model, it is often preferable to parameterize the log-normal distribution by the expected rate:

$$\mathbb{E}[r_i] = e^{\mu + \frac{\sigma^2}{2}},$$

thus allowing for a more straightforward interpretation of the hyperparameters.

This model is perhaps the most commonly applied prior on among-branch rate variation because of its availability in the widely used software package, BEAST (Drummond and Rambaut, 2007; Drummond et al., 2012). The implementation of the UCLN is described by Drummond et al. (2006). Specifically, the underlying log-normal distribution is discretized into a total of $2N - 2$ branch-rate bins and branches are assigned to each of these rate bins. The discrete bins are indexed $1, \ldots, 2N - 2$ along the horizontal axis of the log-normal distribution, the rate for category c, if $c \in (1, \ldots, 2N - 2)$, is determined by the q_c quantile of the log-normal distribution:

$$q_c = \frac{c - 0.5}{2N - 2}.$$

The original implementation of the UCLN model (Drummond et al., 2006; Drummond and Rambaut, 2007) assumed that branches were uniquely assigned to discrete rate categories. This implementation therefore forces branch rates to span the entire range of the log-normal distribution (including the tails), which may be problematic for small trees. More recent versions of BEAST (Drummond et al., 2012), however, allow multiple branches to occupy a single rate category, while other rate categories are allowed to remain empty. Given the discretization, it is important to employ a hierarchical Bayesian framework to accommodate uncertainty in the hyperparameters of the UCLN model; i.e., the mean and variance of the log-normal distribution are assigned hyperpriors. Accordingly, these hyperparameters are random variables and the MCMC algorithm samples from a mixture of discretized log-normal distributions.

Like their autocorrelated counterparts, different independent relaxed-clock models are inherently modular: alternative models could be (and have been) specified by replacing the log-normal distribution with a different distribution. Alternative independent relaxed-clock models have been proposed that adopt the gamma (the UCG model: Lepage et al., 2007; Heath et al., 2012) and exponential distributions (the UCE model: Drummond et al., 2006). Different parametric distributions correspond to different assumptions about the process of rate variation, although those assumptions are often distressingly unclear. The UCE model assumes that the branch rates are independently sampled from an underlying exponential distribution, $r_i \sim \text{Exp}(\lambda)$, with a mean and standard deviation of λ^{-1}. Note that the hyperparameter, λ, reflects both the mean and variance of the branch rates. Accordingly, under UCE—as in the other independent relaxed-clock models—substitution rates are effectively assumed to change in a punctuated manner; i.e., changes in rate are concentrated at speciation events.

The UCG model assumes that the branch rates are i.i.d. draws from an underlying gamma distribution, $r_i \sim \text{Gamma}(\alpha, \lambda)$, where α is the shape parameter of the gamma distribution and λ is the rate. Under this parameterization, the expectation of the gamma distribution is: $\mathbb{E}[r_i] = \alpha/\lambda$ and the variance is equal to α/λ^2. The UCG collapses to the UCE model, with rate λ, when $\alpha = 1$.

As noted previously, most independent-rate relaxed-clock models describe the variation in substitution rate over the tree in a "branch-wise" manner. That is, they are based on branch-specific rates, r_i, that are typically defined as the average rate over each branch, rather than explicitly modeling the change in rate as a continuous function of time, $r(t)$, along each branch. Consequently, even independent relaxed-clock models entail a latent source of autocorrelation: under these branch-wise models, the substitution rate is independent *between* branches but is nevertheless autocorrelated *over the duration* of individual branches. To expunge this residual source of rate autocorrelation, Lepage et al. (2007) proposed the independent-gamma rates (IGR) model, in which the substitution rate is described as a pure "white-noise" process. The IGR model is very similar to the UCG: rates are assumed to be drawn from an underlying gamma distribution. The main difference between the IGR and the UCG models is the variance of the two processes: it scales linearly with time under the IGR, but quadratically under the UCG.

Simulation studies and comparative empirical analyses have begun the arduous but important work of exploring the statistical behavior of the independent relaxed-clock models (e.g., Drummond et al., 2006; Lepage et al., 2007; Heath et al., 2012). Although our understanding of these models is currently incomplete, our sense is that these models span a range of increasingly episodic variation in substitution rate in the following (approximate) sequence: UCLN $<$ UCG $<$ UCE $<$ IGR. Accordingly, the various properties of these models and their biological implications should be carefully considered in divergence-time studies.

13.2.3 Local molecular clock models

Local molecular clock models postulate that, although substitution rates may not be constant across the *entire* tree, substitution rates may nevertheless be constant over *parts* of the tree (Hasegawa et al., 1989; Kishino et al., 1990; Yoder and Yang, 2000a; Yang and Yoder, 2003b). Inference under these models thus involves identifying subsets of related branches that conform to a strict (albeit local) molecular clock. Typically, substitution rates are assumed to change along a branch or at a node, and branches descended from a change point identically inherit the new rate. For example, a single rate change on a tree with N tips can occur on any of the $2N - 2$ branches (Drummond and Suchard, 2010); each rate-shift location corresponds to a unique local-clock model. Accordingly, the state space of local-clock models quickly becomes vast as the size of the tree increases. If we assume that every branch can experience either one or zero rate shifts, there are 2^{2N-2} possible local-clock models (Drummond and Suchard, 2010).

Narrowing the vast pool of candidate local-clock models variously relies on leveraging prior information (but see, e.g., Yang, 2004; Aris-Brosou, 2007). The so-called ad hoc local-clock (AHLC, Yoder and Yang, 2000a; Yang and Yoder, 2003b) model, for example, uses prior biological information to partition the branches of the tree into K rate categories, each with a parameter, r_K, that acts as a substitution-rate multiplier for each partition. Imagine, for instance, that we have prior knowledge that rates of substitution in herbaceous flowering plants are typically higher than those of their woody relatives (e.g., Smith and Donoghue, 2008; Gaut et al., 2011). Accordingly, we might infer the history of this trait on our tree to identify partitions of woody and herbaceous branches. (Note that we ignore uncertainty in this inference, and the possibility that an estimate of divergence times might be required to reliably infer the history of the trait.) Maximum likelihood is then used to estimate the divergence times and substitution rates in each of the K local-clock partitions of the phylogeny. For many problems, however, we will lack sufficient knowledge to form strong prior beliefs about the distribution of substitution rates across branches. Accordingly, recently developed Bayesian local-clock models rely on other (probabilistic) approaches to specify the prior probabilities for partitions of branches to rate categories.

Random-local clock (RLC) model. Drummond and Suchard (2010) introduced a method for efficiently sampling the state space of possible local-clock models. This model specifies a (relative) rate for each branch, r_i, such that $r_i = \phi_i r_{\sigma(i)}$, where $r_{\sigma(i)}$ is the rate of the branch ancestral to branch i and ϕ_i is a branch-specific rate multiplier drawn from a mean-one gamma distribution. The variance of the gamma prior on ϕ_i, when $\delta_i = 1$, is controlled by an additional parameter, ψ:

$$\phi_i \sim \text{Gamma}(1/\psi\delta_i, 1/\psi\delta_i).$$

Under the RLC model, each branch either (1) experienced one or more rate shifts, or (2) experienced zero rate shifts (Drummond and Suchard, 2010). The presence of a rate change on each of the $2N-2$ branches is signified by a vector of indicator variables: $\boldsymbol{\delta} = (\delta_1, \ldots, \delta_{2N-2})$. For a given branch i, if $\delta_i = 0$, no rate change occurred along the branch, so it retains the rate of its ancestor, $r_i = r_{\sigma(i)}$ (the variance of the prior density on ϕ_i collapses to 0, such that the rate multiplier $\phi_i = 1$). Conversely, when $\delta_i = 1$, one or more rate changes occurred along the branch, so the branch rate is the product of the ancestral rate and the branch-specific multiplier: $r_i = \phi_i r_{\sigma(i)}$. Because the prior on ϕ_i has an expectation of 1, some degree of rate autocorrelation can be accommodated, depending on the variance (ψ).

The total number of rate-change events over the tree, K, is the sum of the indicator variables:

$$K = \sum_{i=1}^{2N-2} \delta_i.$$

Drummond and Suchard (2010) assume that each δ_i is a single, independent Bernoulli trial with a very low probability of success, χ, therefore $K \sim \text{Binomial}(2N-2, \chi)$. Because the binomial distribution converges to the Poisson distribution as the number of trials approaches infinity, a Poisson prior on K is a convenient approximation when χ is very small (and $2N-2$ is sufficiently large). Thus,

$$K \sim \text{Poisson}(\lambda),$$

where $\lambda = \chi(2N-2)$.

This approach provides an efficient and flexible means for sampling random local-clock models (Drummond and Suchard, 2010). Specifically, when $K = 0$ the RLC model collapses to the global molecular clock. At the other extreme, when $K = 2N-2$, the RLC model converges to a type of autocorrelated-gamma model (except the variance does not scale with the branch duration). In their implementation, Drummond and Suchard (2010) assume that the variation in substitution rates can be explained by relatively few branch-rate changes, thus, they place substantial prior probability on $K = 0$. For example, the default setting for the Poisson parameter, $\lambda = \log(2)$, places 50% of the prior probability on $K = 0$.

Conveniently, this parameterization of the random local-clock model provides a straightforward approach for assessing support for the global-clock model using Bayes factors (Drummond and Suchard, 2010). This involves calculating the ratio of the posterior odds (odds = probability/(1 − probability)) to the prior odds for the global-clock model (Kass and Raftery, 1995). Specifically, the relative support in favor of $K = 0$ can be determined by dividing the posterior odds (for a given set of sequences X) by the prior odds:

$$\text{BF}(K=0, K>0) = \frac{\text{P}(K=0 \mid X, \lambda)}{1-\text{P}(K=0 \mid X, \lambda)} \div \frac{\text{P}(K=0 \mid \lambda)}{1-\text{P}(K=0 \mid \lambda)}.$$

The prior probability of $K = 0$ is simply given by the Poisson distribution on K:

$$\text{P}(K = 0 \mid \lambda) = e^{-\lambda}.$$

The posterior probability of the global-clock model, $\text{P}(K = 0 \mid X, \lambda)$, is simply the proportion of MCMC samples where $K = 0$. Very small values of this Bayes factor comparison indicate strong support for a model where $K > 0$. The ability to compute Bayes factors directly from the MCMC samples is a major advantage of the RLC model, as most approaches for comparing models via Bayes factors require computationally expensive methods to estimate the marginal likelihoods of the candidate models (Suchard et al., 2001; Lartillot and Philippe, 2006b; Drummond and Suchard, 2010; Xie et al., 2011; Fan et al., 2011).

13.2.4 Mixture models on branch rates

Similar to local-clock models, mixture models adopt the perspective that accommodating variation in substitution rates across branches is fundamentally a partitioning problem; i.e., the task is to assign subsets of branches to a number of substitution-rate categories. Unlike the random local-clock model, however, branches that share a common rate need not be adjacent on the tree. This aspect of mixture models dramatically increases the state space. Specifically, the total number of partitions (branch-rate models) for n elements is described by the Bell numbers (Bell, 1934). The Bell number for n elements is the sum of the Stirling numbers of the second kind:

$$\mathcal{B}(n) = \sum_{k=1}^{n} \mathcal{S}_2(n, k).$$

The Stirling number of the second kind, $\mathcal{S}_2(n, k)$, for n elements and k subsets (corresponding here to the number of branches and rate categories, respectively) is given by the following equation:

$$\mathcal{S}_2(n, k) = \frac{1}{k!} \sum_{i=0}^{k-1} (-1)^i \binom{k}{i} (k - i)^n.$$

The state space of possible branch-rate partitions quickly becomes large, even for small trees. For example, the simple tree in Figure 13.1 has $2^{2N-2} = 1{,}024$ possible random local-clock models (Drummond and Suchard, 2010), compared to $\mathcal{B}(10) = 115,975$ possible branch-rate partitions.

A finite mixture model on branch rates (FMM). Finite-dimensional mixture models specify the number of partitions (substitution-rate categories) *a priori*. To model substitution rate variation across branches, a finite mixture model would require that we specify a fixed number of rate categories, k. The problem thus reduces to evaluating each of the $n = 2N - 2$ branches over the pre-specified k rate categories, and estimating the value of each rate category. A finite-mixture relaxed-clock model will have the following parameters:

k	the fixed number of rate categories
$\boldsymbol{\pi} = (\pi_1, \ldots, \pi_k)$	the value of the rate for each category
$\boldsymbol{z} = (z_1, \ldots, z_{2N-2})$	vector of rate-class assignments for each branch
$\boldsymbol{\phi} = (\phi_1, \ldots, \phi_k)$	vector of probabilities for each class, $\sum_{i \in k} \phi_i = 1$

A fixed-k mixture model on branch rates assumes that branch i is placed in rate category j with probability ϕ_j, where $j = z_i$ and $j \in (1, \ldots, k)$. Accordingly, the rate of each branch is a weighted average over the set of rate categories, where the weights are specified by the vector of rate-class probabilities. Special cases of mixture models can include the global molecular clock ($k = 1$), as well as independent-rate models ($k = 2N - 2$). Additionally, depending on the value of k, a mixture model can identify latent local molecular clocks. Moreover, because the clustering of branches to rate categories is not restricted by the tree topology, a mixture model can identify distantly related lineages that share the same rate.

Although a finite-mixture model has not been applied to the problem of substitution-rate variation among branches, Foster (2004) described a "non-stationary" substitution model in which nucleotide composition may vary across branches. Under this model, the number of nucleotide-composition classes, k, is fixed, which immediately raises the question of how to specify the appropriate number of classes. Green (1995) described an extension of MCMC—reversible-jump MCMC—that provides a means of sampling models with different numbers of parameters. Gowri-Shankar and Rattray (2007) successfully adopted this approach to allow inferences under the "non-stationary" substitution model to be averaged over the number of compositional classes. This approach might also be used to implement a finite-mixture model of substitution rate variation across branches. However, formulating and tuning good proposal distributions for reversible-jump MCMC solutions can pose a significant challenge (Brooks et al., 2003).

An infinite mixture model on branch rates: the Dirichlet process prior (DPP). Like finite mixture models, both the values of the rate categories and the assignment of branches to rate categories are treated as random variables under an infinite mixture model. Unlike the finite mixture models, however, the *number* of rate categories is also a random variable in an infinite mixture model. These so-called "nonparametric" Bayesian models eliminate the need to specify the number of mixture components by modeling a countably infinite number of parameters, thus allowing the data to determine number of parameter categories (Müller and Quintana, 2004; Orbanz and Teh, 2010).

Heath et al. (2012) described an infinite mixture model on branch rates based on the nonparametric Dirichlet process prior model. The Dirichlet process prior model (DPP: Ferguson, 1973b; Antoniak, 1974b) assumes that data elements (branches) can be clustered into distinct parameter classes (substitution-rate categories). When applied as a prior on branch-rate categories, the DPP assumes branches are assignable to an infinite number of latent rate categories and relies on two parameters:

$$\boldsymbol{r} \sim \mathrm{DPP}(\alpha, G_0),$$

where α is the concentration parameter and G_0 is the base distribution of the DPP model. The concentration parameter, α, controls the intensity of clustering: smaller α values place more prior mass on fewer rate categories; larger α values induce greater partitioning and larger numbers of rate categories. The substitution rate associated with each category is drawn from the base distribution, G_0. When used as a prior on branch rates, the base distribution must adequately represent the range of realistic substitution-rate values. Heath et al. (2012) specified a gamma distribution for G_0, with a shape parameter σ_{G_0} and a rate parameter γ_{G_0}: $\pi_i \sim \mathrm{Gamma}(\sigma_{G_0}, \gamma_{G_0})$, where π_i is the rate value associated with substitution-rate category i, for i in k rate categories. The gamma distribution is parameterized such that the expectation of the rate for each category is

$$\mathbb{E}[\pi_i] = \frac{\sigma_{G_0}}{\gamma_{G_0}}.$$

Because it is unclear what values for the concentration parameter, α, are biologically reasonable, Heath et al. (2012) chose to treat it as a random variable and estimate α from the data. Specifically, their hierarchical Bayesian approach specifies a gamma-distributed hyperprior on α: $\alpha \sim \mathrm{Gamma}(\sigma_\alpha, \gamma_\alpha)$. This hierarchical approach accommodates uncertainty in the value of the concentration parameter and leads to full conditional distributions that allow for efficient MCMC sampling (Escobar and West, 1995; Dorazio, 2009).

The DPP model allows calculation of the prior probability for all possible values of k. The state space includes the configuration in which all branches belong to a single rate category ($k = 1$, the global-clock model), the configuration in which each branch belongs to a unique rate category ($k = 2N - 2$, the "uncorrelated-gamma" model), and the range of k values between these two extremes. The prior probability of a given number of rate categories, $\mathrm{P}(k)$, depends on the number of branches in the tree and the α-concentration parameter:

$$\mathrm{P}(k \mid \alpha, 2N-2) = \frac{\mathcal{S}_1(2N-2, k)\alpha^k}{\prod_{i=1}^{2N-2} (\alpha + i - 1)},$$

where $\mathcal{S}_1(\cdot, \cdot)$ is the Stirling number of the first kind, which specifies the possible permutations of $2N - 2$ branches among k rate categories:

$$\mathcal{S}_1(n, k) = (-1)^{n-k} \binom{n}{k}.$$

Furthermore, it is possible to calculate the prior probability that two branches belong to the same rate category, which is conditional only on the α-concentration parameter:

$$P(r_i = r_j \mid \alpha) = \frac{1}{1+\alpha}.$$

MCMC is used to average over the prior probability distribution of branch-rate configurations to approximate the joint posterior probability density of rates and times. Specifically, a Gibbs sampling method (Neal, 2000, algorithm 8) proposes updates to the values for each rate category and reassigns branches to rate categories. When a new rate-class assignment is proposed for branch i, it is placed in any of the k existing classes with a probability proportional to the number of branches occupying each class, or it can be placed in any of the β temporary auxiliary categories with a probability proportional to the concentration parameter, where β is the pre-specified number of auxiliary categories. Heath et al. (2012) fixed $\beta = 4$ in their implementation of the DPP model, which enabled adequate exploration of parameter space without excessive computational costs.

Using sequence data simulated under a range of different branch-rate models (strict clock, local clock, ACLN, ACPP, UCG, and DPP), Heath et al. (2012) evaluated the performance of divergence-time estimation under the DPP relative to estimation under a strict molecular clock ($k = 1$) and an uncorrelated-gamma model ($k = 2N - 2$). Overall, their results indicate that node-age estimates under a DPP model of branch-rate variation were more accurate that the two alternative models. Furthermore, the greater accuracy under the DPP model was not associated with a significant loss in precision. Divergence-time estimation under a DPP model also yielded accurate estimates of the numbers of branch-rate partitions, k, when the data were generated under a local molecular clock model or under the DPP (Heath et al., 2012).

Similar to the random local-clock model (Drummond and Suchard, 2010), the ability to calculate the prior probability of the number of rate categories, k, under the DPP provides an efficient means of comparing certain models or model configurations using Bayes factors (provided that the value of α is constant). For instance, the relative support in favor of the global molecular clock model ($k = 1$) can be determined by computing the ratio of the posterior and prior odds (Kass and Raftery, 1995):

$$\mathrm{BF}(k=1, k>1) = \frac{P(k=1 \mid X, \alpha, 2N-2)}{1 - P(k=1 \mid X, \alpha, 2N-2)} \div \frac{P(k=1 \mid \alpha, 2N-2)}{1 - P(k=1 \mid \alpha, 2N-2)}.$$

This approach can also be applied to determine the support for a model in which branch i and branch j have equivalent rates:

$$\mathrm{BF}(r_i = r_j, r_i \neq r_j) = \frac{P(r_i = r_j \mid X)}{1 - P(r_i = r_j \mid X)} \times \alpha.$$

In general, the flexibility and accuracy of the DPP model make it an appealing prior for accommodating among-lineage substitution rate variation. The inherent flexibility of this nonparametric model may be more amenable to the disparate processes that cause variation in substitution rates. Moreover, when MCMC is used to sample parameters under the DPP, branch-rate partition configurations can be summarized through calculation of partition distances (Gusfield, 2002). This approach identifies the "mean partition," which is the set of branch-rate category assignments that minimizes the squared distances to all other partition configurations sampled by MCMC (Huelsenbeck and Suchard, 2007; Heath et al., 2012). The mean partition helps to uncover latent substitution-rate classes present in the data, and therefore lineages that may have undergone similar evolutionary processes.

13.3 Priors on node times

In order to estimate divergence times from molecular datasets that depart from the molecular clock, the priors on substitution rates must be combined with priors on node ages. As an integral component of the model, the prior on node ages may have a substantial impact on divergence-time estimates (e.g., Aris-Brosou and Yang, 2003; Welch et al., 2005; Lepage et al., 2007). Priors on node ages include those that are exclusively tailored to species-level inference (where a single sequence is sampled for each species in the study group), those tailored to population-level inference (where sequences are sampled from multiple individuals per population/species), and other priors that are relatively generic in nature (making no explicit mechanistic assumptions about the underlying biological processes that gave rise to the data: Figure 13.2).

Our focus here is on estimating divergence times among species. Models appropriate for the analysis of population-level sequence data are described elsewhere in this book (see Chapters 4 and 11).

13.3.1 Generic priors on node ages

The simplest model for specifying a prior distribution of node ages assumes interior nodes are uniformly distributed between the terminal and root nodes. This model makes no explicit assumptions about the underlying biological process of lineage diversification, with the goal of providing non-(or minimally) informative node-age priors (e.g., Lepage et al., 2007; Ronquist et al., 2012a). Given a fixed topology, priors on node ages are generated in a postorder (tip-to-root) traversal of the tree. If the terminal node represents an extant species, the node-age prior density is a point mass on zero. If the terminal node corresponds to a serially sampled taxon (e.g., a fossil or pathogen sampled at some point in the past), the prior density describing the age of the extinct specimen is used (e.g., Drummond et al., 2003; Ronquist et al., 2012a). The age of the root node is sampled from a separate prior probability density. The prior age for an interior node is sampled from a uniform probability density bounded by the age of its immediately ancestral and descendant nodes. Repeating this process for each of the $N - 1$ internal nodes provides a joint prior probability distribution of uniform order statistics with a variable (but finite) time span (Lepage et al., 2007).

Another generic prior applies a single-parameter Dirichlet distribution to the ages of internal nodes (Kishino et al., 2001; Thorne and Kishino, 2002a). In their implementation of the Dirichlet prior, Kishino et al. (2001) apply a gamma prior on the age of the root node for a fixed tree topology of extant taxa. Ages of the interior nodes on a single path spanning the age of the root node to one of the tip nodes are sampled from a flat Dirichlet prior density. The first path is selected from a set of optimal paths that traverse the largest number of internal nodes. This procedure continues recursively until every node has been evaluated, with each path beginning at an internal node that has been visited during a previous path.

Simple, generic node-age models like the uniform or Dirichlet priors are nominally noninformative since they do not attempt to model the biological processes involved in lineage diversification. In principle, the use of vague/noninformative prior densities for parameters is attractive because the corresponding marginal posterior densities for those parameters should be dominated by the information in the data via the likelihood function. In practice, however, what appear to be vague priors may in fact be highly informative, exerting a substantial impact on the posterior by virtue of spreading the prior mass over regions of low likelihood (e.g., Yang and Rannala, 2005). Moreover, empirical phylogenies have been generated by branching processes, so it may be more appropriate to derive priors on node ages using these more biologically reasonable models.

13.3.2 Branching-process priors on node ages

Stochastic branching models describe lineage diversification as a continuous-time process. At any single instant during the forward-time process, all extant lineages have some probability of birth/speciation (giving rise to two descendant species) and, if allowed in the model, some probability of death/extinction (leaving no descendants).

Based on observations of the numbers of described species for different genera, Yule (1924) outlined a stochastic process describing speciation of lineages over time. The Yule-process model (also called the linear-birth or pure-birth process) is the simplest branching-process model (Aldous, 2001). This model simply assumes that at any point in time, every extant lineage has the same rate of speciation, λ, and this rate is constant through time. Thus, there is an exponential waiting time between speciation events, such that the rate of the exponential distribution is equal to the product of the number of extant lineages (n) at a given point in time and the speciation rate: $n\lambda$ (Yule, 1924; Aldous, 2001; Hartmann et al., 2010). Similar to population-level coalescent models, the pure-birth branching model does not allow for lineage extinction. This restrictive assumption may render the Yule model inappropriate for many species-level datasets that involve extinction.

For most clades in the tree of life, it is therefore useful to incorporate lineage extinction in the branching-process model. Kendall (1948) described the constant-rate birth–death process, which extends the Yule model by adding a parameter, μ, for the rate of extinction. As a prior on node ages, the birth–death model assumes that at any point in time, all extant lineages may speciate with rate λ and go extinct with rate μ (Kendall, 1948; Thompson, 1975; Nee et al., 1994; Rannala and Yang, 1996). In their pioneering paper on Bayesian phylogenetic inference, Rannala and Yang (1996) applied the birth–death model as a prior on the branching times and tree topology.

Building on previous work on continuous-time branching models (Yule, 1924; Kendall, 1948; Thompson, 1975; Nee et al., 1994; Rannala and Yang, 1996; Yang and Rannala, 1997b; Popovic, 2004; Aldous and Popovic, 2005), Gernhard (2008) derived the probability density of a phylogeny and individual speciation events under the birth–death process. Because phylogenies represent the relationships of N extant species living in the present time, the probability density of the tree is conditioned on the number of extant tips in the tree. Furthermore, the model also depends on the age of the most recent common ancestor (MRCA), $a_{\sigma(2N-1)}$ (Popovic, 2004; Aldous and Popovic, 2005; Gernhard, 2008). Given that the age of the root node is unknown, an improper uniform (a_{2N-1}, ∞) distribution is assumed for $a_{\sigma(2N-1)}$, leading to the density for the internal nodes including the root $(a_{N+1}, \ldots, a_{2N-1})$:

$$f(a_{N+1}, \ldots, a_{2N-1} \mid \lambda, \mu, N) = \\ N!(\lambda - \mu)\lambda^{N-1} \frac{e^{-(\lambda-\mu)a_{2N-1}}}{\lambda - \mu e^{-(\lambda-\mu)a_{2N-1}}} \prod_{i=N+1}^{2N-1} \frac{(\lambda-\mu)^2 e^{-(\lambda-\mu)a_i}}{(\lambda - \mu e^{-(\lambda-\mu)a_i})^2}. \tag{13.3}$$

The constant-rate birth–death model simplifies to the Yule process when $\mu = 0$, with a probability density equal to:

$$f(a_{N+1}, \ldots, a_{2N-1} \mid \lambda, N) = N!\lambda^{N-1} e^{-\lambda a_{2N-1}} \prod_{i=N+1}^{2N-1} e^{-\lambda a_i}.$$

The probability density $f(\cdot \mid \lambda, \mu, N)$ serves as the prior on node ages, $f(\boldsymbol{a} \mid \theta_a)$, for many implementations of Bayesian divergence-time estimation methods (Rannala and Yang, 2007; Lartillot et al., 2009; Drummond et al., 2012; Ronquist et al., 2012b; Heath et al., 2012).

The constant-rate birth–death process is a convenient and simple model for describing the distribution of species divergence times. However, biological data typically violate many of the assumptions of this simple model. As a result, extensions to the general model have been proposed.

Because it is common for phylogenetic datasets to include only a subsample of species for a given monophyletic clade, a logical extension of the reconstructed birth–death process is to accommodate incomplete sampling of extant taxa. Yang and Rannala (1997b) and Stadler (2009) derived the probability of the node ages under constant rates of speciation and extinction while accounting for incomplete sampling. This model includes an additional parameter, ρ, that represents the probability of randomly sampling an extant taxon. Under this so-called birth–death sampling branching process, the probability density of the tree and speciation times is conditioned on sampling N species (Stadler, 2009):

$$f(\boldsymbol{a} \mid \lambda, \mu, \rho, N) = N!(\lambda - \mu)(\lambda\rho)^{N-1} \frac{e^{-(\lambda-\mu)a_{2N-1}}}{\rho\lambda + (\lambda(1-\rho) - \mu)e^{-(\lambda-\mu)a_{2N-1}}}$$
$$\times \prod_{i=N+1}^{2N-1} \frac{(\lambda-\mu)^2 e^{-(\lambda-\mu)a_i}}{(\rho\lambda + (\lambda(1-\rho) - \mu)e^{-(\lambda-\mu)a_i})^2}.$$

The sampling probability interacts with the lineage speciation rate; note that when $\rho = 1$, the density $f(\boldsymbol{a} \mid \lambda, \mu, \rho = 1, N)$ simplifies to (13.3). Although straightforward to implement, complex models of lineage diversification present significant challenges when it comes to specifying the values of the various hyperparameters (λ, μ, ρ). Therefore, it is necessary to adopt a hierarchical Bayesian framework for these models that specifies separate prior probability densities for the values of the parameters of the birth–death model. This hierarchical approach therefore introduces additional parameters into the joint probability density (13.2): $f(\lambda \mid \theta_\lambda)f(\mu \mid \theta_\mu)f(\rho \mid \theta_\rho)$, where θ_λ, θ_μ, and θ_ρ represent the hyper-hyperparameters associated with the hyperpriors on λ, μ, and ρ. Notably, because representative species are rarely sampled uniformly from the pool of extant taxa, an extension of the birth–death sampling model was described to account for nonrandom sampling in which species are selected to maximize taxonomic diversity (Höhna et al., 2011).

A key assumption of the constant-rate birth–death process—that lineages diversify under constant rates of extinction and/or speciation—is often violated by empirical data. Models that accommodate variation in rates of lineage diversification have been proposed and applied in simulation (e.g., Moore et al., 2004; Heath et al., 2008; Rabosky, 2009) and for inference using model-comparison methods such as the likelihood-ratio test (Chan and Moore, 2005; Stadler, 2011a; Meredith et al., 2011; Etienne et al., 2012). Complex models of variable birth and death rates—often assuming separate speciation/extinction rates for every branch or for groups of branches—are not typically applied as priors on node ages (but see Stadler et al., 2013). Instead, variable diversification rate models are often applied to a pre-specified tree with a fixed, dated tree topology and evaluated using frequentist approaches. These methods typically ignore (the often considerable) uncertainty in the estimates of node ages and topological relationships. Understanding the evolutionary processes responsible for the differential patterns of lineage diversification observed in nature are central to the study of evolution, which is likely to motivate the development of node-age priors that accommodate variation in diversification rates.

Stadler (2011a) proposed a temporal-shift birth–death process model, describing tree-wide changes in diversification rates through time. Under this model, diversification rates change in a step-wise fashion through time: all lineages share the same diversification rate between diversification-rate shifts, and after a shift, all lineages diversify under the new rate. Stadler (2011a) applied this birth–death shift model to a dated phylogeny of extant mammals and used likelihood-ratio tests to estimate the number and magnitude of diversification-rate shifts. This analysis revealed a significant increase in diversification rates approximately 33 Mya, primarily driven by increased diversification rates in rodents, marsupials, cetaceans, and artiodactyls. If used to derive priors on node ages, complex stochastic branching-process models such as these may have the potential to accommodate a greater range of biological processes (Stadler et al., 2013). The effects of different tree priors on species divergence-time estimates has not yet been rigorously explored, however, so it remains unknown whether such complex models provide more robust estimates of node ages.

13.4 Priors for calibrating divergence times

Estimating divergence times from sequence data alone permits inference of relative—but not absolute—node ages and substitution rates. External information must be incorporated to scale relative ages and rates to absolute, geological time.

There are three basic sources of information for calibrating trees. Biogeographic events—such as island formation associated with an episode of endemic speciation, mountain orogeny associated with an episode of vicariant speciation—may provide indirect evidence to calibrate the absolute age of the associated node (e.g., Weir and Schluter, 2008). An obvious concern with biogeographic calibrations is related to the phenomenon of pseudo-congruence (*sensu* Donoghue and Moore, 2003). To illustrate this point, imagine that the phylogeny for our study group specifies the topology $(x, (y, z))$, and that these three species occur in areas A, B, and C, respectively. Further, suppose that geological evidence indicates that previously contiguous areas B and C were subdivided at some point in the past, t_w. The topological congruence between the organismal phylogeny and the geographic history may in fact be causal: specifically, that w, the MRCA of y and z, speciated in response to the fragmentation of areas B and C at time t_w, in which case this vicariance event could serve as a valid biogeographic calibration for estimating divergence times in this group. On the other hand, it is also possible that the apparent topological agreement between phylogeny and geography stems not from a causal and contemporaneous history, but instead may have been achieved by means of organismal dispersal between disjunct areas B and C subsequent to the episode of geographic fragmentation at t_w. Moreover, if the primary objective for estimating absolute divergence times is to explore correlations between biogeographic events and diversification, biogeographic calibrations can lead to a circular line of reasoning.

The second source of calibration data comes from serial sampling, where sequences are sampled at different time horizons—typically restricted to viral data (e.g., Drummond et al., 2003)—but ancient DNA or morphological data may also allow (sub)fossils to be treated as serial samples (e.g., Shapiro et al., 2004; Pyron, 2011; Ronquist et al., 2012a). Serially sampled data impose temporal structure on the tree by providing ages for non-contemporaneous tips.

The final and most common source of calibration information comes—either directly or indirectly—from fossil data. Fossils provide *indirect* information either through *legacy-rate* calibrations or *legacy-date* calibrations. Legacy-rate calibrations are ultimately based on previous estimates from a "comparable" study (i.e., involving gene regions and/or taxonomic groups that are "similar" to the study group) where direct fossil evidence was used. As a basis for calibrating divergence-time estimates under relaxed-clock models, however, legacy-rate calibrations are problematic: presumably, the decision to adopt a relaxed-clock model is motivated by variation in substitution rates across lineages *within* the sampled sequences, which makes it quite likely that rates have varied *between* the sampled sequences and the reference sequences from which the absolute rates were estimated. Similarly, legacy-date calibrations are typically based on previous studies at a broader phylogenetic scale that included some members of the study group. Error propagation is a concern for both legacy-date and legacy-rate calibrations. *Direct* fossil evidence is by far the most common and reliable approach for estimating absolute node ages (Benton and Donoghue, 2007; Marshall, 2008; Parham et al., 2012) and we will focus our discussion on this method.

Direct evidence from the fossil record provides temporal constraints on the ages of nodes in the phylogeny of their extant relatives. Ideally, fossils are assigned to nodes by means of a formal analysis of discrete morphological characters sampled from both extant and fossil species (Figure 13.3a).

Commonly, however, fossils are assigned to nodes based on an informal evaluation of diagnostic (i.e., "synapomorphic") characters believed to indicate the taxonomic affiliation of the fossil. Once placed in the phylogeny, fossil calibrations impose a *minimum* age for the MRCA of the living and fossil species. Accordingly, when multiple fossils have been assigned to the same

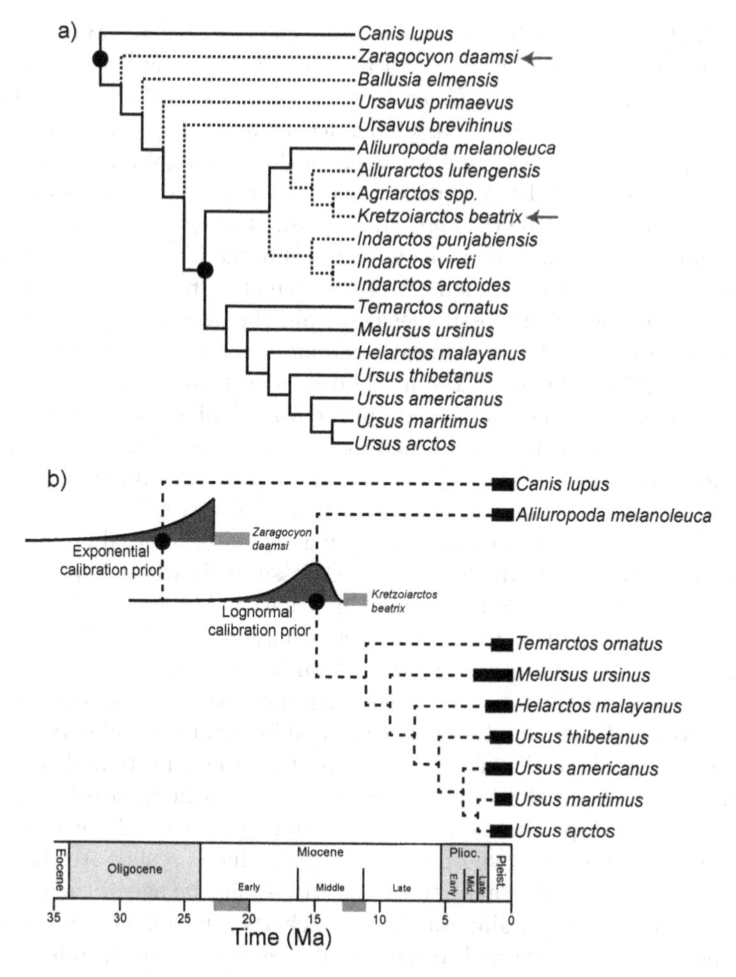

Figure 13.3: a) A cladistic analysis of modern (darker lines) and fossil (dotted lines) bears (outgroup *C. lupus*) using morphological characters from Abella et al. (2012). Because of their phylogenetic placement and stratigraphic ages, only two extinct species are applicable as calibration fossils (arrows): *Z. daamsi* and *K. beatrix*. These fossils calibrate two nodes (filled circles) in the tree of extant species. b) Calibration prior densities applied to the Ursidae phylogeny. Stratigraphic ranges are shown with solid darker bars for each extant taxon and lighter bars for fossil species. We illustrate node calibration using two different parametric densities as calibration priors: an exponential density on the root of the tree is offset by the maximum age of *Z. daamsi*, and a log-normal calibration prior is applied to the MRCA of all bears with a zero-offset based on the maximum age of the *K. beatrix* fossil.

node, only the oldest can be used to calibrate the tree (but see Marshall, 1990, 1994, 1997). For example, in Figure 13.3a, four fossil species—*Z. daamsi, B. elmensis, U. primaevus,* and *U. brevihinus*—branched off the *stem* lineage leading to the MRCA of the bear clade: accordingly, they are referred to as "stem" fossils. Stem fossils are not members of the *crown* group, which includes all descendants of the MRCA of extant species within the study group (*sensu* Doyle and Donoghue, 1993). The oldest fossil, *Zaragocyon daamsi* (family Hemicyonidae), has a date range in the Early Miocene (Abella et al., 2012) and imposes a minimum age on the root of the tree (Figure 13.3a). Additionally, despite the existence of six fossil species within the ursid crown group, only a single fossil can be used as a calibration in an analysis of extant bears. With an age in the Middle Miocene, the newly described fossil panda, *Kretzoiarctos beatrix,* is the oldest known descendant of the MRCA of all living bears (Abella et al., 2012), and provides a minimum age for that node (Figure 13.3). Since calibration fossils impose strong constraints on absolute node ages, correct identification and phylogenetic placement is a critical step.

Provided that fossil species are correctly assigned to internal nodes—where the node is truly older than the fossil age—fossil calibrations impose reliable minimum-age constraints. Stratigraphic information or some bracketing methods (e.g., Marshall, 1990, 1994, 1997, 2008) may provide suitable maximum-age constraints, but the necessary resolution of the fossil record means that maximum-age constraints are much less common. Bayesian methods are well suited to exploring these kinds of problems, and in the absence of maximum-age constraints, parametric distributions are specified to calibrate nodes. Calibration prior densities are offset by the age of the corresponding fossil and do not necessarily require specification of a maximum age (Figure 13.3b). However, calibration densities have hyperparameters (e.g., shape, scale, rate) that must be specified to adequately characterize uncertainty in the age of the calibrated node. Furthermore, depending on the researcher's prior understanding of the age of the calibrated node with respect to its descendant fossil, different parametric distributions can be used as priors. Commonly, log-normal, gamma, or exponential distributions are used as priors for calibrating nodes. Each of these distributions reflects different assumptions about the age of the calibrated node. For example, the mode of the exponential distribution is zero; using the exponential as a calibration prior therefore assumes there is a significant probability density that the age of the node is equal to the age of the fossil. The exponential calibration prior can be made more diffuse (e.g., by decreasing the rate parameter to spread substantial probability over a wider range of values), but the highest prior density for the node age will always be equal to the age of the fossil.

Fundamentally, the prior density applied to a calibrated node is intended to describe the calibration difference, which is the interval between the age of the calibrated node and the age of its oldest fossil descendant. Calibration prior densities are particularly sensitive to the specified values of their hyperparameters and different calibration priors can lead to different expected ages for calibrated nodes (Yang and Rannala, 2006b; Ho, 2007; Ho and Phillips, 2009; Heath, 2012; dos Reis and Yang, 2012). For many molecular biologists, specifying the hyperparameters of calibration priors is particularly difficult. Here, the user is tasked with describing their prior beliefs/knowledge regarding the calibration difference, which, in most cases, is best reflected by a vague prior density. Parameterization of adequately diffuse calibration priors—particularly when there are many calibrated nodes—presents a significant challenge. Accordingly, a hierarchical Bayesian framework provides an objective means of accommodating uncertainty in the values of calibration hyperparameters.

13.4.1 Hierarchical models for calibrating node ages

Imagine that we have a total of F fossils available to calibrate a subset of the nodes in our study phylogeny. The calibrated nodes are contained within the vector $\boldsymbol{C}$. Assume that each node in $\boldsymbol{C}$ is described by an exponential calibration prior density with a single hyperparameter, λ, that specifies the rate of the exponential distribution. Given that divergence-time estimates are sensitive to the chosen specification of these calibration hyperpriors, how should we proceed? One possibility is that all of the exponential calibration priors share the same λ-rate value, or perhaps each of the exponential calibration priors has a unique hyperparameter value, or there might be two or more subsets of exponential calibration priors that share the same λ-rate values, etc. The combinatorics are such that, even for small values of F (where each fossil calibrates one node), the number of possible specification scenarios quickly becomes daunting.

In a recent study, Heath (2012) proposed a nonparametric solution to this problem. The objective is to identify subsets of exponential calibration priors that share the same λ-rate values, where the number of subsets, the assignment of calibrated nodes to subsets, and the values of the hyperparameters for those subsets are treated as random variables under a Dirichlet process prior model. MCMC is then used to sample the calibration distances from a mixture of exponential distributions, while quantifying uncertainty in the λ-hyperparameters, effectively automating the process of parameterizing calibration priors.

For each node in the vector $\boldsymbol{C}$, the exponential calibration prior is offset by the minimum-age constraint for the corresponding fossil. Therefore, this framework induces a vector of exponential-rate parameters: $\boldsymbol{\lambda} = (\lambda_{C_1}, \ldots, \lambda_{C_F})$, and the hyperprior assumed for $\boldsymbol{\lambda}$ is a Dirichlet process: $\boldsymbol{\lambda} \sim \text{DPP}(\alpha, G_0)$. As in the DPP on lineage-specific substitution rates, the concentration parameter for the Dirichlet process hyperprior governs the prior probability that calibration nodes are assigned to the same λ-rate category. In addition, the λ-rate values for each category are drawn from the base distribution $l_i \sim G_0$, where l_i is the λ-rate value corresponding to rate-class i, for i in k rate classes. A gamma distribution is used as the prior for the base distribution, G_0, which has a shape parameter σ_{G_0} and a rate parameter γ_{G_0}: $l_i \sim \text{Gamma}(\sigma_{G_0}, \gamma_{G_0})$.

MCMC samples under this Dirichlet process hyperprior model are based on a variant of a Gibbs sampler (algorithm 8; Neal, 2000).

13.4.1.1 Evaluating hierarchical calibration priors

Using simulated data and artificially generated sets of calibrations, Heath (2012) showed that the Dirichlet process hyperprior on calibration densities leads to more accurate estimates of node ages compared with fixed-hyperparameter calibration priors. In this section, we will expand on those simulations to demonstrate the utility of hierarchical Bayesian models for node calibration even when sets of calibrating fossils are generated from a biologically motivated model of fossilization, preservation, and sampling.

Simulations. We generated 100 replicate datasets, each composed of a tree topology of 20 extant species, 20 corresponding DNA sequences, and a set of 18 fossil calibrations (each associated with a node and a minimum age). Every dataset was generated by first simulating a complete tree topology and branching times under a constant-rate birth–death process ($\lambda = 0.03$, $\mu = 0.015$) using the general sampling approach (Hartmann et al., 2010; Stadler, 2011b). Next, fossilization events were added along each branch in the complete (consisting of both extinct and extant lineages) tree according to a Poisson process with a rate of 0.1. Thus, our approach for generating a complete tree and a complete set of fossils corresponds to the *serially sampled birth–death process* described by Stadler (2010). A preservation/sampling rate was simulated for each fossil "specimen" according to a geometric Brownian-motion process, corresponding to the autocorrelated-log-normal model for lineage substitution rates, with a variance parameter equal to 0.05. Under this model, closely related fossils will therefore have similar sampling rates. For each of the 100 simulated datasets, we sampled a set of 18 "recovered" fossils in proportion to the sampling rate. Figure 13.4 shows an example of the tree and fossil generation for a single realization of the simulation. Typically, the recovered set of fossils includes some calibration nodes with more than one fossil descendant. In these cases, only the oldest fossil is selected to calibrate the node, thus each set of 18 observed fossils was reduced for every simulation replicate. Across our simulations, the sets of calibration fossils ranged in size from 5 to 11, with a median of 8 (i.e., out of 100 replicates, no dataset included all 18 recovered fossils). A (relative) clock rate (r) was drawn for each replicate from a Gamma(5,10) distribution ($\mathbb{E}(r) = 0.5$), thus each replicate dataset was simulated under the global molecular clock model. The sequence data were generated on each tree under its given clock rate and a GTR+Γ substitution model (with exchangeability rates, base frequencies, and Γ-shape parameters matching those used in Heath, 2012).

Analysis. We estimated the ages of internal nodes under a strict molecular clock model using four separate approaches for calibrating node ages in the program DPPDiv (Heath et al., 2012; Heath, 2012; Darriba et al., 2013). Each

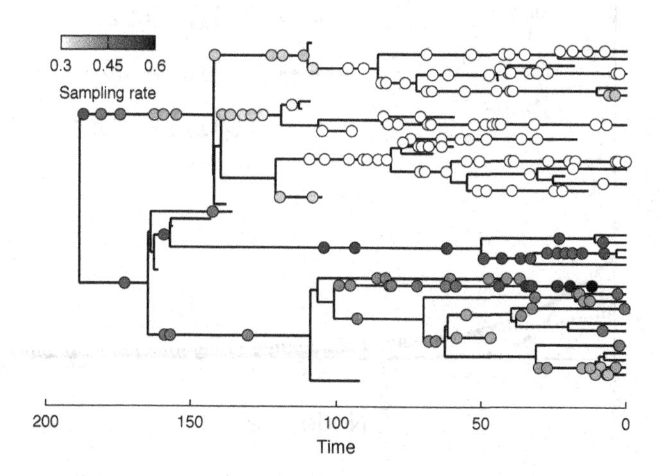

Figure 13.4: An example of the model used to generate tree topologies, branch times, and sets of calibration points. The tree (20 extant lineages) was simulated according to a constant-rate birth–death process. Fossilization events (circles) were drawn from a Poisson process on the complete phylogeny with both extant and extinct lineages. Then, a sampling rate was simulated for each fossil according to an autocorrelated, geometric Brownian-motion process (the gradient ranges from white to black indicating low to high sampling rates, respectively). For each simulation replicate, we sampled 18 fossils from the set of fossilization events in proportion their sampling rate. Only the oldest fossil assigned to a node provides a node-age constraint.

replicate was analyzed under the Dirichlet process hyperprior (DPP-HP) and three different exponential densities with fixed λ-hyperparameters. Figure 13.5 illustrates the fixed-exponential calibration priors. Each fossil calibration imposes a minimum age (A_F) on the age of the calibrated node. For each fossil in each simulation replicate, fixed exponential calibration densities were specified using the following hyperparameters:

$$\lambda_I = (A_F * 0.1)^{-1},$$
$$\lambda_V = (A_F * 0.3333)^{-1},$$
$$\lambda_T = (A_T - A_F)^{-1},$$

where A_T represents the true age of the calibration node (Figure 13.5). Thus, for each simulated alignment and set of fossils, we compared node-age estimates under the DPP-HP to estimates under an informative exponential prior (Fixed-λ_I), a vague prior (Fixed-λ_V), and a fixed-rate exponential prior with an expectation equal to the true age of the calibrated node (Fixed-λ_T).

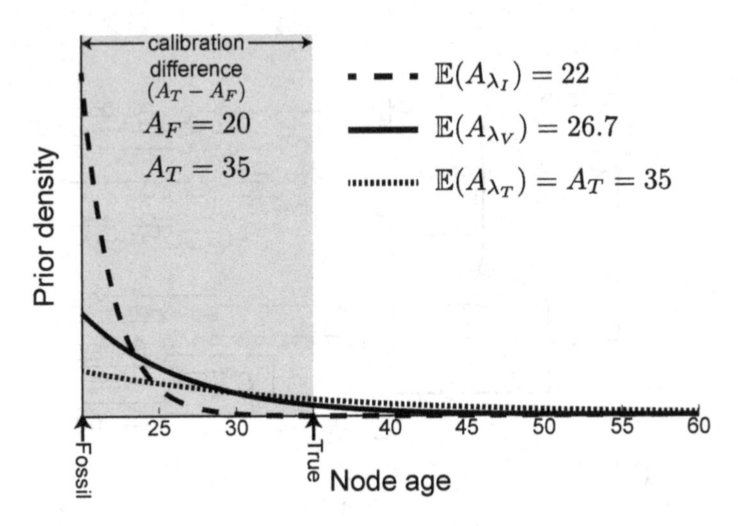

Figure 13.5: Examples of the three fixed-parameter exponential priors used to calibrate analyses of simulated data: Fixed-λ_I (dashed line), Fixed-λ_V (solid line), and Fixed-λ_T (dotted line). In this example, the age of the fossil, A_F, is 20 time units and the true node age is $A_T = 35$. The legend shows expected ages of the node under each prior: $\mathbb{E}[A_{\lambda_*}]$.

For each analysis under a given calibration prior, we fixed the tree topology to match that of the true tree and assumed the generating substitution (GTR+Γ) and clock (strict) models, where the clock rate was assumed to be Gamma-distributed. Accuracy of node-age estimates was evaluated using the coverage probability: i.e., the proportion of times that the true node age falls within the 95% credible interval. We also measured the precision of each estimator by evaluating the widths of the 95% credible intervals.

Results. Divergence-time estimates under the DPP-HP approach were, on average, more accurate (higher coverage probability) than those based on fixed calibration priors (Table 13.1). These results are consistent with previous simulations where sets of calibration models were artificially generated under a non-biological model (Heath, 2012). The present results demonstrate that when calibration fossils are simulated and sampled according to a complex diversification model that accounts for taphonomic processes, the DPP-HP leads to robust estimates of node ages.

We examined the effect of the true node age—for all internal nodes—on the accuracy and precision of divergence-time estimates under each of the different priors (Figure 13.6a,b). Similar to previous results (Heath, 2012), we show that application of overly informative calibration priors results in decreased accuracy in the age estimates of older nodes (Figure 13.6a). The effect of this excessively informative prior is manifest in the relatively precise (narrow) credible intervals

Table 13.1: The coverage probability (the proportion of estimates where the true value falls within the 95% credible interval) for estimates of node ages under the Dirichlet process hyperprior (DPP-HP), the calibration prior centered on the true node age (Fixed-λ_T), the vague fixed exponential prior (Fixed-λ_V), and the informative fixed-calibration prior (Fixed-λ_I).

Calibration Prior	**Coverage probability**
DPP-HP	0.926
Fixed-λ_T	0.917
Fixed-λ_V	0.843
Fixed-λ_I	0.550

on node-age estimates (Figure 13.6b). By contrast, the coverage probabilities for node-age estimates under the DPP-HP approach have greater accuracy and are very similar to the estimates under exponential prior densities centered on the *true* node ages (Figure 13.6a). This increased accuracy, however, comes at the expense of decreased precision: the credible intervals under the DPP-HP are wider than those from the Fixed-λ_T calibration prior (Figure 13.6b).

Despite the slight decrease in the precision of node-age estimates under the DPP-HP compared with the Fixed-λ_T prior, divergence-time estimation using a hierarchical calibration prior offers significant advantages. Importantly, the DPP-HP yields robust and reliable results without requiring the user to specify a calibration prior centered on the true (unknown) age of the calibrated node. The use of parametric prior densities is a common practice

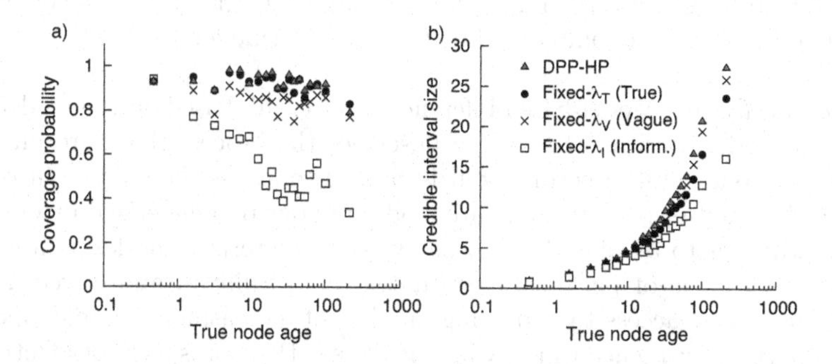

Figure 13.6: The coverage probabilities and credible interval widths of node age estimates under the Dirichlet process hyperprior (DPP-HP; gray triangles), the fixed exponential prior centered on the true node age (Fixed-λ_T; filled circles), the vague fixed exponential (Fixed-λ_V; crosses), and the informative exponential prior (Fixed-λ_I; boxes). For both graphs, the true node ages were binned so that each bin contained 100 nodes (log scale) and the (a) coverage probability or the (b) average credible interval size was calculated for each bin.

for calibrating Bayesian divergence-time estimates. Specifying details of the calibration density (values for the hyperparameters, etc.) is difficult and often entails a considerable degree of subjectivity, and these choices are known to have a large impact on the resulting divergence-time estimates. Our simulations and those conducted by Heath (2012) illustrate the benefit of applying a hierarchical model to calibration prior distributions, as this approach automates the process of parameterizing calibration prior densities and, more importantly, accommodates uncertainty in the values of calibration hyperparameters.

13.5 Practical issues for estimating divergence times

Bayesian methods for dating species divergences have transformed the field of evolutionary biology by allowing researchers to investigate a wide range of questions about the timing and rates of biological processes. Like any statistical approach, however, these methods are not without distinct challenges. In this section we outline three potentially problematic issues associated with the application of Bayesian divergence-time methods.

13.5.1 Model selection and model uncertainty

The burgeoning toolbox of models for estimating divergence times raises new challenges for assessing the fit of our data to this swelling pool of candidate models (model selection) and/or accommodating uncertainty when many models may provide a comparable fit to our data (model averaging).

Model selection. Model-based inference is, after all, based on the model: our choice of model is critical because it describes the process that (presumably) gave rise to our data. Accordingly, it is critical to assess the relative fit of our data to the relevant pool of candidate models: under-parameterized models will cause estimates to be biased, whereas over-parameterized models will inflate the error variance of parameter estimates (e.g., Sullivan and Joyce, 2005). Researchers have access to a growing number of increasingly complex models for estimating divergence times, which increases their prospects of identifying a model that provides a reasonable fit to a given dataset. On the other hand, the expanding list of candidate models will require a commensurate increase in effort to reliably assess their fit to a given dataset.

For example, there are at least 75 candidate relaxed-clock models for estimating divergence times from species-level datasets (i.e., that can be derived from all possible combinations of priors on rates and ages; *cf*, Figure 13.2). Despite the diversity of available models, however, the vast majority of empirical studies have simply assumed the UCLN model of rate variation and a birth–death prior on node ages, without assessing relative fit of the data to other candidate relaxed-clock models. The prevalence of this combination of models in the literature is, in part, due to their availability in the widely used and well-designed software package, BEAST (Drummond and Rambaut, 2007; Drummond et al., 2012).

This emerging convention might not be a concern if, for example, all other candidate models were special cases of the UCLN+BD relaxed-clock model (i.e., if the model could neatly collapse to all other, nested models if dictated by the information in the data), or if the model generally provided the best fit to all (or most) empirical datasets. Unfortunately, the evidence (albeit limited) suggests reasons for concern. The study by Lepage et al. (2007), for example, assessed the fit of several empirical datasets to a number of relaxed-clock models using rigorous model-comparison methods: i.e., the marginal likelihoods of candidate models were estimated using thermodynamic integration (Lartillot and Philippe, 2006b) and compared using Bayes factors (Suchard et al., 2001). Not surprisingly, this survey found that no single relaxed-clock model emerged as a panacea. Instead, the "best" relaxed-clock model was different for different datasets, and, in fact, most of the datasets had a better fit to autocorrelated relaxed-clock models (Lepage et al., 2007). Moreover, the chosen prior model of substitution-rate variation (and to a lesser extent the prior on node ages) is known to dramatically impact divergence-time estimates (e.g., Aris-Brosou and Yang, 2003; Welch et al., 2005; Lepage et al., 2007).

Clearly, it would be preferable to assess the fit of each dataset to the relevant pool of candidate relaxed-clock models. It is also evident that reliable Bayesian model selection will require substantially increased computational effort. Bayesian model selection is based on comparing the marginal likelihoods of candidate models. Relatively efficient methods for estimating marginal likelihoods have been developed that are based on a standard MCMC sample—these include the harmonic mean estimator (Newton and Raftery, 1994a), the stabilized harmonic mean estimator (Suchard et al., 2005), the Bayesian analogue of the Akaike information criterion (AICM) based on posterior samples (Raftery et al., 2007b), and posterior Bayes factor (Aitkin, 1991). These efficient estimators, however, are known to be strongly biased toward the selection of overly complex models (e.g., Baele et al., 2012b, 2013c). Fortunately, more reliable estimators have recently been proposed, such as path sampling techniques (e.g., Lartillot and Philippe, 2006b) and stepping-stone sampling approaches (e.g., Fan et al., 2011; Xie et al., 2011). These approaches require separate MCMC analyses to simulate from the joint prior to the joint posterior of each candidate model, or between the posteriors spanning two candidate models. Accordingly, these new model-selection methods are a mixed blessing—in principle they enable reliable model selection; in practice, however, the increased computational burden they entail imposes severe limits on the number of models that can be evaluated. Accordingly, efficient methods that offer reliable selection among candidate models are urgently needed, and it appears that important progress on this front is imminent (see Chapter 3).

Model uncertainty. Model selection entails ranking a set of candidate models according to their relative fit to the data at hand. Even if we have exhaustively evaluated all possible candidate models and accurately assessed the relative fit of each model to a given dataset, it may nevertheless be unwise to condition our inference on the best model. This relates to the issue of model uncertainty. Imagine, for example, that several (possibly many) alternative models provide a similarly good fit to a given dataset. Under these circumstances, conditioning inference on *any single model* (even the "best") ignores uncertainty in the chosen model and will cause estimates to be biased. This scenario is apt to become increasingly plausible as the pool of candidate models continues to swell.

Model uncertainty can be addressed by means of *model averaging*—parameters are estimated under each of the candidate models, and then summarized as the weighted average over the candidate models, where the weighting is based on the probability of each candidate model. The Bayesian framework provides a natural approach for accommodating model uncertainty; we simply adopt the perspective that the models (like the parameters within each model) are themselves random variables.

Bayesian model averaging has been implemented for various phylogenetic problems using reversible-jump MCMC, where the chain integrates over the joint prior probability density of a given model in the usual manner, but may also jump between models, visiting each model in proportion to its marginal probability. For example, rjMCMC has been used to average over the pool of 203 substitution models corresponding to all possible combinations of the six exchangeability parameters (Huelsenbeck et al., 2004), but this approach has not (to our knowledge) been used to average over relaxed-clock models. An alternative approach for Bayesian model averaging was recently proposed by Li and Drummond (2012): typically, many relaxed-clock models have considerable marginal posterior probability (i.e., the credible set of models contains several models), and averaging over models better captures uncertainty in divergence-time estimates (Li and Drummond, 2012).

Non-parametric Bayesian models also have considerable potential as a means of model averaging. For example, the application of the Dirichlet process prior model to accommodate variation in substitution rates across branches (Heath et al., 2012) may be viewed as a form of model averaging. From this perspective, each unique configuration—the number of substitution-rate categories, the assignment of branches to those rate categories, and the rates of each category—corresponds to a unique relaxed-clock model. The DPP specifies the prior probability of each possible model, and MCMC provides an efficient means of averaging the inference over this often vast model space. The beauty of these non-parametric models is their apparent ability to shrink down on the data:

the dimensions of the model are not specified *a priori*, but rather are defined by the patterns in the data.

13.5.2 (Hyper)prior sensitivity

Model selection/uncertainty is closely related to the issue of prior sensitivity. In the former case, we are concerned with uncertainty in the choice of candidate models; in the latter we are concerned with uncertainty in the choice and parameterization of priors used to describe each of the parameters in the candidate models (Berger, 1994). Bayesian inference adopts the perspective that parameters are random variables. This perspective requires that we specify a prior probability distribution for each parameter. The joint prior probability density is essentially the product of the prior densities for all of the parameters in the model, which collectively describe our expectations about the inference before evaluating the data at hand. The joint posterior probability density is the updated version of the joint prior probability density: it is updated by the information in the data via the likelihood function. The current convention in Bayesian phylogenetic inference is to specify "vague" prior probability densities that reflect a high degree of ignorance about the expected parameter values. Specifying diffuse priors is intended to reduce their influence on the corresponding posterior estimate, i.e., the marginal posterior density for a parameter with a minimally informative prior should be dominated by the information in the data.

The dimensionality of the joint prior probability density increases with the complexity of the corresponding model, while the amount of information in the data remains fixed. Accordingly, the specified prior will exert a greater influence on parameter estimates under more complex models; that is, the estimates exhibit prior sensitivity. This issue is of particular concern for divergence-time estimation, which often invokes quite complex models. One solution is to adopt a hierarchical Bayesian approach, in which we avoid committing to specific values for hyperparameters, and instead assume they are also random variables. The inference is then integrated over the joint prior probability density that includes dimensions for the newly specified second-order hyperpriors. This is a successful strategy for many model parameters. For example, applications of the Dirichlet process prior model typically specify a second-order gamma hyperprior on the concentration parameter, α, which can often be reliably estimated from the data (e.g., Gelfand et al., 2005; Dorazio, 2009). In particular, the number of parameter categories, k, is often robust to the parameterization of the hyperprior on the concentration parameter. In their analysis of a primate mitochondrial dataset, Heath et al. (2012) compared the marginal posterior and prior probabilities of k under four different parameterizations of the gamma hyperprior on α, such that $\mathbb{E}[k]$ was 3, 6, 18, or 59. This comparison revealed strong support in the data for branch-rate partitions, where the posterior median value of k ranged from 4 to 9 under the different parameterizations of the hyperprior. Thus, even when $\mathbb{E}[k] = 59$ under the hyperprior on α, the information in the data supported $k = 9$, with a 95% credible interval of $[3, 16]$.

Despite the benefits of the hierarchical Bayesian approach, however, this essentially *defers* rather than *resolves* the issue of prior sensitivity: at some point we need to "put our money down" on specific (hyper-)parameter values. Importantly, there are many cases in which the chosen hyperprior values are strongly informative. For example, estimates of divergence times may also be sensitive to the second-order hyperpriors for the general birth–death prior on node ages. This sensitivity was vividly demonstrated in an analysis of metazoan divergence times: the conclusion that this group arose via a "Cambrian explosion" or arose gradually throughout the pre-Cambrian hinged on the upper bound specified for the uniform second-order hyperprior for the sampling fraction, ρ (e.g., Aris-Brosou and Yang, 2003; Welch et al., 2005).

Accordingly, robust inference demands that we assess the sensitivity of estimates to the chosen (hyper)priors, especially for complex models. Unfortunately, protocols and best practices for assessing prior sensitivity are not well established. Prior sensitivity essentially entails performing complete MCMC simulations in which a single (hyper)prior value at a time is modified. The combinatorics of this problem conspire to make this a daunting prospect for complex phylogenetic models.

13.5.3 Limitations and challenges of fossil calibration

The common practice of applying parametric densities to calibrated nodes is a less-than-ideal approach for estimating absolute divergence times (Heled and Drummond, 2012; Warnock et al., 2012; Heath, 2012; dos Reis and Yang, 2012). In principle, numerous parametric distributions provide infinite flexibility for describing our beliefs (and uncertainty) about the age of calibrated nodes. In practice, however, it is often difficult to summarize these beliefs as parametric probability densities. Moreover, unlike most priors—which are updated (and typically dominated) by the information in the data—calibration priors are essentially 'pseudo data'. That is, there is no information in the sequence data (or elsewhere) to update the prior calibration densities. Consequently, posterior samples of absolute times/rates are reliant on, and highly sensitive to, the choice and parameterization of prior calibration densities.

Furthermore, calibration priors on nodes often entail (cryptic) interactions with other node-age priors (Heled and Drummond, 2012; Warnock et al., 2012). Regardless of the minimum-age constraint imposed by a fossil calibration, all nodes in the tree must be younger than their ancestors and older than their descendant nodes. Thus, calibrated node ages are affected by the ages of surrounding nodes. This property can yield a marginal posterior density for the age of a calibrated node that is different from the corresponding prior density (Warnock et al., 2012). In addition, under the multiplicative construction of the tree-wide age prior in the program BEAST (Drummond and Rambaut,

2007; Drummond et al., 2012), the "induced" prior density of a calibrated node is the product of the calibration prior density and the corresponding prior density for the age of that node generated under the chosen tree-wide age prior (e.g., the birth–death prior on node ages). Importantly, Heled and Drummond (2012) have developed an alternative tree prior that eliminates this problem when a single calibration is used in conjunction with the Yule model. Regardless of the priors applied, any analysis of divergence times that relies on calibration densities must carefully inspect the "induced" marginal calibration priors rendered by node-age parameter interactions. Induced calibration priors can be visualized by first targeting the joint prior probability density (e.g., by running the MCMC without data) and then examining the marginal prior probability density for each of the calibrated nodes (dos Reis et al., 2012; Heled and Drummond, 2012; Warnock et al., 2012; Heath, 2012).

Perhaps the most significant source of error in divergence-time estimation is the misidentification and/or misplacement of fossils used for calibration. Presumably, a misplaced fossil would not have an adverse impact unless the fossil was truly *older* than the node it is being used to calibrate. In this case, the hard minimum-age bound enforced by the fossil calibration results in a calibration density that has zero prior probability density on the true node age, with the potential to bias the divergence-time estimates of other nodes in the tree. Yang and Rannala (2006b) described an approach for constructing calibration densities that allows the calibrated node to "escape" hard bounds. These flexible, "soft-bound," calibration priors place a small (but non-zero) prior mass on node ages outside of the hard bound, a prudent approach for any application of uniform calibration priors.

Fossil tip-dating. Ideally, we would prefer to directly incorporate information from fossil specimens, such that divergence-time estimates are conditioned on having observed a set of sequences *and* a set of fossils. This approach can mitigate problems related to induced calibration priors and accommodate uncertainty in the position of calibrated nodes (when jointly estimating the topology of extant and fossil species). Furthermore, treating fossils in this way provides a means for using *all* of the available fossils—not just the oldest specimens for each calibrated node—when estimating divergence times.

Several recent studies (Lee et al., 2009; Pyron, 2011; Ronquist et al., 2012a) have investigated approaches for the combined analysis of extant and fossil species for estimating divergence times. This so-called tip-dating approach treats fossils as serially sampled taxa, such that the sampling times of the fossil specimens impose temporal structure on the tree. These methods rely on data matrices comprising molecular sequence data sampled from the extant species *and* discrete morphological characters for both extant and extinct species. Joint estimation of phylogeny and divergence times from these combined molecular and morphological datasets effectively accommodates uncertainty in the placement of fossil taxa. The conventional "node-based" approach is susceptible to error propagation and only uses a subset of the available fossil evidence. That is, a fossil is first assigned to a node (by some means) and a calibration prior is then specified for that node (by some means) reflecting the age of the fossil and our beliefs about the reliability of the fossil record. This approach therefore ignores uncertainty associated with the phylogenetic position of the fossil. Moreover, because the calibration prior imposes a minimum-age constraint, only the oldest fossil assignable to a node is relevant under this approach. Ronquist et al. (2012a) analyzed a wasp dataset using the tip-dating approach and compared the resulting divergence-time estimates to those obtained using conventional node-based calibration. Their results suggest that tip dating produces divergence-time estimates that are both more precise (i.e., narrower credible intervals) and also less sensitive to prior inputs (e.g., the age of the root-node prior calibration) compared with node dating.

The development of tip-dating approaches represents an important direction for the field of Bayesian divergence-time estimation, but nevertheless presents some challenges. Tip dating relies on combining molecular sequence data for extant species and discrete morphological data for both extant and fossil taxa. Because molecular data are typically unavailable for fossil taxa, these data matrices will usually exhibit a considerable degree of non-randomly distributed missing data. High levels of missing data can confound phylogeny estimation by magnifying systematic errors, particularly when the pattern of missing states is non-random (Lemmon et al., 2009; Ronquist et al., 2012a; Roure et al., 2013). Although this artifact may not be problematic for topological inference if the data are highly informative regarding the tree topology (Wiens et al., 2010; Wiens and Morrill, 2011), systematically incomplete data matrices may nevertheless significantly impact estimates of branch lengths (and therefore divergence times). Missing data is also an issue within the morphological partition: the fragmentary nature of fossil preservation imposes a hard lower bound on the level of missing data that is impervious to our efforts to sample additional fossil taxa and/or score additional morphological characters.

Furthermore, Bayesian analysis of morphological characters assumes that the data conform to a (relaxed) "morphological clock" (Lewis, 2001; Ronquist et al., 2012a). This may be a reasonable assumption, given that the accumulation of morphological disparity is generally accepted to be correlated with absolute time. However, even if morphological evolution may be appropriately described by a (relaxed) morphological clock, the pattern of morphological-rate variation across lineages may be quite different from the corresponding pattern for the molecular sequence data; i.e., the morphological and molecular clocks may be largely "asynchronous." Moreover, it is reasonable to expect that—given their relative exposure to selection—morphological traits may be prone to "heterotachy" (i.e., variation in the rate of morphological change across traits and across lineages). Thus, the common approach of assigning a uniform model to the morphological data—and specifying a common relaxed-clock model for the morphological and molecular partitions—may lead to systematic errors in estimates of divergence times, rates, and tree topology.

Fully integrative analysis of extant and extinct taxa requires the development of new models for lineage diversification that account for serially sampled species as well as taphonomic processes—rates of fossilization, preservation, and recovery. Previous applications of tip dating have used inappropriate branching-process priors such as the Yule (pure birth) or reconstructed birth–death prior (Pyron, 2011; Wood et al., 2013). Both the Yule and reconstructed birth–death priors assume that all tips were sampled at the same time, which is clearly violated in the case of serial sampling (Gernhard, 2008; Stadler, 2010; Stadler and Yang, 2013). Moreover, the Yule prior assumes that the rate of extinction is zero; unless all sampled fossils are direct ancestors of living taxa, the assumptions of the Yule process are seriously violated. Ronquist et al. (2012a) avoided these pitfalls by constructing a putatively vague uniform prior on node times, but this generic prior does not provide a biologically meaningful description of the processes responsible for generating the tree.

In principle, the processes of lineage diversification, fossilization/preservation, and sampling should all be components of the prior on node and tip ages. Stadler (2010) described a general model for datasets where some of the species represent extinct fossil taxa (or pathogens) sampled at various time horizons. When applied to species-level datasets that include discrete morphological data for both extant and fossil specimens, this model can be used to specify the prior distribution of speciation times and species-sampling events through time. As a prior on node ages, this serially sampled birth–death model incorporates an additional parameter, ψ, that represents the rate of sampling lineages in the past (Stadler, 2010; Stadler and Yang, 2013). Essentially, when applied to species-level datasets that include fossils, this parameter summarizes the various taphonomic processes—fossilization, preservation, recovery, etc.—that constitute the sampling rate of extinct lineages. When plugged into (13.2), the probability density of $\boldsymbol{a}$ is: $f(\boldsymbol{a} \mid \lambda, \mu, \rho, \psi)$, with additional hyperpriors on λ, μ, ρ, and ψ: $f(\lambda \mid \theta_\lambda)f(\mu \mid \theta_\mu)f(\rho \mid \theta_\rho)f(\psi \mid \theta_\psi)$. The serially sampled birth–death process can also be combined with the birth–death shift model of (Stadler, 2011a), which would allow ψ to change episodically over time (Stadler et al., 2013). Therefore, this complex model could account for serially sampled fossil species where the rate of fossil sampling changes through time to provide a more realistic description of the variation in preservation observed in the stratigraphic record.

Analysis of combined molecular and morphological datasets comprising both fossil and extant species requires complex models of lineage diversification and rate variation. Application of these methods should also be coupled with model comparisons using Bayes factors (Ronquist et al., 2012a) and evaluation of prior sensitivity (Warnock et al., 2012). Despite the challenges associated with integrative approaches and parameter-rich models, methods for directly incorporating fossil data in divergence-time estimation is a promising direction with the potential to enhance our ability to explore a wide range of questions about macroevolution.

13.6 Summary and prospectus

Advances in the development of stochastic models for estimating species divergence times is one of the major success stories in phylogenetic systematics. Rapid progress on this problem over the last decade is due in large measure to the adoption of a Bayesian statistical framework, as it provides a natural and flexible environment for incorporating prior beliefs/assumptions regarding the nature of substitution rate variation and the distribution of node ages. Moreover, the MCMC algorithms used in Bayesian inference permit the use of increasingly complex (and hopefully realistic) models that would otherwise be prohibitive, and provide a natural means of estimating and accommodating uncertainty. Our enhanced ability to more accurately estimate species divergence times has, in turn, advanced research in a vast and growing number of scientific disciplines.

Although adopting a Bayesian framework has revolutionized estimation of species divergence times, the revolution is incomplete: further progress is needed in the application, evaluation, and elaboration of current methods. Conventions have emerged as "standard practice" in comparative studies that do not fully embrace the inherent benefits of Bayesian divergence-time estimation. Although these methods estimate a marginal posterior probability distribution of dated trees, comparative studies are typically based on summaries of these posterior distributions (using point estimates of the tree topology, divergence times, rate estimates, etc.), which ignores associated uncertainty that may impact the comparative inferences. For example, studies of trait evolution (and historical biogeography, diversification rates, host-parasite co-evolution, etc.) often proceed by first estimating divergence times from molecular sequence data, then summarize the resulting marginal distribution of dated phylogenies (e.g., as mean or median branch times on a majority-rule consensus, maximum-clade credibility, or MAP tree), and finally condition inference of trait history on that point estimate. This approach treats phylogenies as "observations" that are known perfectly and without error. As estimates from data, however, divergence times are typically associated with (sometimes considerable) uncertainty. If a relative/absolute time scale is crucial for addressing a question, it is equally important to accommodate uncertainty in divergence-time estimates. It is well established that failure to accommodate phylogenetic uncertainty can dramatically impact the conclusions of comparative studies (e.g., Huelsenbeck et al., 2000d; Huelsenbeck and Bollback, 2001; Huelsenbeck et al., 2001a; Huelsenbeck and Imennov, 2002).

A Bayesian framework provides a natural approach for accommodating uncertainty. The comparative inference is simply averaged over the marginal posterior probability distribution of dated trees. In principle, the comparative inference need not be Bayesian: e.g., "quasi-Bayesian" estimates of trait histories (or historical biogeography, diversification rates, host-parasite co-evolution, etc.) might be performed on a posterior sample of trees using parsimony, maximum likelihood, or some other method. Such comparative methods are not strictly Bayesian as point—rather than marginal—estimates are generated for each sampled tree: although uncertainty in the phylogeny is accommodated under this approach, uncertainty associated with the comparative inference on each tree is nevertheless ignored. A preferable, "sequential-Bayesian" approach would involve first estimating the marginal posterior probability density of dated trees (e.g., from molecular sequence data) and then performing marginal comparative inference on a posterior sample of trees. Alternatively, a fully integrative "hierarchical-Bayesian" approach to evolutionary studies would simultaneously estimate the joint posterior probability density of divergence times and the comparative inference.

In theory, it should not matter whether we adopt a fully hierarchical or serial Bayesian approach for comparative inferences. For example, we might examine the evolution of a discrete morphological trait in one of two ways. Under the serial-inference scenario, we would first estimate the marginal posterior probability distribution of dated trees from a molecular sequence dataset using one of the models described in this chapter, and subsequently integrate the inference of trait history over the distribution of resulting trees using a stochastic model of morphological evolution (e.g., Pagel and Meade, 2004, 2006). In the second (hierarchical-inference) scenario, we would combine the sequence dataset and the morphological trait in a single data matrix, and estimate the joint posterior probability density of trees and trait histories simultaneously under a composite model (e.g., Ronquist et al., 2012a). The inference of trait history under either scenario should—in theory—be identical (assuming conditional independence of the molecular and morphological data). In practice, however, estimates under these two approaches may differ substantially. Technical issues, for example, may make it difficult to reliably mix over the marginal posterior probability distribution of trees or other parameters. These issues motivate the development of fully hierarchical Bayesian approaches that jointly estimate divergence times and the comparative inference as the gold standard for evolutionary studies; however, this approach is currently only possible for a small subset of comparative methods (e.g., Lartillot and Poujol, 2011; Lartillot and Delsuc, 2012).

Although there is clearly room for the elaboration of hierarchical Bayesian methods, we also desperately need to thoroughly evaluate the performance of existing methods via the analysis of simulated and empirical data. Simulation studies have begun the important work of evaluating the statistical behavior of existing models (e.g., Drummond et al., 2006; Battistuzzi et al., 2010; Lepage et al., 2006; Heath et al., 2012; Baele et al., 2013c), but further investigation in this area is critically important. The simplicity conferred by simulation studies is both an advantage and a liability: these studies can clearly assess the ability of the inference models to recover the true (known) parameters of the process used to generate the data. The success of a model on simulated data, however, may not provide a reliable guide to its behavior when applied to real data, which is more directly assessed by measuring the statistical fit of the model to the data at hand. Accordingly, it is crucial that we complement simulation studies with careful empirical analyses—using robust Bayesian model-selection methods—to assess the performance of available models on real datasets (e.g., Lepage et al., 2007; Linder et al., 2011; Baele et al., 2013c).

The multitude of available models for estimating species divergence times constitutes an embarrassment of riches: the considerable and growing pool of candidate models improves our odds of providing a reasonable fit to—and therefore provide reliable inference from—a given empirical dataset. On the other hand, exploring the fit of a given dataset to all (or even most) available divergence-time methods represents a considerable computational investment. Perhaps for this reason, this ideal has not been embraced as standard practice in empirical studies. However, the benefits of adopting more rigorous model-selection standards are potentially huge. Studies that rigorously assess the fit of empirical datasets to a broad range of existing models will realize the full potential of existing methods for estimating divergence times: thorough model selection is needed to identify the existing model that provides the best fit to—and therefore renders the most reliable estimates from—individual datasets. Comprehensive model selection practices would also reveal the relative importance of the three model components—associated with the prior models on rates, node ages, and calibrations—and thereby identify priorities for refining existing and developing new Bayesian divergence-time estimation methods.

We are optimistic that future extensions of the Bayesian framework for estimating species divergence times—coupled with rigorous evaluation and careful application of these methods—will continue to enhance our ability to answer a broad range of fundamental questions in biology.

Acknowledgments

We are grateful to the editors for the invitation to contribute this chapter and to Jeff Thorne and Paul Lewis for their insightful comments. TAH was supported by NSF grant DEB-1256993 and NIH grants GM-069801 and GM-086887. BRM was supported by NSF grants DEB-0842181 and DEB-0919529.

Bibliography

Abecasis, A. B., Lemey, P., Vidal, N., de Oliveira, T., Peeters, M., Camacho, R., Shapiro, B., Rambaut, A., and Vandamme, A.-M. (2007). Recombination confounds the early evolutionary history of human immunodeficiency virus type 1: subtype G is a circulating recombinant form. *Journal of Virology* **81**, 8543–8551.

Abella, J., Alba, D. M., Robles, J. M., Valenciano, A., Rotgers, C., Carmona, R., Montoya, P., and Morales, J. (2012). *Kretzoiarctos* gen. nov., the oldest member of the giant panda clade. *PLoS One* **17**, e48985.

Adams, R. P., Murray, I., and MacKay, D. J. (2009). Tractable nonparametric Bayesian inference in Poisson processes with Gaussian process intensities. *Proceedings of the 26th Annual International Conference on Machine Learning* pages 9–16.

Aitichinson, J. (1984). *The statistical analysis of compositional data.* London: Chapman and Hall.

Aitkin, M. N. (1991). Posterior Bayes factor. *Journal of the Royal Statistical Society, Series B* **53**, 111–142.

Akaike, H. (1973). Information theory and an extension of the maximum likelihood principle. In *Second international symposium on information theory*, volume 1, pages 267–281. Berlin: Springer.

Al-Hussaini, A. N. and Elliot, R. J. (1989). Markov bridges and enlarged filtrations. *Canadian Journal of Statistics* **17**, 329–332.

Al-Mohy, A. H. and Higham, N. J. (2009). A new scaling and squaring algorithm for the matrix exponential. *SIAM Journal on Matrix Analysis and Applications* **31**, 970–989.

Aldous, D. (1996). Probability distributions on cladograms in random discrete structures. In D. Aldous and R. Pemantle, editors, *Random discrete structures. IMA Volumes in Mathematics and its Applications 76*, pages 1–18. Berlin: Springer.

Aldous, D. (2001). Stochastic models and descriptive statistics for phylogenetic trees, from Yule to today. *Statistical Science* **16**, 23–34.

Aldous, D. and Popovic, L. (2005). A critical branching process model for biodiversity. *Advances in Applied Probability* **37**, 1094–1115.

Alfaro, M. E. and Holder, M. T. (2006). The posterior and the prior in Bayesian phylogenetics. *Annual Review of Ecology Evolution and Systematics* **37**, 19–42.

Altekar, G., Dwarkadas, S., Huelsenbeck, J. P., and Ronquist, F. (2004). Parallel Metropolis coupled Markov chain Monte Carlo for Bayesian phylogenetic inference. *Bioinformatics* **20**, 407–415.

Andersen, P. K., Borgan, O., Gill, R. D., and Keiding, N. (1995). *Statistical models based on counting processes.* Berlin: Springer, second edition.

Andrieu, C., Doucet, A., and Holenstein, R. (2010). Particle Markov chain Monte Carlo methods. *Journal of the Royal Statistical Society, Series B* **72**, 269–342.

Andrieu, C., Moulines, É., and Priouret, P. (2005). Stability of stochastic approximation under verifiable conditions. *SIAM Journal on Control and Optimization* **44**, 283–312.

Antoniak, C. E. (1974a). Mixtures of Dirichlet processes with applications to Bayesian nonparametric problems. *Annals of Statistics* **2**, 1152–1174.

Antoniak, C. E. (1974b). Mixtures of Dirichlet processes with applications to non-parametric problems. *Annals of Statistics* **2**, 1152–1174.

Ardia, D., Basturk, N., Hoogerheide, L., and van Dijk, H. (2012). A comparative study of Monte Carlo methods for efficient evaluation of marginal likelihood. *Computational Statistics and Data Analysis* **56**, 3398–3414.

Arenas, M. and Posada, D. (2007). Recodon: coalescent simulation of coding DNA sequences with recombination, migration, and demography. *BMC Bioinformatics* **8**, 458.

Arenas, M., Valiente, G., and Posada, D. (2008). Characterization of reticulate networks based on the coalescent with recombination. *Molecular Biology and Evolution* **25**, 2517–2520.

Arima, S. (2009). Bayesian tools for complex statistical models in genetics, (PhD thesis).

Arima, S. and Tardella, L. (2012). Improved harmonic mean estimator for phylogenetic model evidence. *Journal of Computational Biology* **19**, 418–438.

Aris-Brosou, S. (2007). Dating phylogenies with hybrid local molecular clocks. *PLoS ONE* **2**, e879.

Aris-Brosou, S. and Yang, Z. (2002). Effects of models of rate evolution on estimation of divergence dates with special reference to the metazoan 18S ribosomal RNA phylogeny. *Systematic Biology* **51**, 703–714.

Aris-Brosou, S. and Yang, Z. (2003). Bayesian models of episodic evolution support a late precambrian explosive diversification of the Metazoa. *Molecular Biology and Evolution* **20**, 1947–1954.

Arjas, E. and Heikkinen, J. (1997). An algorithm for nonparametric Bayesian estimation of a Poisson intensity. *Computational Statistics* **12**, 385–402.

Arndt, P. F., Burge, C. B., and Hwa, T. (2003). DNA sequence evolution with neighbor-dependent mutation. *Journal of Computational Biology* **10**, 313–322.

Asmussen, S. and Hobolth, A. (2008). Bisection ideas in end-point conditioned Markov process simulation. In G. Rubino and B. Tuffin, editors, *Proceedings of the 7th International Workshop on Rare Event Simulation*, pages 121–130. Rennes: RESIM 2008.

Asmussen, S. and Hobolth, A. (2012). Markov bridges, bisection and variance reduction. In L. Plaskota and H. Wozniakowski, editors, *Monte Carlo and quasi-Monte Carlo methods 2010,* Springer Proceedings in Mathematics and Statistics 23, pages 3–22. Berlin: Springer.

Baele, G., Lemey, P., Bedford, T., Rambaut, A., Suchard, M. A., and Alekseyenko, A. V. (2012a). Improving the accuracy of demographic and molecular clock model comparison while accommodating phylogenetic uncertainty. *Molecular Biology and Evolution* **29**, 2157–2167.

Baele, G., Lemey, P., Bedford, T., Rambaut, A., Suchard, M. A., and Alekseyenko, A. V. (2012b). Improving the accuracy of demographic and molecular clock model comparison while accommodating phylogenetic uncertainty. *Molecular Biology and Evolution* **30**, 2157–2167.

Baele, G., Lemey, P., and Vansteelandt, S. (2013a). Make the most of your samples: Bayes factor estimators for high-dimensional models of sequence evolution. *BMC Bioinformatics* **14**, 1–18.

Baele, G., Li, W. L. S., Drummond, A. J., Suchard, M. A., and Lemey, P. (2013b). Accurate model selection of relaxed molecular clocks in Bayesian phylogenetics. *Molecular Biology and Evolution* **30**, 239–243.

Baele, G., Li, W. L. S., Drummond, A. J., Suchard, M. A., and Lemey, P. (2013c). Accurate model selection of relaxed molecular clocks in Bayesian phylogenetics. *Molecular Biology and Evolution* **30**, 239–243.

Baele, G., Peer, Y. V. d., and Vansteelandt, S. (2008). A model-based approach to study nearest-neighbor influences reveals complex substitution patterns in non-coding sequences. *Systematic Biology* **57**, 675–692.

Barry, D. and Hartigan, J. (1987). Statistical analysis of hominoid molecular evolution. *Statistical Science* **2**, 191–207.

Bartolucci, F., Scaccia, L., and Mira, A. (2006). Efficient Bayes factor estimation from the reversible jump output. *Biometrika* **93**, 41–52.

Battistuzzi, F. U., Filipski, A., Hedges, S. B., and Kumar, S. (2010). Performance of relaxed-clock methods in estimating evolutionary divergence times and their credibility intervals. *Molecular Biology and Evolution* **27**, 1289–1300.

Bayes, T. (1763). An essay towards solving a problem in the doctrine of chances. *Philosophical Transactions of the Royal Society of London* **53**, 370–416.

Beaumont, M. A., Zhang, W. Y., and Balding, D. J. (2002). Approximate Bayesian computation in population genetics. *Genetics* **162**, 2025–2035.

Beerli, P. (2004). Effect of unsampled populations on the estimation of population sizes and migration rates between sampled populations. *Molecular Ecology* **13**, 827–836.

Beerli, P. (2006). Comparison of Bayesian and maximum likelihood inference of population genetic parameters. *Bioinformatics* **22**, 341–345.

Beerli, P. and Felsenstein, J. (1999). Maximum likelihood estimation of migration rates and effective population numbers in two populations using a coalescent approach. *Genetics* **152**, 763–773.

Beerli, P. and Felsenstein, J. (2001a). Maximum likelihood estimation of a migration matrix and effective population size in n subpopulations by using a coalescent approach. *Proceedings of the National Academy of Science USA* **98**, 4563–4568.

Beerli, P. and Felsenstein, J. (2001b). Maximum likelihood estimation of a migration matrix and effective population sizes in n subpopulations by using a coalescent approach. *Proceedings of the National Academy of Sciences USA* **98**, 4563–4568.

Beerli, P. and Palczewski, M. (2010). Unified framework to evaluate panmixia and migration direction among multiple sampling locations. *Genetics* **185**, 313–326.

Bell, E. T. (1934). Exponential numbers. *American Mathematics Monthly* **41**, 411–419.

Bennett, S. N., Drummond, A. J., Kapan, D. D., Suchard, M. A., Muñoz Jordán, J. L., Pybus, O. G., Holmes, E. C., and Gubler, D. J. (2010). Epidemic dynamics revealed in dengue evolution. *Molecular Biology and Evolution* **27**, 811–818.

Benton, M. J. and Donoghue, P. C. J. (2007). Paleontological evidence to date the tree of life. *Molecular Biology and Evolution* **24**, 26–53.

Berger, J. (1994). An overview of robust Bayesian analysis. *Test* **3**, 5–124.

Berger, J. O. and Bernardo, J. M. (1992). *On the development of reference priors (with Discussion)*, volume 4, pages 35–60. London: Oxford University Press.

Berger, J. O., Bernardo, J. M., and Sun, D. (2009). The formal definition of reference priors. *Annals of Statistics* **37**, 905–938.

Bernardo, J. M. (1979). Reference posterior distributions for Bayesian inference. *Journal of the Royal Statistical Society, Series B* **41**, 113–147.

Bernardo, J. M. (2005). Reference analysis. In D. K. Dey and C. R. Rao, editors, *Handbook of Statistics*, volume 25, pages 17–90. Amsterdam: Elsevier.

Besag, J. and Green, P. J. (1993). Spatial statistics and Bayesian computation (with discussion). *Journal of the Royal Statistical Society, Series B* **55**, 25–37.

Besag, J., Green, P. J., Higdon, D., and Mengersen, K. (1995). Bayesian computation and stochastic systems. *Statistical Science* **10**, 3–66.

Bininda-Emonds, O. R. P., editor (2004). *Phylogenetic supertrees: combining information to reveal the tree of life.* Dordrecht: Kluwer.

Bishop, C. M. (2006). *Pattern recognition and machine learning.* Berlin: Springer.

Bladt, M. and Sørensen, M. (2005). Statistical inference for discretely observed Markov jump processes. *Journal of the Royal Statistical Society, Series B* **67**, 395–410.

Blanquart, S. and Lartillot, N. (2006). A Bayesian compound stochastic process for modeling nonstationary and nonhomogeneous sequence evolution. *Molecular Biology and Evolution* **23**, 2058–2071.

Blanquart, S. and Lartillot, N. (2008). A site- and time-heterogeneous model of amino acid replacement. *Molecular Biology and Evolution* **25**, 842–858.

Bolhuis, P. G., Chandler, D., Dellago, C., and Geissler, P. L. (2002). Transition path sampling: throwing ropes over rough mountain passes, in the dark. *Annual Review of Physical Chemistry* **53**, 291–318.

Bouchard-Côté, A., Sankararaman, S., and Jordan, M. I. (2012). Phylogenetic inference via sequential Monte Carlo. *Systematic Biology* **61**, 579–593.

Brandes, U., Eiglsperger, M., Herman, I., Himsolt, M., and Marshall, M. S. (2002). GraphML progress report: structural layer proposal. In P. Mutzel, M. JÃijnger, and S. Leipert, editors, *Proceedings of the 9th International Symposium on Graph Drawing (Springer LNCS 2265)*, pages 501–512. Berlin: Springer.

Brandley, M. C., Leache, A. D., Warren, D. L., and McGuire, J. A. (2006). Are unequal clade priors problematic for Bayesian phylogenetics? *Systematic Biology* **55**, 138–146.

Brooks, S. P., Giudici, P., and Roberts, G. O. (2003). Efficient construction of reversible jump Markov chain Monte Carlo proposal distributions. *Journal of the Royal Statistical Society, Series B* **65**, 3–39.

Brown, J. M., Hedtke, S. M., Lemmon, A. R., and Lemmon, E. M. (2010). When trees grow too long: investigating the causes of highly inaccurate Bayesian branch-length estimates. *Systematic Biology* **59**, 145–161.

Brown, W. M., Prager, E. M., Wang, A., and Wilson, A. C. (1982). Mitochondrial DNA sequences of primates: tempo and mode of evolution. *Journal of Molecular Evolution* **18**, 225–239.

Bryant, D. and Steel, M. (2009). Computing the distribution of a tree metric. *IEEE IEEE/ACM Transactions on Computational Biology and Bioinformatics* **6**, 420–426.

Buonocore, A., Pirozzi, E., and Caputo, L. (2009). A note on the sum of uniform random variables. *Statistics and Probability Letters* **79**, 2092–2097.

Calderhead, B. and Girolami, M. (2009). Estimating Bayes factors via thermodynamic integration and population MCMC. *Computational Statistics and Data Analysis* **53**, 4028–4045.

Campos, P. F., Willerslev, E., Sher, A., Orlando, L., Axelsson, E., Tikhonov, A., Aaris-Sørensen, K., Greenwood, A. D., Kahlke, R., Kosintsev, P., Krakhmalnaya, T., Kuznetsova, T., Lemey, P., MacPhee, R., Norris, C. A., Shepherd, K., Suchard, M. A., Zazula, G. D., Shapiro, B., and Gilbert, M. T. P. (2010). Ancient DNA analyses exclude humans as the driving force behind late Pleistocene musk ox (*Ovibos moschatus*) population dynamics. *Proceedings of the National Academy of Sciences USA* **107**, 5675–5680.

Cappé, O., Guillin, A., Marin, J., and Robert, C. (2004). Population Monte Carlo. *Journal of Computational and Graphical Statistics* **13**, 907–929.

Carlin, B. P. and Chib, S. (1995). Bayesian model choice via Markov Chain Monte Carlo methods. *Journal of the Royal Statistical Society, Series B* **57**, 473–484.

Carling, M. D. and Brumfield, R. T. (2007). Gene sampling strategies for multi-locus population estimates of genetic diversity (θ). *PLoS One* **2**, 160.

Carpenter, J., Clifford, P., and Fearnhead, P. (1999). An improved particle filter for non-linear problems. *IEEE Proceedings: Radar, Sonar and Navigation* **146**, 2–7.

Cavalli-Sforza, L. and Edwards, A. (1967). Phylogenetic analysis: models and estimation procedures. *Evolution* **21**, 550–570.

Chan, K. M. A. and Moore, B. R. (2005). SymmeTREE: whole-tree analysis of differential diversification rates. *Bioinformatics* **21**, 1709–1710.

Chen, H. F. (2002). *Stochastic approximation and its applications.* Dordrecht: Kluwer Academic Publishers.

Chen, M. and Shao, Q. (1997). On Monte Carlo methods for estimating ratios of normalizing constants. *The Annals of Statistics* **25**, 1563–1594.

Cheon, S. and Liang, F. (2008). Phylogenetic tree construction using sequential stochastic approximation Monte Carlo. *BioSystems* **91**, 94–107.

Cheon, S. and Liang, F. (2009). Bayesian phylogeny analysis via stochastic approximation Monte Carlo. *Molecular Phylogenetics and Evolution* **53**, 394–403.

Chib, S. (1995). Marginal likelihood from the Gibbs output. *Journal of the American Statistical Association* **90**, 1313–1321.

Choi, S. C., Hobolth, A., Robinson, D. M., Kishino, H., and Thorne, J. L. (2007). Quantifying the impact of protein tertiary structure on molecular evolution. *Molecular Biology and Evolution* **24**, 1769–1782.

Choi, S. C., Redelings, B. D., and Thorne, J. L. (2008). Basing population genetic inferences and models of molecular evolution upon desired stationary distributions of DNA or protein sequences. *Philosophical Transactions of the Royal Society of London, Series B* **363**, 3931–3939.

Cowles, M. K. and Carlin, B. P. (1996). Markov chain Monte Carlo convergence diagnostics: a comparative review. *Journal of the American Statistical Society* **91**, 883–904.

Crawford, F. W. and Suchard, M. A. (2012). Transition probabilities for general birth-death processes with applications in ecology, genetics, and evolution. *Journal of Mathematical Biology* **65**, 553–580.

Cressie, N. A. (1993). *Statistics for spatial data.* New York: Wiley-Interscience.

Crisan, D. and Doucet, A. (2002). A survey of convergence results on particle filtering for practitioners. *IEEE Transactions on Signal Processing* **50**, 736–746.

Darriba, D., Aberer, A., Flouri, T., Heath, T., Izquierdo-Carrasco, F., and Stamatakis, A. (2013). Boosting the performance of Bayesian divergence time estimation with the Phylogenetic Likelihood Library. *12th IEEE International Workshop on High Performance Computational Biology.*

Datta, G. S. and Ghosh, M. (1996). On the invariance of noninformative priors. *Annals of Statistics* **24**, 141–159.

de Koning, A. P. J., Gu, W., and Pollock, D. D. (2010). Rapid likelihood analysis on large phylogenies using partial sampling of substitution histories. *Molecular Biology and Evolution* **27**, 249–265.

Dey, D. K., Ghosh, S., and Mallick, B. K., editors (2011). *Bayesian modeling in bioinformatics.* Chapman and Hall/CRC.

Donoghue, M. J. and Moore, B. R. (2003). Toward an integrative historical biogeography. *Integrative and Comparative Biology* **43**, 261–270.

Dorazio, R. M. (2009). On selecting a prior for the precision parameter of the Dirichlet process mixture models. *Journal of Statistical Planning and Inference* **139**, 3384–3390.

dos Reis, M., Inoue, J., Hasegawa, M., Asher, R., Donoghue, P., and Yang, Z. (2012). Phylogenomic datasets provide both precision and accuracy in estimating the timescale of placental mammal phylogeny. *Proceedings of the Royal Society, Series B* **279**, 3491–3500.

dos Reis, M. and Yang, Z. (2012). The unbearable uncertainty of Bayesian divergence time estimation. *Journal of Systematics and Evolution* **51**, 30–43.

Douc, R., Cappé, O., and Moulines, E. (2005). Comparison of resampling schemes for particle filtering. In S. Lončarić, H. Babić, and M. Bellanger, editors, *Proceedings of the 4th International Symposium on Image and Signal Processing and Analysis*, pages 64–69. Zagreb: Faculty of Electrical Engineering and Computing.

Doucet, A. (1997). *Monte Carlo methods for Bayesian estimation of hidden Markov models. Application to radiation signals.* Ph.D. thesis, University Paris-Sud Orsay.

Doucet, A., Freitas, N. D., and Gordon, N., editors (2001). *Sequential Monte Carlo methods in practice.* Berlin: Springer.

Doucet, A. and Johansen, A. M. (2009). A tutorial on particle filtering and smoothing: fifteen years later. In D. Crisan and B. Rozovsky, editors, *Oxford Handbook of Nonlinear Filtering.* Cambridge: Cambridge University Press.

Doyle, J. A. and Donoghue, M. J. (1993). Phylogenies and angiosperm diversification. *Paleobiology* **19**, 141–167.

Drozdek, A. and Simon, D. (1995). *Data structure in C.* Boston: PWS Publishing Company.

Drummond, A., Pybus, O., Rambaut, A., Forsberg, R., and Rodrigo, A. (2003). Measurably evolving populations. *Trends in Ecology and Evolution* **18**, 481–488.

Drummond, A. J., Ho, S. Y. W., Phillips, M. J., and Rambaut, A. (2006). Relaxed phylogenetics and dating with confidence. *PLoS Biology* **4**, e88.

Drummond, A. J., Nicholls, G. K., Rodrigo, A. G., and Solomon, W. (2002). Estimating mutation parameters, population history and genealogy simultaneously from temporally spaced sequence data. *Genetics* **161**, 1307–1320.

Drummond, A. J. and Rambaut, A. (2007). BEAST: Bayesian evolutionary analysis by sampling trees. *BMC Evolutionary Biology* **7**, 214.

Drummond, A. J., Rambaut, A., Shapiro, B., and Pybus, O. G. (2005). Bayesian coalescent inference of past population dynamics from molecular sequences. *Molecular Biology and Evolution* **22**, 1185–1192.

Drummond, A. J. and Suchard, M. A. (2010). Bayesian random local clocks, or one rate to rule them all. *BMC Biology* **8**, 114.

Drummond, A. J., Suchard, M. A., Xie, D., and Rambaut, A. (2012). Bayesian phylogenetics with BEAUti and the BEAST 1.7. *Molecular Biology and Evolution* **29**, 1969–1973.

Durbin, R., Eddy, S., Krogh, A., and Mitchison, G. (1998). *Biological sequence analysis: probabilistic models of proteins and nucleic acids.* Cambridge: Cambridge University Press.

Duret, L. and Galtier, N. (2000). The covariation between TpA deficiency, CpG deficiency, and G + C content of human isochores is due to a mathematical artifact. *Molecular Biology and Evolution* **17**, 1620–1625.

Edwards, S. V. (2009). Is a new and general theory of molecular systematics emerging? *Evolution* **63**, 1–19.

Escobar, M. D. and West, M. (1995). Bayesian density estimation and inference using mixtures. *Journal of the American Statistical Association* **90**, 577–588.

Etienne, R., Haegeman, B., Stadler, T., Aze, T., Pearson, P., Purvis, A., and Phillimore, A. (2012). Diversity-dependence brings molecular phylogenies closer to agreement with the fossil record. *Proceedings of the Royal Society, Series B* **279**, 1300–1309.

Ewing, G. and Rodrigo, A. (2006a). Coalescent-based estimation of population parameters when the number of demes changes over time. *Molecular Biology and Evolution* **23**, 988–996.

Ewing, G. and Rodrigo, A. (2006b). Estimating population parameters using the structured serial coalescent with Bayesian MCMC inference when some demes are hidden. *Evolutionary Bioinformatics* **2**, 227–235.

Eyre-Walker, A. and Awadalla, P. (2001). Does human mtDNA recombine? *Journal of Molecular Evolution* **53**, 430–435.

Fan, Y., Wu, R., Chen, M.-H., Kuo, L., and Lewis, P. O. (2011). Choosing among partition models in Bayesian phylogenetics. *Molecular Biology and Evolution* **28**, 523–532.

Fearnhead, P. and Donnelly, P. (2001). Estimating recombination rates from population genetic data. *Genetics* **159**, 1299–1318.

Fearnhead, P. and Sherlock, C. (2006). An exact Gibbs sampler for the Markov modulated Poisson processes. *Journal of the Royal Statistical Society, Series B* **68**, 767–ø784.

Felsenstein, J. (1973). Maximum-likelihood estimation of evolutionary trees from continuous characters. *American Journal of Human Genetics* **25**, 471–492.

Felsenstein, J. (1978a). Cases in which parsimony or compatibility methods will be positively misleading. *Systematic Zoology* **27**, 401–410.

Felsenstein, J. (1978b). The number of evolutionary trees. *Systematic Zoology* **27**, 27–33.

Felsenstein, J. (1981). Evolutionary trees from DNA sequences: a maximum likelihood approach. *Journal of Molecular Evolution* **17**, 368–376.

Felsenstein, J. (1983). Statistical inference of phylogenies. *Journal of the Royal Statistical Society, Series A* **146**, 246–272.

Felsenstein, J. (1985). Confidence limits on phylogenies: an approach using the bootstrap. *Evolution* **39**, 783–791.

Felsenstein, J. (1989). Phylogenetic inference package (PHYLIP), version 3.2. *Cladistics* **5**, 164–166.

Felsenstein, J. (1992a). Estimating effective population size from samples of sequences: a bootstrap Monte Carlo approach. *Genetical Research* **60**, 209–220.

Felsenstein, J. (1992b). Estimating effective population size from samples of sequences: inefficiency of pairwise and segregating sites as compared to phylogenetic estimates. *Genetical Research* **59**, 139–147.

Felsenstein, J. (1993). *PHYLIP (phylogenetic inference package), version 3.5.* Seattle: University of Washington.

Felsenstein, J. (2004). *Inferring phylogenies.* Sunderland: Sinauer Associates.

Felsenstein, J. (2005). Accuracy of coalescent likelihood estimates: do we need more sites, more sequences, or more loci? *Molecular Biology and Evolution* **23**, 691–700.

Felsenstein, J. and Churchill, G. A. (1996). A hidden Markov model approach to variation among sites in rate of evolution. *Molecular Biology and Evolution* **13**, 93–104.

Felsenstein, J. and Rodrigo, A. G. (1999). Coalescent approaches to HIV population genetics. In *The evolution of HIV*, pages 233–272. Baltimore: Johns Hopkins University Press.

Feng, X., Buell, D. A., Rose, J. R., and Waddell, P. J. (2003). Parallel algorithms for Bayesian phylogenetic inference. *Journal of Parallel and Distributed Computing* **63**, 707–718.

Ferguson, T. S. (1973a). A Bayesian analysis of some nonparametric problems. *The Annals of Statistics* **1**, 209–230.

Ferguson, T. S. (1973b). A Bayesian analysis of some nonparametric problems. *Annals of Statistics* **1**, 209–230.

Ferreira, M. A. R. and Suchard, M. A. (2008). Bayesian analysis of elapsed times in continuous-time Markov chains. *Canadian Journal of Statistics* **36**, 355–368.

Fitch, W. and Margoliash, E. (1967). Construction of phylogenetic trees. *Science* **155**, 279–284.

Fitch, W. M. (1971). Toward defining the course of evolution: minimal change for a specific tree topology. *Journal of Parallel and Distributed Computing* **20**, 406–416.

Foster, I. (1995). *Designing and building parallel programs.* London: Addison-Wesley.

Foster, P. G. (2004). Modeling compositional heterogeneity. *Systematic Biology* **53**, 485–495.

Frank, C., Mohamed, M., Strickland, G., Lavanchy, D., Arthur, R., Magder, L., El Khoby, T., Abdel-Wahab, Y., Ohn, E., Anwar, W., and Sallam, I. (2000). The role of parenteral antischistosomal therapy in the spread of hepatitis C virus in Egypt. *Lancet* **355**, 887–891.

Friedland, B. (2005). *Control system design: an introduction to state space methods.* Mineola: Dover Publications.

Friel, N. and Pettitt, A. N. (2008). Marginal likelihood estimation via power posteriors. *Journal of the Royal Statistical Society, Series B* **70**, 589–607.

Frost, S. D. W. and Volz, E. M. (2010). Viral phylodynamics and the search for an 'effective number of infections'. *Philosophical Transactions of the Royal Society of London, Series B* **365**, 1879–1890.

Fu, Y. (1994). Estimating effective population size or mutation rate using the frequencies of mutations of various classes in a sample of DNA sequences. *Genetics* **138**, 1375–1386.

Galtier, N., Gascuel, O., and Jean-Marie, A. (2005). Markov models in molecular evolution. In R. Nielsen, editor, *Statistical Methods in Molecular Evolution*, pages 3–24. Berlin: Springer.

Galtier, N. and Gouy, M. (1998). Inferring pattern and process: maximum-likelihood implementation of a nonhomogeneous model of DNA sequence evolution for phylogenetic analysis. *Molecular Biology and Evolution* **15**, 871–879.

Gao, F., Robertson, D. L., Morrison, S. G., Hui, H., Craig, S., Decker, J., Fultz, P. N., Girard, M., Shaw, G. M., Hahn, B. H., and Sharp, P. M. (1996). The heterosexual human immunodeficiency virus type 1 epidemic in Thailand is caused by an intersubtype (A/E) recombinant of African origin. *Journal of Virology* **70**, 7013–29.

Gaut, B., Yang, L., Takuno, S., and Eguiarte, L. E. (2011). The patterns and causes of variation in plant nucleotide substitution rates. *Annual Review of Ecology, Evolution, and Systematics* **42**, 245–266.

Gelfand, A. and Dey, D. (1994). Bayesian model choice: asymptotics and exact calculations. *Journal of the Royal Statistical Society, Series B: Statistical Methodology* **56**, 501–514.

Gelfand, A. E., Kottas, A., and MacEachern, S. N. (2005). Bayesian non-parametric spatial modeling with Dirichlet process mixing. *Journal of the American Statistical Association* **100**, 1021–1035.

Gelfand, A. E. and Smith, A. F. M. (1990). Sampling-based approaches to calculating marginal densities. *Journal Of The American Statistical Association* **85**, 398–409.

Gelman, A., Carlin, J. B., Stern, H. S., and Rubin, D. B. (2004). *Bayesian data analysis.* London: Chapman and Hall/CRC.

Gelman, A. and Meng, X.-L. (1998). Simulating normalizing constants: from importance sampling to bridge sampling to path sampling. *Statistical Science* **13**, 163–185.

Geman, S. and Geman, D. (1984). Stochastic relaxation, gibbs distributions, and the bayesian restoration of images. *IEEE transactions on pattern analysis and machine intelligence* **6**, 721–741.

Gene Ontology Consortium (2000). Gene ontology: tool for the unification of biology. *Nature Genetics* **25**, 25–29.

Gernhard, T. (2008). The conditioned reconstructed process. *Journal of Theoretical Biology* **253**, 769–778.

Geweke, J. (1989). Bayesian inference in econometric models using Monte Carlo integration. *Econometrica* **57**, 1317–1339.

Geyer, C. J. (1991). Estimating normalizing constants and reweighting mixtures in Markov chain Monte Carlo. Technical Report 568, School of Statistics, University of Minnesota. `http://purl.umn.edu/58433`.

Geyer, C. J. (1992a). Markov chain Monte Carlo maximum likelihood. In E. M. Keramigas and S. M. Kaufman, editors, *Computing Science and Statistics: Proceedings of the 23rd Symposium on the Interface*, pages 156–163. Fairfax, Virginia: Interface Foundation of North America.

Geyer, C. J. (1992b). Practical markov chain monte carlo (with discussion). *Statistical Science* **7**, 437–511.

Geyer, C. J. and Thompson, E. A. (1995). Annealing Markov chain Monte Carlo with applications to ancestral inference. *Journal of the American Statistical Association* **90**, 909–920.

Ghosh, M. (2011). Objective priors: an introduction for frequentists. *Statistical Science* **26**, 187–202.

Gilks, W., Richardson, S., and Spiegalhalter, D. (1996a). *Markov chain Monte Carlo in practice.* London: Chapman and Hall.

Gilks, W. R., Spiegelhalter, D. J., and Richardson, S. (1996b). *Markov chain Monto Carlo in practice.* Boca Raton: CRC Press.

Gill, M. S., Lemey, P., Faria, N. R., Rambaut, A., Shapiro, B., and Suchard, M. A. (2012). Improving Bayesian population dynamics inference: a coalescent-based model for multiple loci. *Molecular Biology and Evolution* **30**, 713–724.

Goldman, N. and Yang, Z. (1994). A codon-based model of nucleotide substitution for protein-coding DNA sequences. *Molecular Biology and Evolution* **11**, 725–736.

Goodreau, S. M. (2006). Assessing the effects of human mixing patterns on human immunodeficiency virus-1 interhost phylogenetics through social network simulation. *Genetics* **172**, 2033–2045.

Görür, D., Boyles, L., and Welling, M. (2012). Scalable inference on Kingman's coalescent using pair similarity. *Journal of Machine Learning Research* **22**, 440–448.

Görür, D. and Teh, Y. W. (2008). An efficient sequential Monte Carlo algorithm for coalescent clustering. In *Advances in Neural Information Processing*, pages 521 – 528. Red Hook: Curran Associates.

Gowri-Shankar, V. (2006). RNA phylogenetic inference with heterogeneous substitution models. Ph.D. thesis, University of Manchester.

Gowri-Shankar, V. and Rattray, M. (2007). A reversible jump method for Bayesian phylogenetic inference with a nonhomogeneous substitution model. *Molecular Biology and Evolution* **24**, 1286–1299.

Green, P. J. (1995). Reversible jump Markov chain Monte Carlo computation and Bayesian model determination. *Biometrika* **82**, 711–732.

Green, P. J. and Richardson, S. (2001). Modelling heterogeneity with and without the Dirichlet process. *Scandinavian Journal of Statistics* **28**, 355–375.

Grenfell, B. T., Pybus, O. G., Gog, J. R., Wood, J. L. N., Daly, J. M., Mumford, J. A., and Holmes, E. C. (2004). Unifying the epidemiological and evolutionary dynamics of pathogens. *Science* **303**, 327–332.

Griffiths, R. C. and Marjoram, P. (1996). Ancestral inference from samples of dna sequences with recombination. *Journal of Computational Biology* **3**, 479–502.

Griffiths, R. C. and Marjoram, P. (1997). An ancestral recombination graph. In *Progress in Population Genetics and Human Evolution*, IMA Volumes in Mathematics and its Applications, Volume 87, pages 257–270. Berlin: Springer.

Griffiths, R. C. and Tavaré, S. (1994). Sampling theory for neutral alleles in a varying environment. *Philosophical Transactions of the Royal Society of London, Series B* **344**, 403–410.

Gusfield, D. (2002). Partition-distance: a problem and class of perfect graphs arising in clustering. *Information Processing Letters* **82**, 159–164.

Hackett, S. J. et al. (2008). A phylogenomic study of birds reveals their evolutionary history. *Science* **320**, 1763–1768.

Hall, B. G. (2004). *Phylogenetic trees made easy.* Sunderland: Sinauer.

Halpern, A. L. and Bruno, W. J. (1998). Evolutionary distances for protein-coding sequences: modeling site-specific residue frequencies. *Molecular Biology and Evolution* **15**, 910–917.

Hartmann, K., Wong, D., and Stadler, T. (2010). Sampling trees from evolutionary models. *Systematic Biology* **59**, 465–476.

Hasegawa, M., Kishino, H., and Yano, T. (1989). Estimation of branching dates among primates by molecular clocks of nuclear DNA which slowed down in Hominoidea. *Journal of Human Evolution* **18**, 461–476.

Hasegawa, M., Kishino, H., and Yano, T. A. (1985). Dating of the human-ape splitting by a molecular clock of mitochondrial DNA. *Journal of Molecular Evolution* **22**, 160–174.

Hasegawa, M., Yano, T., and Kishino, H. (1984). A new molecular clock of mitochondrial-DNA and the evolution of hominoids. *Proceedings of the Japan Academy, Series B* **60**, 95–98.

Hastings, W. K. (1970). Monte Carlo sampling methods using Markov chains and their applications. *Biometrika* **57**, 97–109.

Heath, T. A. (2012). A hierarchical Bayesian model for calibrating estimates of species divergence times. *Systematic Biology* **61**, 793–809.

Heath, T. A., Holder, M. T., and Huelsenbeck, J. P. (2012). A Dirichlet process prior for estimating lineage-specific substitution rates. *Molecular Biology and Evolution* **29**, 939–255.

Heath, T. A., Zwickl, D. J., Kim, J., and Hillis, D. M. (2008). Taxon sampling affects inferences of macroevolutionary processes from phylogenetic trees. *Systematic Biology* **57**, 160–166.

Hedtke, S. M., Stanger-Hall, K., Baker, R. J., and Hillis, D. M. (2008). All-male asexuality: origin and maintenance of androgenesis in the Asian clam *Corbicula*. *Evolution* **62**, 1119–1136.

Hein, J. (1993). A heuristic method to reconstruct the history of sequences subject to recombination. *Journal of Molecular Evolution* **36**, 396–405.

Hein, J., Schierup, M. H., and Wiuf, C. (2005). *Gene genealogies, variation and evolution: a primer in coalescent theory.* Oxford: Oxford University Press, first edition.

Heled, J. and Drummond, A. (2008). Bayesian inference of population size history from multiple loci. *BMC Evolutionary Biology* **8**, 289.

Heled, J. and Drummond, A. J. (2010). Bayesian inference of species trees from multilocus data. *Molecular Biology and Evolution* **27**, 570–580.

Heled, J. and Drummond, A. J. (2012). Calibrated tree priors for relaxed phylogenetics and divergence time estimation. *Systematic Biology* **61**, 138–149.

Hellenthal, G. and Stephens, M. (2007). msHOT: modifying Hudson's ms simulator to incorporate crossover and gene conversion hotspots. *Bioinformatics* **23**, 520–521.

Hey, J. and Nielsen, R. (2004). Multilocus methods for estimating population sizes, migration rates and divergence time, with applications to the divergence of *Drosophila pseudoobscura* and *D. persimilis*. *Genetics* **167**, 747–760.

Ho, S. Y. W. (2007). Calibrating molecular estimates of substitution rates and divergence times in birds. *Journal of Avian Biology* **38**, 409–414.

Ho, S. Y. W. and Phillips, M. J. (2009). Accounting for calibration uncertainty in phylogenetic estimation of evolutionary divergence times. *Systematic Biology* **58**, 367–380.

Ho, S. Y. W. and Shapiro, B. (2011). Skyline-plot methods for estimating demographic history from nucleotide sequences. *Molecular Ecology Resources* **11**, 423–434.

Hobolth, A. (2008). A Markov chain Monte Carlo expectation maximization algorithm for statistical analysis of DNA sequence evolution with neighbor-dependent substitution rates. *Journal of Computational and Graphical Statistics* **17**, 138–162.

Hobolth, A. and Jensen, J. (2011). Summary statistics for endpoint-conditioned continuous-time Markov chains. *Journal of Applied Probability* **48**, 911–924.

Hobolth, A. and Jensen, J. L. (2005). Statistical inference in evolutionary models of DNA sequences via the EM algorithm. *Statistical Applications in Genetics and Molecular Biology* **4**, Article 18.

Hobolth, A. and Stone, E. A. (2009a). Simulation from endpoint-conditioned, continuous-time Markov chains on a finite state space, with applications to molecular evolution. *Annals of Applied Statistics* **3**, 1204–1231.

Hobolth, A. and Stone, E. A. (2009b). Simulation from endpoint-conditioned, continuous-time Markov chains on a finite state space, with applications to molecular evolution. *Annals of Applied Statistics* **3**, 1204.

Hodgkinson, A. and Eyre-Walker, A. (2011). Variation in the mutation rate across mammalian genomes. *Nature Reviews Genetics* **12**, 756–766.

Hoeting, J. A., Madigan, D., Raftery, A. E., and Volinsky, C. T. (1999). Bayesian model averaging: a tutorial. *Statistical Science* **14**, 382–401.

Höhna, S. and Drummond, A. J. (2011). Guided tree topology proposals for Bayesian phylogenetic inference. *Systematic Biology* **61**, 1–11.

Höhna, S., Stadler, T., Ronquist, F., and Britton, T. (2011). Inferring speciation and extinction rates under different sampling schemes. *Molecular Biology and Evolution* **28**, 2577–2589.

Holmes, I. and Rubin, G. M. (2002). An expectation maximization algorithm for training hidden substitution models. *Journal of Molecular Biology* **317**, 753–764.

Hoogerheide, L., Kaashoek, J., and Van Dijk, H. (2007). On the shape of posterior densities and credible sets in instrumental variable regression models with reduced rank: An application of flexible sampling methods using neural networks. *Journal of Econometrics* **139**, 154–180.

Hudelot, C., Gowri-Shankar, V., Jow, H., Rattray, M., and Higgs, P. (2003). RNA-based phylogenetic methods: application to mammalian mitochondrial RNA sequences. *Molecular Phylogenetics and Evolution* **28**, 241–252.

Hudson, R. R. (1983). Properties of a neutral allele model with intragenic recombination. *Theoretical Population Biology* **23**, 183–201.

Hudson, R. R. (1985). Statistical properties of the number of recombination events in the history of a sample of DNA sequences. *Genetics* **111**, 147–164.

Hudson, R. R. (1990). Gene genealogies and the coalescent process. *Oxford Surveys in Evolutionary Biology* **7**, 1–44.

Hudson, R. R. (2001). Two-locus sampling distributions and their application. *Genetics* **159**, 1805–1817.

Hudson, R. R. (2002). Generating samples under a Wright-Fisher neutral model of genetic variation. *Bioinformatics* **18**, 337–338.

Huelsenbeck, J., Larget, B., and Alfaro, E. (2004). Bayesian phylogenetic model selection using reversible jump Markov chain Monte Carlo. *Molecular Biology and Evolution* **21**, 1123–1133.

Huelsenbeck, J. P. and Bollback, J. P. (2001). Empirical and hierarchical Bayesian estimation of ancestral states. *Systematic Biology* **50**, 351–366.

Huelsenbeck, J. P. and Dyer, K. (2004). Bayesian estimation of positively selected sites. *Journal Of Molecular Evolution* **58**, 661–672.

Huelsenbeck, J. P. and Imennov, N. S. (2002). Geographic origin of human mitochondrial DNA: accommodating phylogenetic uncertainty and model comparison. *Systematic Biology* **51**, 155–165.

Huelsenbeck, J. P., Jain, S., Frost, S., and Pond, S. (2006). A Dirichlet process model for detecting positive selection in protein-coding DNA sequences. *Proceedings of the National Academy of Sciences USA* **103**, 6263–6268.

Huelsenbeck, J. P., Larget, B., and Swofford, D. (2000a). A compound Poisson process for relaxing the molecular clock. *Genetics* **154**, 1879–1892.

Huelsenbeck, J. P., Larget, B., and Swofford, D. L. (2000b). A compound Poisson process for relaxing the molecular clock. *Genetics* **154**, 1879–1892.

Huelsenbeck, J. P., Nielsen, R., and Bollback, J. (2003). Stochastic mapping of morphological characters. *Systematic Biology* **52**, 131–158.

Huelsenbeck, J. P. and Rannala, B. (2003). Detecting correlation between characters in a comparative analysis with uncertain phylogeny. *Evolution* **57**, 1237–1247.

Huelsenbeck, J. P., Rannala, B., and Larget, B. (2000c). A Bayesian framework for the analysis of cospeciation. *Evolution* **54**, 352–364.

Huelsenbeck, J. P., Rannala, B., and Masly, J. P. (2000d). Accommodating phylogenetic uncertainty in evolutionary studies. *Science* **288**, 2349–2350.

Huelsenbeck, J. P. and Ronquist, F. (2001). MRBAYES: Bayesian inference of phylogenetic trees. *Bioinformatics* **17**, 754–755.

Huelsenbeck, J. P., Ronquist, F., Nielsen, R., and Bollback, J. (2001a). Bayesian inference of phylogeny and its impact on evolutionary biology. *Science* **294**, 2310–2314.

Huelsenbeck, J. P., Ronquist, F., Nielsen, R., and Bollback, J. P. (2001b). Bayesian inference of phylogeny and its impact on evolutionary biology. *Science* **294**, 2310–2314.

Huelsenbeck, J. P. and Suchard, M. A. (2007). A nonparametric method for accommodating and testing across-site rate variation. *Systematic Biology* **56**, 975–987.

Hukushima, K. and Nemoto, K. (1996). Exchange Monte Carlo method and application to spin glass simulations. *Journal of the Physical Society of Japan* **65**, 1604–1608.

Huson, D. H. and Bryant, D. (2006). Application of phylogenetic networks in evolutionary studies. *Molecular Biology and Evolution* **23**, 254–267.

Hwang, D. G. and Green, P. (2004). Bayesian Markov chain Monte Carlo sequence analysis reveals varying neutral substitution patterns in mammalian evolution. *Proceedings of the National Academy of Sciences USA* **101**, 13994–14001.

Jeffreys, H. (1935). Some tests of significance treated by theory of probability. *Proceedings of the Cambridge Philosophical Society* **31**, 203–222.

Jeffreys, H. (1961a). *The theory of probability.* Oxford: Oxford University Press, third edition.

Jeffreys, H. (1961b). *Theory of Probability.* Oxford: Oxford University Press.

Jensen, A. (1953). Markoff chains as an aid in the study of Markoff processes. *Scandinavian Actuarial Journal* **36**, 87–91.

Jensen, J. L. and Pedersen, A.-M. K. (2000). Probabilistic models of DNA sequence evolution with context dependent rates of substitution. *Advances in Applied Probability* **32**, 499–517.

Jorgensen, B. (1982). *Statistical properties of the generalized inverse Gaussian distribution*, volume 9 of *Lecture Notes in Statistics*. Berlin: Springer.

Jukes, T. and Cantor, C. (1969a). Evolution of protein molecules. In H. Munro, editor, *Mammalian Protein Metabolism*. Academic Press.

Jukes, T. H. and Cantor, C. R. (1969b). Evolution of protein molecules. In H. N. Munro, editor, *Mammalian Protein Metabolism*, pages 21–123. New York: Academic Press.

Jung, A., Maier, R., Vartanian, J.-P., Bocharov, G., Jung, V., Fischer, U., Meese, E., Wain-Hobson, S., and Meyerhans, A. (2002). Recombination: multiply infected spleen cells in HIV patients. *Nature* **418**.

Kadane, J. B. and Lazar, N. A. (2004). Methods and criteria for model selection. *Journal of the American Statistical Association* **99**, 279–290.

Kang, C. J. (2008). The full likelihood approach for inferring variation of recombination rate using a Markov chain Monte Carlo method with a hidden Markov model. Ph.D. thesis, University of Washington, Seattle, Washington.

Kass, R. E. and Raftery, A. E. (1995). Bayes factors. *Journal of the American Statistical Association* **90**, 773–795.

Kass, R. E. and Wasserman, L. (1996). The selection of prior distributions by formal rules. *Journal of the American Statistical Association* **91**, 1343–1370.

Kendall, D. G. (1948). On the generalized "birth-and-death" process. *Annals of Mathematical Statistics* **19**, 1–15.

Kidd, K. K. and Sgaramella-Zonta, L. A. (1971). Phylogenetic analysis: concepts and methods. *The American Journal of Human Genetics* **23**, 235–252.

Kimura, M. (1980). A simple method for estimating evolutionary rates of base substitutions through comparative studies of nucleotide-sequences. *Journal of Molecular Evolution* **16**, 111–120.

Kingman, J. (1982a). The coalescent. *Stochastic Processes and Their Applications* **13**, 235–248.

Kingman, J. F. C. (1982b). On the genealogy of large populations. *Journal of Applied Probability* **19**, 27–43.

Kishino, H. and Hasegawa, M. (1990). Converting distance to time: application to human-evolution. *Methods in Enzymology* **183**, 550–570.

Kishino, H., Miyata, T., and Hasegawa, M. (1990). Maximum likelihood inference of protein phylogeny and the origin of chloroplasts. *Journal of Molecular Evolution* **31**, 151–160.

Kishino, H., Thorne, J. L., and Bruno, W. J. (2001). Performance of a divergence time estimation method under a probabilistic model of rate evolution. *Molecular Biology and Evolution* **18**, 352–361.

Knowles, L. L. and Kubatko, L. S., editors (2011). *Estimating species trees: practical and theoretical aspects.* Hoboken: Wiley-Blackwell.

Kocher, T. D., Conroy, J. A., McKaye, K. R., Stauffer, J. R., and Lockwood, S. F. (1995). Evolution of NADH dehydrogenase subunit 2 in East African cichlid fish. *Molecular Phylogenetics and Evolution* **4**, 420–432.

Kotecha, J., Jianqui, Z., Yufei, H., Ghirmai, T., Bugallo, M., and Miguez, J. (2003). Particle filtering. *Signal Processing Magazine, IEEE* **20**, 19–38.

Kuhner, M. and Smith, L. (2007). Comparing likelihood and Bayesian coalescent estimation of population parameters. *Genetics* **175**, 155–165.

Kuhner, M., Yamato, J., and Felsenstein, J. (1995). Estimating effective population size and mutation rate from sequence data using Metropolis-Hastings sampling. *Genetics* **140**, 1421–1430.

Kuhner, M. K. (2006). LAMARC 2.0: maximum likelihood and Bayesian estimation of population parameters. *Bioinformatics* **22**, 768–770.

Kuhner, M. K., Yamato, J., and Felsenstein, J. (1998). Maximum likelihood estimation of population growth rates based on the coalescent. *Genetics* **149**, 429–434.

Kuhner, M. K., Yamato, J., and Felsenstein, J. (2000). Maximum likelihood estimation of recombination rates from population data. *Genetics* **156**, 1393–1401.

Kühnert, D., Wu, C.-H., and Drummond, A. J. (2011). Phylogenetic and epidemic modeling of rapidly evolving infectious diseases. *Infection, Genetics and Evolution* **11**, 1825–1841.

Lanave, C., Preparata, G., Saccone, C., and Serio, G. (1984). A new method for calculating evolutionary substitution rates. *Journal of Molecular Evolution* **20**, 86–93.

Laplace, P. (1812). Theorie analytique des probabilites. *Courcier.*

Lapointe, F. J. and Legendre, P. (1991). The generation of random ultrametric matrices representing dendrograms. *Journal of Classification* **8**, 177–200.

Larget, B. (2013). The estimation of tree posterior probabilities using conditional clade probability distributions. *Systematic Biology* **62**, 501–511.

Larget, B. and Simon, D. (1999). Markov chain Monte Carlo algorithms for the Bayesian analysis of phylogenetic trees. *Molecular Biology and Evolution* **16**, 750–759.

Lartillot, N. (2006). Conjugate Gibbs sampling for Bayesian phylogenetic models. *Journal of Computational Biology* **13**, 1701–1722.

Lartillot, N., Brinkmann, H., and Philippe, H. (2007). Suppression of long branch attraction artefacts in the animal phylogeny using site-heterogeneous model. *BMC Evolutionary Biology* **7**, 1–14.

Lartillot, N. and Delsuc, F. (2012). Joint reconstruction of divergence times and life-history evolution in placental mammals using a phylogenetic covariance model. *Evolution* **66**, 1773–1787.

Lartillot, N., Lepage, T., and Blanquart, S. (2009). PhyloBayes 3: a Bayesian software package for phylogenetic reconstruction and molecular dating. *Bioinformatics* **25**, 2286–1288.

Lartillot, N. and Philippe, H. (2004). A Bayesian mixture model for across-site heterogeneities in the amino-acid replacement process. *Molecular Biology and Evolution* **21**, 1095–1109.

Lartillot, N. and Philippe, H. (2006a). Computing Bayes factors using thermodynamic integration. *Systematic Biology* **55**, 195–207.

Lartillot, N. and Philippe, H. (2006b). Computing Bayes factors using thermodynamic integration. *Systematic Biology* **55**, 195–207.

Lartillot, N. and Poujol, R. (2011). A phylogenetic model for investigating correlated evolution of substitution rates and continuous phenotypic characters. *Molecular Biology and Evolution* **28**, 729–744.

Lee, M. S. Y., Oliver, P. M., and Hutchinson, M. N. (2009). Phylogenetic uncertainty and molecular clock calibrations: a case study of legless lizards (pygopodidae, gekkota). *Molecular Phylogenetics and Evolution* **50**, 661–666.

Lefebvre, G., Steele, R., and Vandal, A. C. (2010). A path sampling identity for computing the Kullback-Leibler and J divergences. *Computational Statistics and Data Analysis* **54**, 1719–1731.

Lemey, P., Salemi, M., and vandamme, A.-M., editors (2009). *The phylogenetic handbook: a practical approach to phylogenetic analysis and hypothesis testing.* Cambridge: Cambridge University Press, second edition.

Lemmon, A. R., Brown, J. M., Stanger-Hall, K., and Lemmon, E. M. (2009). The effect of ambiguous data on phylogenetic estimates obtained by maximum likelihood and Bayesian inference. *Systematic Biology* **58**, 130–145.

Lenk, P. (2009). Simulation pseudo-bias correction to the harmonic mean estimator of integrated likelihoods. *Journal of Computational and Graphical Statistics* **18**, 941–960.

Lepage, T., Bryant, D., Philippe, H., and Lartillot, N. (2007). A general comparison of relaxed molecular clock models. *Molecular Biology and Evolution* **24**, 2669–2680.

Lepage, T., Lawi, S., Tupper, P., and Bryant, D. (2006). Continuous and tractable models for the variation of evolutionary rates. *Mathematical Biosciences* **199**, 216–233.

Leventhal, G. E., Kouyos, R., Stadler, T., von Wyl, V., Yerly, S., Böni, J., Cellerai, C., Klimkait, T., Günthard, H. F., and Bonhoeffer, S. (2012). Inferring epidemic contact structure from phylogenetic trees. *PLoS Computational Biology* **8**, 1–10.

Lewis, L. A. and Trainor, F. R. (2011). Survival of *Protosiphon botryoides* (chlorophyceae, chlorophyta) from a connecticut soil dried for 43 years. *Phycologia* **51**, 662–665.

Lewis, P. O. (2001). A likelihood approach to estimating phylogeny from discrete morphological character data. *Systematic Biology* **50**, 913–925.

Lewis, P. O., Holder, M. T., and Holsinger, K. E. (2005). Polytomies and Bayesian phylogenetic inference. *Systematic Biology* **54**, 241–253.

Lewis, P. O., Holder, M. T., and Swofford., D. L. (2008). Phycas: software for phylogenetic analysis. `http://www.phycas.org/`.

Li, H. and Durbin, R. (2011). Inference of human population history from individual whole-genome sequences. *Nature* **475**, 493–496.

Li, J. Z., Absher, D. M., Tang, H., Southwick, A. M., Casto, A. M., Ramachandran, S., Cann, H. M., Barsh, G. S., Feldman, M., Cavalli-Sforza, L. L., and Myers, R. M. (2008). Worldwide human relationships inferred from genome-wide patterns of variation. *Science* **319**, 1100–1104.

Li, S., Pearl, D., and Doss, H. (2000). Phylogenetic tree construction using Markov chain Monte Carlo. *Journal of the American Statistical Association* **95**, 493–508.

Li, W. H. (1997). *Molecular evolution.* Sunderland: Sinauer Associates.

Li, W. L. S. and Drummond, A. J. (2012). Model averaging and Bayes factor calculation of relaxed molecular clocks in Bayesian phylogenetics. *Molecular Biology and Evolution* **29**, 751–761.

Liang, F. (2003). Use of sequential structure in simulation from high dimensional systems. *Physical Review E* **67**, 56101–56107.

Liang, F., Liu, C., and Carroll, R. J. (2007). Stochastic approximation in Monte Carlo computation. *Journal of the American Statistical Association* **102**, 305–320.

Liang, F. and Wong, W. H. (2000). Evolutionary Monte Carlo applications to c_p model sampling and change-point problems. *Statistica Sinica* **10**, 317–342.

Liang, L., Zollner, S., and Abecasis, G. R. (2006). GENOME: a rapid coalescent-based whole genome simulator. *Bioinformatics* **23**, 1565–1567.

Linder, M., Britton, T., and Sennblad, B. (2011). Evaluation of Bayesian models of substitution rate evolution — parental guidance versus mutual independence. *Systematic Biology* **60**, 329–342.

Liu, J. S. (2001). *Monte Carlo strategies in scientific computing.* Berlin: Springer.

Liu, L. and Pearl, D. K. (2007). Species trees from gene trees: reconstructing Bayesian posterior distributions of a species phylogeny using estimated gene tree distributions. *Systematic Biology* **56**, 504–514.

Maddison, D. R. (1991). The discovery and importance of multiple islands of most parsimonious trees. *Systematic Zoology* **40**, 315–328.

Maddison, D. R., Swofford, D. L., and Maddison, W. P. (1997). Nexus: an extensible file format for systematic information. *Systematic Biology* **46**, 590–621.

Madigan, D. and Raftery, A. E. (1994). Model selection and accounting for model uncertainty in graphical models using Occam's window. *Journal of the American Statistical Association* **89**, 1535–1546.

Mailund, T., Schierup, M. H., Pederson, C. N. S., Madsen, J. N., Hein, J., and Schauser, L. (2006). GeneREcon: a coalescent based tool for fine-scale association mapping. *Bioinformatics* **22**, 2317–2318.

Mardis, E. (2011). A decade's perspective on DNA sequencing technology. *Nature* **470**, 198–203.

Marinari, E. and Parisi, G. (1992). Simulated tempering: a new Monte Carlo scheme. *Europhysics Letters* **19**, 451–458.

Marjoram, P. and Wall, J. D. (2006). Fast "coalescent" simulation. *BMC Genetics* **7**, 16.

Marshall, C. R. (1990). Confidence intervals on stratigraphic ranges. *Paleobiology* **16**, 1–10.

Marshall, C. R. (1994). Confidence intervals on stratigraphic ranges: partial relaxation of the assumption of randomly distributed fossil horizons. *Paleobiology* **20**, 459–469.

Marshall, C. R. (1997). Confidence intervals on stratigraphic ranges with nonrandom distributions of fossil horizons. *Paleobiology* **23**, 165–173.

Marshall, C. R. (2008). A simple method for bracketing absolute divergence times on molecular phylogenies using multiple fossil calibration points. *The American Naturalist* **171**, 726–742.

Marshall, D. C. (2010). Cryptic failure of partitioned Bayesian phylogenetic analyses: lost in the land of long trees. *Systematic Biology* **59**, 108–117.

Mateiu, L. and Rannala, B. (2006). Inferring complex DNA substitution processes on phylogenies using uniformization and data augmentation. *Systematic Biology* **55**, 259–269.

Mau, B. and Newton, M. A. (1997). Phylogenetic inference for binary data on dendrograms using Markov chain Monte Carlo. *Journal of Computational and Graphical Statistics* **6**, 122–131.

Mau, B., Newton, M. A., and Larget, B. (1999). Bayesian phylogenetic inference via Markov chain Monte Carlo. *Biometrics* **55**, 1–12.

McDade, L. (1992). Hybrids and phylogenetic systematics II: the impact of hybrids on cladistic analysis. *Evolution* **46**, 1329–1346.

McGill, J. R., Walkup, E. A., and Kuhner, M. K. (2013). GraphML specializations to codify ancestral recombination graphs. *Frontiers in Genetics* **4**, 146.

McVean, G. A. T., Awadalla, P., and Fearnhead, P. (2002). A coalescent-based method for detecting and estimating recombination from gene sequences. *Genetics* **160**, 1231–1241.

McVean, G. A. T. and Cardin, N. J. (2005). Approximating the coalescent with recombination. *Philosophical Transactions of the Royal Society of London, Series B* **360**, 1387–1393.

Meng, X.-L. and Wong, W. H. (1996). Simulating ratios of normalizing constants via simple identity: a theoretical exploration. *Statistica Sinica* **6**, 831–860.

Meredith, R. W., Janečka, J. E., Gatesy, J., Ryder, O. A., Fisher, C. A., Teeling, E. C., Goodbla, A., Eizirik, E., Simão, T. L. L., Stadler, T., Rabosky, D. L., Honeycutt, R. L., Flynn, J. J., Ingram, C. M., Steiner, C., Williams, T. L., Robinson, T. J., Burk-Herrick, A., Westerman, M., Ayoub, N. A., Springer, M. S., and Murphy, W. J. (2011). Impacts of the cretaceous terrestrial revolution and KPg extinction on mammal diversification. *Science* **334**, 521–524.

Message Passing Interface Forum (1994). MPI: a message passing interface standard. *The International Journal of Supercomputer Applications* **8**, 157–416.

Metropolis, N., Rosenbluth, A. W., Rosenbluth, M. N., Teller, A. H., and Teller, E. (1953a). Equation of state calculations by fast computing machines. *Journal of Chemical Physics* **21**, 1087–1091.

Metropolis, N., Rosenbluth, A. W., Rosenbluth, M. N., Teller, A. H., and Teller, E. (1953b). Equations of state calculations by fast computing machines. *Journal of Chemical Physics* **21**, 1087–1092.

Metropolis, N., Rosenbluth, A. W., Rosenbluth, N., Teller, A. H., and Teller, E. (1953c). Equation of state calculation by fast computing machines. *Journal of Chemical Physics* **21**, 1087–1092.

Miklós, I., Lunter, G. A., and Holmes, I. (2004). A "long indel" model for evolutionary sequence alignment. *Molecular Biology and Evolution* **21**, 529–540.

Minin, V., Abdo, Z., Joyce, P., and Sullivan, J. (2003). Performance-based selection of likelihood models for phylogeny estimation. *Systematic Biology* **52**, 674–683.

Minin, V. N., Bloomquist, E. W., and Suchard, M. A. (2008). Smooth skyride through a rough skyline: Bayesian coalescent-based inference of population dynamics. *Molecular Biology and Evolution* **25**, 1459–1471.

Moler, C. and Loan, C. V. (2003). Nineteen dubious ways to compute the exponential of a matrix, twenty-five years later. *SIAM Review* **45**, 3–49.

Møller, J., Syversveen, A. R., and Waagepetersen, R. P. (1998). Log Gaussian Cox processes. *Scandinavian Journal of Statistics* **25**, 451–482.

Moore, B. R., Chan, K. M. A., and Donoghue, M. J. (2004). Detecting diversification rate variation in supertrees. In O. R. P. Bininda-Emonds, editor, *Phylogenetic supertrees: combining information to reveal the tree of life*, pages 487–533. Dordrecht: Kluwer Academic Publishers.

Moral, P. D. (2004). *Feynman-Kac formulae.* Berlin: Springer.

Moral, P. D., Doucet, A., and Jasra, A. (2006). Sequential Monte Carlo samplers. *Journal of the Royal Statistical Society, Series B* **68**, 411–436.

Morris, A. P., Whittaker, J. C., and Balding, D. J. (2002). Fine-scale mapping of disease loci via shattered coalescent modeling of genealogies. *American Journal of Human Genetics* **70**, 686–707.

Morrison, D. (2009). Why would phylogeneticists ignore computerized sequence alignment? *Systematic Biology* **58 (1)**, 150–158.

Müller, P. and Quintana, F. A. (2004). Nonparametric Bayesian data analysis. *Statistical Science* **19**, pp. 95–110.

Nakleh, L. (2010). Evolutionary phylogenetic networks: models and issues. In N. Ramakrishnan, editor, *The problem solving handbook for computational biology and bioinformatics*, pages 125–158. Berlin: Springer.

Nasrallah, C. A., Mathews, D., and Huelsenbeck, J. (2011). Quantifying the impact of dependent evolution among sites in phylogenetic inference. *Systematic Biology* **60(1)**, 60–73.

Neal, R. (2008). The harmonic mean of the likelihood: worst Monte Carlo method ever. `http://radfordneal.wordpress.com/2008/08/17/the-harmonic-mean-of-the-likelihood-worst-monte-carlo-method-ever/`.

Neal, R. M. (2000). Markov chain sampling methods for Dirichlet process mixture models. *Journal of Computational and Graphical Statistics* **9**, 249–265.

Nee, S., May, R. M., and Harvey, P. H. (1994). The reconstructed evolutionary process. *Philosophical Transactions of the Royal Society B* **344**, 305–311.

Newton, M. and Raftery, A. E. (1994a). Approximate Bayesian inference by the weighted likelihood bootstrap. *Journal of the Royal Statistical Society, Series B* **56**, 3–48.

Newton, M. A. (1996). Bootstrapping phylogenies: large deviations and dispersion effects. *Biometrika* **83**, 315–328.

Newton, M. A., Mau, B., and Larget, B. (1999). Markov chain Monte Carlo for the Bayesian analysis of evolutionary trees from aligned molecular sequences. In F. Seillier-Moseiwitch, editor, *IMS Lecture Notes-Monograph Series: Statistics in Molecular Biology and Genetics*, 33, pages 143–162. Providence: American Mathematical Society.

Newton, M. A. and Raftery, A. E. (1994b). Approximate Bayesian inference by the weighted likelihood bootstrap. *Journal of the Royal Statistical Society, Series B* **56**, 3–48.

Newton, M. A. and Raftery, A. E. (1994c). Approximating Bayesian inference with the weighted likelihood bootstrap. *Journal of the Royal Statistical Society, Series B* **56**, 3–48.

Nielsen, R. (2002). Mapping mutations on phylogenies. *Systematic Biology* **51**, 729–739.

Nielsen, R., editor (2005). *Statistical methods in molecular evolution.* Berlin: Springer.

Nielsen, R. and Yang, Z. (2003). Estimating the distribution of selection coefficients from phylogenetic data with applications to mitochondrial and viral DNA. *Molecular Biology and Evolution* **20**, 1231–1239.

Nordborg, M. (2001). Coalescent theory. In D. Balding, M. Bishop, and C. Cannings, editors, *Handbook of Statistical Genetics*, pages 179–212. Chichester: John Wiley and Sons.

Nordborg, M. and Innan, H. (2003). The genealogy of sequences containing multiple sites subject to strong selection in a subdivided population. *Genetics* **163**, 1201–1213.

Nylander, J. A., Ronquist, F., Huelsenbeck, J. P., and Nieves-Aldrey, J. L. (2004). Bayesian phylogenetic analysis of combined data. *Systematic Biology* **53**, 47–67.

O'Dea, E. B. and Wilke, C. O. (2010). Contact heterogeneity and phylodynamics: how contact networks shape parasite evolutionary trees. *Interdisciplinary Perspectives on Infectious Diseases* **2011**.

O'Fallon, B. (2011). ACG: analysis of recombinant genealogies. `http://arup.utah.edu/ACG/`.

O'Fallon, B., Seger, J., and Adler, F. R. (2010). A continuous-state coalescent and the impact of weak selection on the structure of gene genealogies. *Molecular Biology and Evolution* **27**, 1162–1172.

Ogata, Y. (1989). A Monte Carlo method for high dimensional integration. *Numerische Mathematik* **55**, 137–157.

Opgen-Rhein, R., Fahrmeir, L., and Strimmer, K. (2005). Inference of demographic history from genealogical trees using reversible jump Markov chain Monte Carlo. *BMC Evolutionary Biology* **5**, 1–6.

Orbanz, P. and Teh, Y. (2010). Bayesian nonparametric models. In C. Sammut and G. I. Webb, editors, *Encyclopedia of Machine Learning*. Berlin: Springer.

Pachter, L. (2007). An introduction to reconstructing ancestral genomes. *Proceedings of Symposia in Applied Mathematics, AMS Short Course Subseries* **64**, 1–20.

Pagel, M. and Meade, A. (2004). A phylogenetic mixture model for detecting pattern-heterogeneity in gene sequence or character-state data. *Systematic Biology* **53**, 571–581.

Pagel, M. and Meade, A. (2006). Bayesian analysis of correlated evolution of discrete characters by reversible-jump Markov chain Monte Carlo. *The American Naturalist* **167**, 808–825.

Palacios, J. A. and Minin, V. N. (2013). Gaussian process-based Bayesian nonparametric inference of population trajectories from gene genealogies. *Biometrics* **69**, 8–18.

Paradis, E., Claude, J., and Strimmer, K. (2004). APE: analyses of phylogenetics and evolution in R language. *Bioinformatics* **20**, 289–290.

Parham, J. F., Donoghue, P. C. J., Bell, C. J., Calway, T. D., Head, J. J., Holroyd, P. A., Inoue, J. G., Irmis, R. B., Joyce, W. G., Ksepka, D. T., Patané, J. S. L., Smith, N. D., Tarver, J. E., van Tuinen, M., Yang, Z., Angielczyk, K. D., Greenwood, J. M., Hipsley, C. A., Jacobs, L., Makovicky, P. J., Müller, J., Smith, K. T., Theodor, J. M., Warnock, R. C. M., and Benton, M. J. (2012). Best practices for justifying fossil calibrations. *Systematic Biology* **61**, 346–359.

Pedersen, A. M. and Jensen, J. L. (2001). A dependent-rates model and an MCMC-based methodology for the maximum-likelihood analysis of sequences with overlapping reading frames. *Molecular Biology and Evolution* **18**, 763–776.

Petris, G. and Tardella, L. (2003). A geometric approach to trandimensional Markov chain Monte Carlo. *Canadian Journal of Statistics* **31**, 469–482.

Petris, G. and Tardella, L. (2007). New perspectives for estimating normalizing constants via posterior simulation. Technical report, Università di Roma "La Sapienza."

Pickett, K. M. and Randle, C. P. (2005). Strange Bayes indeed: uniform topological priors imply non-uniform clade priors. *Molecular Phylogenetics and Evolution* **34**, 203–211.

Pitt, M. and Shephard, N. (1999). Filtering via simulation: auxiliary particle filters. *Journal of the American Statistical Association* **94**, 590–599.

Pluzhnikov, A. and Donnelly, P. (1996). Optimal sequencing strategies for surveying molecular genetic diversity. *Genetics* **144**, 1247–1262.

Pollock, D. D., Taylor, W. R., and Goldman, N. (1999). Coevolving protein residues: maximum likelihood identification and relationship to structure. *Journal of Molecular Biology* **287**, 187–198.

Popovic, L. (2004). Asymptotic genealogy of a critical branching process. *Annals of Applied Probability* **14**, 2120–2148.

Posada, D. and Buckley, T. R. (2004). Model selection and model averaging in phylogenetics: advantages of Akaike Information Criterion and Bayesian approaches over likelihood ratio tests. *Systematic Biology* **53**, 793–808.

Press, W. H., Teukolsky, S. A., Vetterling, W. T., and Flannery, B. P. (1992). *Numerical recipes in C: the art of scientific computing. 2nd ed.* Cambridge: Cambridge University Press, second edition.

Pybus, O. G., Drummond, A. J., Nakano, T., Robertson, B. H., and Rambaut, A. (2003). The epidemiology and iatrogenic transmission of hepatitis C virus in Egypt: a Bayesian coalescent approach. *Molecular Biology and Evolution* **20**, 381–387.

Pybus, O. G. and Rambaut, A. (2002). GENIE: estimating demographic history from molecular phylogenies. *Bioinformatics* **18**, 1404–1405.

Pybus, O. G. and Rambaut, A. (2009). Evolutionary analysis of the dynamics of viral infectious disease. *Nature Reviews Genetics* **10**, 540–550.

Pybus, O. G., Rambaut, A., and Harvey, P. H. (2000a). An integrated framework for the inference of viral population history from reconstructed genealogies. *Genetics* **155**, 1429–1437.

Pybus, O. G., Rambaut, A., and Harvey, P. H. (2000b). An integrated framework for the inference of viral population history from reconstructed genealogies. *Genetics* **155**, 1429–1437.

Pyron, R. A. (2011). Divergence time estimation using fossils as terminal taxa and the origins of Lissamphibia. *Systematic Biology* **60**, 466–481.

R Development Core Team (2008). *R: a language and environment for statistical computing.* R Foundation for Statistical Computing, Vienna. ISBN 3-900051-07-0.

Rabosky, D. L. (2009). Heritability of extinction rates links diversification patterns in molecular phylogenies and fossils. *Systematic Biology* **58**, 629–640.

Raftery, A. and Akman, V. (1986). Bayesian analysis of a Poisson process with a change-point. *Biometrika* **73**, 85–89.

Raftery, A., Newton, M., Satagopan, J., and Krivitsky, P. (2007a). Estimating the integrated likelihood via posterior simulation using the harmonic mean identity. In *Memorial Sloan-Kettering Cancer Center, Dept. of Epidemiology and Biostatistics Working Paper Series. Working Paper 6.* Berkeley: Berkeley Electronic Press.

Raftery, A., Newton, M., Satagopan, J., and Krivitsky, P. (2007b). Estimating the integrated likelihood via posterior simulation using the harmonic mean identity. In J. M. Bernardo, M. J. Bayarri, and J. O. Berger, editors, *Bayesian Statistics*, pages 1–45. Oxford: Oxford University Press.

Rambaut, A. (2000). Estimating the rate of molecular evolution: incorporating non-contemporaneous sequences into maximum likelihood phylogenies. *Bioinformatics* **16**, 395–399.

Rambaut, A. and Bromham, L. (1998). Estimating divergence dates from molecular sequences. *Molecular Biology and Evolution* **15**, 442–448.

Rambaut, A. and Grassly, N. C. (1997). Seq-Gen: an application for the Monte Carlo simulation of DNA sequence evolution along phylogenetic trees. *Computer Applications in the Biosciences* **13**, 235–238.

Rambaut, A., Pybus, O. G., Nelson, M. I., Viboud, C., Taubenberger, J. K., and Holmes, E. C. (2008). The genomic and epidemiological dynamics of human influenza A virus. *Nature* **453**, 615–619.

Rannala, B. (2002). Identifiability of parameters in MCMC Bayesian inference of phylogeny. *Systematic Biology* **51**, 754–760.

Rannala, B. and Yang, Z. (1996). Probability distribution of molecular evolutionary trees: a new method of phylogenetic inference. *Journal of Molecular Evolution* **43**, 304–311.

Rannala, B. and Yang, Z. (2007). Inferring speciation times under an episodic molecular clock. *Systematic Biology* **56**, 453–466.

Rannala, B. and Yang, Z. H. (2003). Bayes estimation of species divergence times and ancestral population sizes using DNA sequences from multiple loci. *Genetics* **164**, 1645–1656.

Rannala, B., Zhu, T., and Yang, Z. (2012). Tail paradox, partial identifiability, and influential priors in Bayesian branch length inference. *Molecular Biology and Evolution* **29**, 325–335.

Rao, V. and Teh, Y. W. (2012). Fast mcmc sampling for markov jump processes and extensions. *arXiv preprint arXiv:1208.4818* .

Rasmussen, C. E. and Williams, C. K. I. (2006). *Gaussian processes for machine learning.* Cambridge: The MIT Press.

Rasmussen, D., Ratmann, O., and Koelle, K. (2011). Inference for nonlinear epidemiological models using genealogies and time series. *PLoS Computational Biology* **7**, e1002136.

Ray, N., Currat, M., Foll, M., and Excoffier, L. (2010). SPLATCHE2: a spatially-explicit simulation framework for complex demography, genetic admixture and recombination. *Bioinformatics* **26**, 2993–2994.

Redelings, B. D. and Suchard, M. A. (2005). Joint Bayesian estimation of alignment and phylogeny. *Systematic Biology* **54**, 401–418.

Reeves, J. H. (1992). Heterogeneity in the substitution process of amino acid sites of proteins coded for by mitochondrial DNA. *Journal of Molecular Evolution* **35**, 17–31.

Ripley, B. D. (1987). *Stochastic simulation.* New York: John Wiley and Sons.

Robbins, H. and Monro, S. (1951). A stochastic approximation method. *The Annals of Mathematical Statistics* **22**, 400–407.

Robert, C. and Wraith, D. (2009). Computational methods for Bayesian model choice. *AIP Proceedings* **1193**, 251–262.

Roberts, G. O. and Tweedie, R. L. (1996). Geometric convergence and central limit theorems for multidimensional hastings and Metropolis algorithms. *Biometrika* **83**, 95–110.

Robertson, A. and Hill, W. G. (1983). Population and quantitative genetics of many linked loci in finite populations. *Proceedings of the Royal Society of London, Series B* **219**, 253–264.

Robinson, D. F. and Foulds, L. R. (1981). Comparison of phylogenetic trees. *Mathematical Biosciences* **53**, 131–147.

Robinson, D. M. (2003). D.R.EVOL: Three-Dimensional Realistic Evolution. Ph.D. thesis, Bioinformatics, North Carolina State University.

Robinson, D. M., Jones, D. T., Kishino, H., Goldman, N., and Thorne, J. L. (2003). Protein evolution with dependence among codons due to tertiary structure. *Molecular Biology and Evolution* **20**, 1692–1704.

Rodrigue, N. (2007). Phylogenetic structural modeling of molecular evolution. Ph.D. thesis, University of Montreal.

Rodrigue, N., Lartillot, N., Bryant, D., and Philippe, H. (2005). Site interdependence attributed to tertiary structure in amino acid sequence evolution. *Gene* **347**, 207–217.

Rodrigue, N. and Philippe, H. (2010). Mechanistic revisions of phenomenological modeling strategies in molecular evolution. *Trends in Genetics* **26**, 248–252.

Rodrigue, N., Philippe, H., and Lartillot, N. (2006). Assessing site-interdependent phylogenetic models of sequence evolution. *Molecular Biology and Evolution* **23**, 1762–1775.

Rodrigue, N., Philippe, H., and Lartillot, N. (2007). Exploring fast computational strategies for probabilistic phylogenetic analysis. *Systematic Biology* **55**, 711–726.

Rodrigue, N., Philippe, H., and Lartillot, N. (2008). Uniformization for sampling realizations of Markov processes: applications to Bayesian implementations of codon substitution models. *Bioinformatics* **24**, 56–62.

Rodrigue, N., Philippe, H., and Lartillot, N. (2010). Mutation-selection models of coding sequence evolution with site-heterogeneous amino acid fitness profiles. *Proceedings of the National Academy of Science USA* **107**, 4629–4634.

Rodriguez, F., Oliver, J. L., Marin, A., and Medina, J. R. (1990). The general stochastic model of nucleotide substitution. *Journal of Theoretical Biology* **142**, 485–501.

Ronquist, F. and Deans, A. (2010). Bayesian phylogenetics and its influence on insect systematics. *Annual Review of Entomology* **55**, 189–206.

Ronquist, F., Klopfstein, S., Vilhelmsen, L., Schulmeister, S., Murray, D. L., and Rasnitsyn, A. P. (2012a). A total-evidence approach to dating with fossils, applied to the early radiation of the Hymenoptera. *Systematic Biology* **61**, 973–999.

Ronquist, F., Teslenko, M., van der Mark, P., Ayres, D. L., Darling, A., Höhna, S., Larget, B., Liu, L., Suchard, M. A., and Huelsenbeck, J. P. (2012b). Mrbayes 3.2: efficient Bayesian phylogenetic inference and model choice across a large model space. *Systematic Biology* **61**, 539–542.

Roure, B., Baurain, D., and Philippe, H. (2013). Impact of missing data on phylogenies inferred from empirical phylogenomic data sets. *Molecular Biology and Evolution* **30**, 197–214.

Rzhetsky, A. and Nei, M. (1992). A simple method for estimating and testing minimum evolution trees. *Molecular Biology and Evolution* **9**, 945–967.

Saitou, N. and Nei, M. (1987). The neighbor-joining method: a new method for reconstructing phylogenetic trees. *Molecular Biology and Evolution* **4**, 406–425.

Salemi, M. and Vandamme, A. M. (2003). *The phylogenetic handbook: a practical approach to DNA and protein phylogeny.* Cambridge: Cambridge University Press.

Salminen, M. O., Carr, J. K., Burke, D. S., and McCutchan, F. E. (1995). Identification of breakpoints in intergenotypic recombinants of HIV type 1 by bootscanning. *AIDS Research and Human Retroviruses* **11**, 1423–1425.

Salter, L. A. and Pearl, D. K. (2001). Stochastic search strategy for estimation of maximum likelihood phylogenetic trees. *Systematic Biology* **50**, 7–17.

Schrider, D. R., Hourmozdi, J. N., and Hahn, M. W. (2011). Pervasive multinucleotide mutational events in eukaryotes. *Current Biology* **21**, 1051–1054.

Schwarz, G. (1978). Estimating the dimension of a model. *Annals of Statistics* **6**, 461–464.

Shah, S. et al. (2009). Mutational evolution in a lobular breast tumour profiled at single nucleotide resolution. *Nature* **461**, 809–813.

Shapiro, B., Drummond, A. J., Rambaut, A., Wilson, M. C., Matheus, P. E., Sher, A. V., Pybus, O. G., Gilbert, M. T. P., Barnes, I., Binladen, J., Willerslev, E., Hansen, A. J., Baryshnikov, G. F., Burns, J. A., Davydov, S., Driver, J. C., Froese, D. G., Harington, C. R., Keddie, G., Kosintsev, P., Kunz, M. L., Martin, L. D., Stephenson, R. O., Storer, J., Tedford, R., Zimov, S., and Cooper, A. (2004). Rise and fall of the Beringian Steppe Bison. *Science* **306**, 1561–1565.

Siepel, A. and Haussler, D. (2004). Phylogenetic estimation of context-dependent substitution rates by maximum likelihood. *Molecular Biology and Evolution* **21**, 468–488.

Simon, D. and Larget, B. (1998). Bayesian analysis in molecular biology and evolution (BAMBE). version 1.01 beta.

Simon, D. and Larget, B. (2001). Bayesian analysis in molecular biology and evolution (BAMBE). version 2.03 beta.

Sinshheimer, J. S., Lake, J. A., and Little, R. J. A. (1996). Bayesian hypothesis testing of four-taxon topologies using molecular sequence data. *Biometrics* **52**, 193–210.

Slatkin, M. and Hudson, R. (1991). Pairwise comparisons of mitochondrial dna sequences in stable and exponentially growing populations. *Genetics* **129**, 555–562.

Smith, L. P. and Kuhner, M. K. (2009). The limits of fine-scale mapping. *Genetic Epidemiology* **33**, 344–356.

Smith, S. A. and Donoghue, M. J. (2008). Rates of molecular evolution are linked to life history in flowering plants. *Science* **322**, 86–89.

Spencer, C. C. A. and Coop, G. (2004). SelSim: a program to simulate population genetic data with natural selection and recombination. *Bioinformatics* **20**, 3673–3675.

Stadler, T. (2009). On incomplete sampling under birth-death models and connections to the sampling-based coalescent. *Journal of Theoretical Biology* **261**, 58–66.

Stadler, T. (2010). Sampling-through-time in birth-death trees. *Journal of Theoretical Biology* **267**, 396–404.

Stadler, T. (2011a). Mammalian phylogeny reveals recent diversification rate shifts. *Proceedings of the National Academy of Sciences USA* **108**, 6187–6192.

Stadler, T. (2011b). Simulating trees on a fixed number of extant species. *Systematic Biology* **60**, 668–675.

Stadler, T., Kühnert, D., Bonhoeffer, S., and Drummond, A. J. (2013). Birth-death skyline plot reveals temporal changes of epidemic spread in HIV and hepatitis C virus (HCV). *Proceedings of the National Academy of Sciences USA* **110**, 228–233.

Stadler, T. and Yang, Z. (2013). Dating phylogenies with sequentially sampled tips. *Systematic Biology* **62**, 674–688.

Steel, M. and Pickett, K. M. (2006). On the impossibility of uniform priors on clades. *Molecular Phylogenetics and Evolution* **39**, 585–586.

Steel, M. A. (2005). Should phylogenetic models be trying to 'fit an elephant'? *Trends in Genetics* **21**, 307–309.

Stephens, M., Smith, N. J., and Donnelly, P. (2001). A new statistical method for haplotype reconstruction from population data. *The American Journal of Human Genetics* **68**, 978–989.

Stets, R., Dwarkadas, S., Hardavellas, N., Hunt, G., Kontothanassis, L., Parthasarathy, S., and Scott, M. L. (1997). Cashmere-2l: software coherent shared memory on a clustered remote-write network. In *Proceedings of the 16th ACM Symposium on Operating Systems Principles*, pages 170–183. New York: Association for Computing Machinery.

Strimmer, K. and Pybus, O. G. (2001). Exploring the demographic history of DNA sequences using the generalized skyline plot. *Molecular Biology and Evolution* **18**, 2298–2305.

Suchard, M., Weiss, R., and Sinsheimer, J. (2005). Models for estimating Bayes factors with applications to phylogeny and tests of monophyly. *Biometrics* **61**, 665–673.

Suchard, M. A. and Rambaut, A. (2009). Many-core algorithms for statistical phylogenetics. *Bioinformatics* **25**, 1370–1376.

Suchard, M. A., Weiss, R. E., and Sinsheimer, J. S. (2001). Bayesian selection of continuous-time Markov chain evolutionary models. *Molecular Biology and Evolution* **18**, 1001–1013.

Sullivan, J. and Joyce, P. (2005). Model selection in phylogenetics. *Annual Review of Ecology, Evolution, and Systematics* **36**, 445–466.

Susko, E. (2008). On the distributions of bootstrap support and posterior distributions for a star tree. *Systematic Biology* **57**, 602–612.

Suzuki, Y., Glazko, G. V., and Nei, M. (2002). Overcredibility of molecular phylogenies obtained by Bayesian phylogenetics. *Proceedings of the National Academy of Sciences USA* **99**, 16138–16143.

Tajima, F. (1983). Evolutionary relationship of DNA sequences in finite populations. *Genetics* **105**, 437–460.

Tamura, K. and Nei, M. (1993). Estimation of the number of nucleotide substitutions in the control region of mitochondrial DNA in humans and chimpanzees. *Molecular Biology and Evolution* **10**, 512–526.

Tanner, M. A. and Wong, W. H. (1987). The calculation of posterior distributions by data augmentation. *Journal of the American Statistical Association* **82**, 528–550.

Tataru, P. and Hobolth, A. (2011). Comparison of methods for calculating conditional expectations of sufficient statistics for continuous time Markov chains. *BMC Bioinformatics* **12**, 465.

Tavaré, S. (1986). Some probabilistic and statistical problems in the analysis of DNA sequences. *Lectures on Mathematics in the Life Sciences (American Mathematical Society)* **17**, 57–86.

Tavaré, S. (2004). Part I: ancestral inference in population genetics. In J. Picard, editor, *Lectures on Probability Theory and Statistics, Ecolé d'Estes Probabilité de Saint-Flour XXXI – 2001*, volume 1837 of *Lecture Notes in Mathematics*, pages 1–188. New York: Springer.

Teh, Y. W., Daume III, H., and Roy, D. M. (2008). Bayesian agglomerative clustering with coalescents. In J. Platt, D. Koller, Y. Singer, and S. Roweis, editors, *Advances in Neural Information Processing 20*, pages 1473–1480. Cambridge: MIT Press.

Teshima, K. M. and Innan, H. (2009). MBS: modifying Hudson's MS software to generate samples of DNA sequences with a biallelic site under selection. *BMC Bioinformatics* **10**, 166.

Thompson, E. A. (1975). *Human evolutionary trees.* Cambridge: Cambridge University Press.

Thorne, J. and Kishino, H. (2002a). Divergence time and evolutionary rate estimation with multilocus data. *Systematic Biology* **51**, 689–702.

Thorne, J. L., Choi, S. C., Yu, J., Higgs, P. G., and Kishino, H. (2007). Population genetics without intraspecific data. *Molecular Biology and Evolution* **24**, 1667–1677.

Thorne, J. L., Goldman, N., and Jones, D. T. (1996). Combining protein evolution and secondary structure. *Molecular Biology and Evolution* **13**, 666–673.

Thorne, J. L. and Kishino, H. (2002b). Divergence time and evolutionary rate estimation with multilocus data. *Systematic Biology* **51**, 689–702.

Thorne, J. L. and Kishino, H. (2005). Estimation of divergence times from molecular sequence data. In R. Nielsen, editor, *Statistical methods in molecular evolution*, pages 235–256. Berlin: Springer.

Thorne, J. L., Kishino, H., and Felsenstein, J. (1991). An evolutionary model for maximum likelihood alignment of DNA sequences. *Journal of Molecular Evolution* **33**, 114–124.

Thorne, J. L., Kishino, H., and Painter, I. S. (1998). Estimating the rate of evolution of the rate of molecular evolution. *Molecular Biology and Evolution* **15**, 1647–1657.

Tiao, G. G. and Cuttman, I. (1965). The inverted Dirichlet distribution with applications. *Journal of the American Statistical Association* **60**, 793–805.

Tillier, E. R. M. (1994). Maximum likelihood with multi-parameter models of substitution. *Journal of Molecular Evolution* **39**, 409–417.

Valentini, A., Pompanon, F., and Taberlet, P. (2009). DNA barcoding for ecologists. *Trends in Ecology and Evolution* **24**, 110–117.

Van Loan, C. (1978). Computing integrals involving the matrix exponential. *IEEE Transactions on Automatic Control* **23**, 395–404.

Verdinelli, I. and Wasserman, L. (1995). Computing Bayes factors using a generalization of the Savage-Dickey density ratio. *Journal of the American Statistical Association* **90**, 614–618.

Volz, E. M. (2012). Complex population dynamics and the coalescent under neutrality. *Genetics* **190**, 187–201.

Volz, E. M., Pond, S. L. K., Ward, M. J., Brown, A. J. K., and Frost, S. D. W. (2009). Phylodynamics of infectious disease epidemics. *Genetics* **183**, 1421–1430.

von Reumont, B., Meusemann, K., Szucsich, N., Dell'Ampio, E., Gowri-Shankar, V., Bartel, D., Simon, S., Letsch, H., Stocsits, R., Luan, Y., Wagele, J., Pass, G., Hadrys, H., and Misof, B. (2009). Can comprehensive background knowledge be incorporated into substitution models to improve phylogenetic analyses? A case study on major arthropod relationships. *BMC Evolutionary Biology* **9**, 1–19.

Wakeley, J. and Sargsyan, O. (2008). Extensions of the coalescent effective population size. *Genetics* **181**, 341–345.

Wall, J. D. (2000). A comparison of estimators of the population recombination rate. *Molecular Biology and Evolution* **17**, 156–163.

Wang, L. (2012). Bayesian phylogenetic inference via Monte Carlo methods. Ph.D. thesis, University of British Columbia.

Wang, Y. and Rannala, B. (2008). Bayesian inference of fine-scale recombination rates using population genomic data. *Philosophical Transactions of the Royal Society of London. Series B* **363**, 3921–3930.

Wang, Y. and Rannala, B. (2009). Population genomic inference of recombination rates and hotspots. *Proceedings of the National Academy of Sciences USA* **106**, 6215–6219.

Wang, Z., Johnston, P., Yang, Z., and Townsend, J. (2009). Evolution of reproductive morphology in leaf endophytes. *PlosOne* **4**, 42–46.

Wapinski, I., Pfeffer, A., Friedman, N., and Regev, A. (2007). Automatic genome-wide reconstruction of phylogenetic gene trees. *Bioinformatics* **23(13)**, i549–i558.

Warnock, R. C. M., Yang, Z., and Donoghue, P. C. J. (2012). Exploring the uncertainty in the calibration of the molecular clock. *Biology Letters* **8**, 156–159.

Watterson, G. (1975). On the number of segregating sites in genetical models without recombination. *Theoretical Population Biology* **7**, 256–276.

Weir, J. and Schluter, D. (2008). Calibrating the avian molecular clock. *Molecular Ecology* **17**, 2321–2328.

Welch, D., Bansal, S., and Hunter, D. R. (2011). Statistical inference to advance network models in epidemiology. *Epidemics* **3**, 38–45.

Welch, J. J., Fontanillas, E., and Bromham, L. (2005). Molecular dates for the "Cambrian Explosion": the influence of prior assumptions. *Systematic Biology* **54**, 672–678.

Wetterstrand, K. A. (2012). DNA sequencing costs: data from the NHGRI large-scale genome sequencing program. `http://www.genome.gov/sequencingcosts`. Accessed: 20/06/2012.

Whelan, S. and Goldman, N. (2004). Estimating the frequency of events that cause multiple-nucleotide changes. *Genetics* **167**, 2027–2043.

Wiens, J. J., Kuczynski, C. A., Townsend, T., Reeder, T. W., Mulcahy, D. G., and Sites, J. W. (2010). Combining phylogenomics and fossils in higher-level squamate reptile phylogeny: molecular data change the placement of fossil taxa. *Systematic Biology* **59**, 674–688.

Wiens, J. J. and Morrill, M. C. (2011). Missing data in phylogenetic analysis: reconciling results from simulations and empirical data. *Systematic Biology* **60**, 719–731.

Wiley, E. O. and Lieberman, B. S. (2011). *Phylogenetics: theory and practice of phylogenetic systematics.* Hoboken: Wiley-Blackwell.

Wiuf, C. and Hein, J. (1999). Recombination as a point process along sequences. *Theoretical Population Biology* **55**, 248–259.

Wong, W. H. and Liang, F. (1997). Dynamic weighting in Monte Carlo and optimization. *Proceedings of the National Academy of Sciences USA* **94**, 14220–14224.

Wood, H. M., Matzke, N. J., Gillespie, R. G., and Griswold, C. E. (2013). Treating fossils as terminal taxa in divergence time estimation reveals ancient vicariance patterns in the palpimanoid spiders. *Systematic Biology* **62**, 264–284.

Worobey, M., Gemmel, M., Teuwen, D. E., Haselkorn, T., Kunstman, K., Bunce, M., Muyembe, J. J., Kabongo, J. M. M., Kalengayi, R. M., Marck, E. V., Thomas, M., Gilbert, P., and Wolinsky, S. M. (2008). Direct evidence of extensive diversity of HIV-1 in Kinshasa by 1960. *Nature* **455**, 661–665.

Wright, S. (1931). Evolution in Mendelian populations. *Genetics* **16**, 97–159.

Xie, W., Lewis, P. O., Fan, Y., Kuo, L., and Chen, M.-H. (2011). Improving marginal likelihood estimation for Bayesian phylogenetic model selection. *Systematic Biology* **60**, 150–160.

Yamanoue, Y., Miya, M., Matsuura, K., Katoh, M., Sakai, H., and Nishida, M. (2008). A new perspective on phylogeny and evolution of tetraodontiform fishes (Pisces: Acanthopterygii) based on whole mitochondrial genome sequences: basal ecological diversification? *SBMC Evolutionary Biology* **8**, 212–226.

Yang, Z. (1993). Maximum likelihood estimation of phylogeny from DNA sequences when substitution rates differ over sites. *Molecular Biology and Evolution* **10**, 1396–1401.

Yang, Z. (1994a). Estimating the pattern of nucleotide substitution. *Journal of Molecular Evolution* **39**, 105–111.

Yang, Z. (1994b). Maximum likelihood phylogenetic estimation from DNA sequences with variable rates over sites: approximate methods. *Journal of Molecular Evolution* **39**, 306–314.

Yang, Z. (1995a). A space-time process model for the evolution of DNA sequences. *Genetics* **139**, 993–1005.

Yang, Z. (1995b). A space-time process model for the evolution of DNA sequences. *Genetics* **139**, 993–1005.

Yang, Z. (1996). Among-site rate variation and its impact on phylogenetic analyses. *Trends in Ecology and Evolution* **11**, 367–372.

Yang, Z. (2000a). Complexity of the simplest phylogenetic estimation problem. *Proceedings of the Royal Society of London, Series B* **267**, 109–116.

Yang, Z. (2000b). Maximum likelihood estimation on large phylogeneis and analysis of adaptive evolution in human influenza virus A. *Journal of Molecular Evolution* **51**, 423–432.

Yang, Z. (2004). A heuristic rate smoothing procedure for maximum likelihood estimation of species divergence times. *Acta Zoologica Sinica* **50**, 645–656.

Yang, Z. (2006). *Computational Molecular Evolution*. Oxford: Oxford University Press.

Yang, Z. (2007). Fair-balance paradox, star-tree paradox, and Bayesian phylogenetics. *Molecular Biology and Evolution* **24**, 1639–1655.

Yang, Z. and Nielsen, R. (2008). Mutation-selection models of codon substitution and their use to estimate selective strengths on codon usage. *Molecular Biology and Evolution* **25**, 568–579.

Yang, Z. and Rannala, B. (1997a). Bayesian phylogenetic inference using DNA sequences: a Markov chain Monte Carlo method. *Molecular Biology and Evolution* **14**, 717–724.

Yang, Z. and Rannala, B. (1997b). Bayesian phylogenetic inference using DNA sequences: a Markov chain Monte Carlo method. *Molecular Biology and Evolution* **14**, 717–724.

Yang, Z. and Rannala, B. (2005). Branch-length prior influences Bayesian posterior probability of phylogeny. *Systematic Biology* **54**, 455–470.

Yang, Z. and Rannala, B. (2006a). Bayesian estimation of species divergence times under a molecular clock using multiple fossil calibrations with soft bounds. *Molecular Biology and Evolution* **23**, 212–226.

Yang, Z. and Rannala, B. (2006b). Bayesian estimation of species divergence times under a molecular clock using multiple fossil calibrations with soft bounds. *Molecular Biology and Evolution* **23**, 212–226.

Yang, Z. and Rannala, B. (2012). Molecular phylogenetics: principles and practice. *Nature Reviews Genetics* **13**, 303–314.

Yang, Z. and Roberts, D. (1995). On the use of nucleic acid sequences to infer early branchings in the tree of life. *Molecular Biology and Evolution* **12**, 451–458.

Yang, Z. and Yoder, A. D. (2003a). Comparison of likelihood and Bayesian methods for estimating divergence times using multiple gene loci and calibration points, with application to a radiation of cute-looking mouse lemur species. *Systematic Biology* **52**, 705–716.

Yang, Z. and Yoder, A. D. (2003b). Comparison of likelihood and Bayesian methods for estimating divergence times using multiple gene loci and calibration points, with application to a radiation of cute-looking mouse lemur species. *Systematic Biology* **52**, 705–716.

Yap, V. and Speed, T. (2004). Modeling DNA base substitution in large genomic regions from two organisms. *Journal of Molecular Evolution* **58**, 12–18.

Yoder, A. D. and Yang, Z. (2000a). Estimation of primate speciation dates using local molecular clocks. *Molecular Biology and Evolution* **17**, 1081–1090.

Yoder, A. D. and Yang, Z. H. (2000b). Estimation of primate speciation dates using local molecular clocks. *Molecular Biology and Evolution* **17**, 1081–1090.

Yu, J. and Thorne, J. L. (2006). Dependence among sites in RNA evolution. *Molecular Biology and Evolution* **23**, 1525–1537.

Yule, G. U. (1924). A mathematical theory of evolution based on the conclusions of Dr. J. C. Willis. *Philosophical Transactions of the Royal Society of London, Series B* **213**, 21–87.

Zhang, C., Rannala, B., and Yang, Z. (2012). Robustness of compound Dirichlet priors for Bayesian inference of branch lengths. *Systematic Biology* **61**, 779–784.

Zhu, T., Korber, B. T., Nahmias, A. J., Hooper, E., Shaper, P. M., and Ho, D. D. (1998). An African HIV-1 sequence from 1959 and implications for the origin of the epidemic. *Nature* **391**, 594–597.

Zollner, S. and Pritchard, J. K. (2005). Coalescent-based association mapping and fine mapping of complex trait loci. *Genetics* **169**, 1071–1092.

Zuckerkandl, E. and Pauling, L. (1962). Molecular disease, evolution, and genetic heterogeneity. In M. Kasha and B. Pullman, editors, *Horizons in Biochemistry*, pages 189–225. New York: Academic Press.

Zuckerkandl, E. and Pauling, L. (1965). Evolutionary divergence and convergence in proteins. In V. Bryson and H. Vogel, editors, *Evolving Genes and Proteins*, pages 97–166. New York: Academic Press.

Zwickl, D. J. and Holder, M. T. (2004). Model parameterization, prior distributions, and the general time-reversible model in Bayesian phylogenetics. *Systematic Biology* **53**, 877–888.

Index

Printed in the United States
by Baker & Taylor Publisher Services